GLASS MICROELECTRODES

Glass Microelectrodes

Edited by

MARC LAVALLÉE
Département de Biophysique, Faculté de Médecine
Université de Sherbrooke, Canada

Research and Development Laboratories
Corning Glass Works, Corning, New York

OTTO F. SCHANNE
Département de Biophysique, Faculté de Médecine
Université de Sherbrooke, Canada

NORMAND C. HÉBERT
Research and Development Laboratories
Corning Glass Works, Corning, New York

Département de Biophysique, Faculté de Médecine
Université de Sherbrooke, Canada

JOHN WILEY & SONS, INC.
New York · London · Sydney · Toronto

Library of Congress Catalog Card Number: 68–9252

SBN 471 51885 9
Printed in the United States of America

Contributors

DANIEL P. AGIN, Department of Physiology, University of Chicago, Chicago, Illinois

ROGER G. BATES, National Bureau of Standards, Washington, D.C.

TUSHAR K. CHOWDHURY, Department of Physiology, George Washington University, Washington, D.C.

EDOUARD CORABŒUF, Laboratoire de Physiologie comparée et de Physiologie cellulaire associé au CNRS, Faculté des Sciences, Orsay, France

GEORGE EISENMAN, Department of Physiology, University of Chicago, Chicago, Illinois

B. FRANKENHAEUSER, The Nobel Institute for Neurophysiology, Karolinska Institutet, Stockholm, Sweden

HARRY GRUNDFEST, Laboratory of Neurophysiology, Department of Neurology, College of Physicians and Surgeons, Columbia University, New York

NORMAND C. HÉBERT, Research and Development Laboratories, Corning Glass Works, Corning, New York

J. A. M. HINKE, Department of Anatomy, University of British Columbia, Vancouver, British Columbia, Canada

YU. D. KHOLODOVA, Institute of Physiology, Kiev, U.S.S.R.

RAJA N. KHURI, Department of Physiology, American University of Beirut, Beirut, Lebanon

P. G. KOSTYUK, Institute of Physiology, Kiev, U.S.S.R.

GREGOR A. KURELLA, Department of Biophysics, Moscow State University, U.S.S.R.

MARC LAVALLÉE, Département de Biophysique, Faculté de Médecine, Université de Sherbrooke, Sherbrooke, Québec, Canada

A. A. LEV, Laboratory of Cell Physiology, Institute of Cytology, Academy of Sciences, Leningrad, U.S.S.R.

FRANK W. ORME, Department of Physiology-Anatomy, University of California, Berkeley, California

OTTO F. SCHANNE, Département de Biophysique, Faculté de Médecine, Université de Sherbrooke, Sherbrooke, Québec, Canada

FRED M. SNELL, Department of Biophysics, State University of New York at Buffalo, Buffalo, New York

Z. A. SOROKINA, Institute of Physiology, Kiev, U.S.S.R.

GABOR SZABO, Département de Biophysique, Faculté de Médecine, Université de Sherbrooke, Sherbrooke, Québec, Canada

TSUNEO TOMITA, Department of Physiology, Keio University School of Medicine, Shinjuku-ku, Tokyo, Japan

A. B. VALLBO, Department of Physiology, University of Umea, Umea, Sweden

WILLIAM J. WHALEN, St. Vincent Charity Hospital, Cleveland, Ohio

Preface

As biology evolved, research interests drifted from whole organisms to systems, organs, and then single cells and cellular components. At the same time, because of a greater awareness of the complexities of biological phenomena and the number of factors that could affect them, more attention was devoted to maintaining a regular environment for the processes under observation. Therefore the improvement of methods for biological investigation was directed toward the elimination of the reactive error, that is, the reduction of the interference of the method with the process being measured, coupled, of course, with a higher degree of precision and stability.

Intracellular glass micropipette electrodes have been used for the measurement of membrane potentials for about 20 years. Their universal acceptance as a research tool is attributed to their ability to probe single living cells without altering their environment. This unique feature of intracellular glass micropipette electrodes has stimulated the development of related electrodes capable of determining the ionic activities of chemical species and other parameters within single living cells.

Despite its numerous applications, the intracellular glass micropipette electrode has remained a poorly defined tool with unpredictable behavior. Most papers that deal with its applications lack descriptive material of the problems encountered in its fabrication, discussion of artifacts originating within the electrode itself, or changes in the cellular environment by impalement of cells. The absence of an accepted theory of the behavior and accomplishments of the glass microelectrode as a most powerful tool inspired the preparation of this book.

Four chapters are devoted to the electrochemistry of open-tip glass microelectrodes. The general conclusions of the four independently done works agree fairly well. However, since they were done with different approaches and dealt with different types of glass, filling solutions, and experimental set-up, the editors have preferred not to remove from each chapter material that may have seemed to overlap at first glance.

Two chapters are devoted to microelectrode glass technology and one chapter to inner reference electrodes.

We have included chapters on special types of microelectrode designed for injecting current or dyes or for measuring hydrogen, potassium, sodium, calcium, chloride activities, and other cellular parameters such as cytoplasmic resistivity and pO_2.

The book does not treat devices such as platinum or tungsten intra-cellular electrodes. Also, the reader should not expect to find discussions of electronic circuits such as high-input impedance preamplifiers or suitable read-out systems.

Neither is it restricted to the design and construction of micro-electrodes, for some chapters deal with their application to the solution of biological problems. These problems are exposed sometimes at length; we thought that their very nature, for example, the nonuniform ionic distribution inside the cell, is of utmost importance in the design of the microelectrode — its size, reproducibility. and speed of response.

We are grateful to Corning Glass Works for sponsoring the Intra-cellular Glass Microelectrodes Conference, held in May 1967 in Montreal, Canada, which enabled the book's contributors to exchange ideas and to present their work to an audience of more than 200 microelectrode workers. It should be pointed out, however, that the book is not the proceedings of this conference. We are grateful also to the Université de Sherbrooke, through its Comité des Cahiers, for its financial aid in the preparation of the manuscript. We acknowledge the previous collaboration of Professor Ronald Sutherland for editing the manuscript for English and of Miss Claudette Gaumond for typing it.

Sherbrooke, Canada   MARC LAVALLÉE
June, 1968   OTTO SCHANNE
                                                       NORMAND C. HÉBERT

Contents

GLASS MICROELECTRODES

CHAPTER 1

Inner Reference Electrodes and Their Characteristics

ROGER G. BATES

National Bureau of Standards

1.1 INTRODUCTION

This chapter is concerned primarily with the selection of electrodes and inner solutions most suited to provide electrical contact with the reference side of the glass membrane. The discussion will be limited to that situation in which the glass electrode is part of a cell designed for the measurement of pH. Particular attention will be directed to the effect of the inner electrode and inner solution on the temperature coefficient of the total cell's electromotive force. The conclusions regarding the effective design of a pH cell will not apply in their entirety to the varied types of cells with glass microelectrodes [1, 2]; it is hoped, nevertheless, that the principles set forth will constitute a useful guide in minimizing errors caused by the selection of unsuitable inner reference elements.

When a metal electrode is employed for the measurement of the activity of a metal ion to which the electrode is reversible, changes in the electrode potential are easily observed by connecting the electrode to the potentiometer or some other measuring instrument by a piece of copper wire or a similar electronic conductor. Because of the nature of the glass membrane, however, the potential of the pH-sensitive surface cannot be measured in this simple fashion; a more complex arrangement is necessary. Electrical contact with the inner surface of the membrane has been accomplished by means of mercury or by films of silver or other metals[3]. Fused salts such as the silver halides have also been found suitable, but the most common means of measuring changes in the electric potential of the responsive surface is through an electrode immersed in a suitable solution which makes contact with the reference ("inner") side of the glass membrane.

Copper wire is a nearly perfect connector—free from instability and drifts. Similarly, the ideal inner cell should have a high thermal stability. Its potential should remain constant from day to day, and there should be no marked hysteresis after changes of temperature.

The inner cell is evidently a far less simple and direct connector than is the copper wire used with metal electrodes. It is subject to many possible difficulties. However, the combination of inner electrode and inner solution offers certain advantages not attainable with the direct metallic connection. For example, the inner cell provides a bias voltage in series with the glass electrode and thus permits the voltage of the total cell to be changed by any desired amount. Furthermore, this arrangement offers the opportunity of compensating in some degree for the unwanted temperature variation of the emf of the total pH cell, thus minimizing the likelihood of chance errors owing to temperature fluctuations and also simplifying the calculation of results. The manner in which this may be accomplished will be discussed in detail in the sections to come.

1.2 GLASS ELECTRODE pH CELLS

The design of a typical pH cell with glass electrode and external reference electrode is shown diagrammatically in Fig. 1.1. Both the inner and

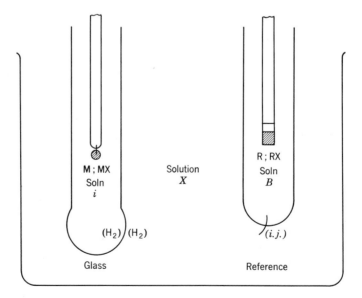

Fig. 1.1. Diagram of a typical pH cell with glass electrode and reference electrode.

outer surfaces of the glass membrane are presumed to be perfectly reversible to hydrogen ions and thus to function as hydrogen gas electrodes.
The electrode within the glass bulb consists of a metal M, often in the
form of a coating on a platinum wire. It may be in equilibrium with an
insoluble salt MX containing the metal M to which the electrode is reversible. Similarly, the metal electrode R within the external reference
electrode may be combined with a salt RX insoluble in solution B. The
inner solution is designated i. Some sort of mechanical device serves to
establish a stable liquid junction between solution B of the reference
electrode and solution X (of unknown pH) with which the solution vessel
is filled.

The total cell may be represented schematically as follows:

$$\underbrace{\text{M;MX, Soln } i, \text{H}_2}_{E_i} \| \underbrace{\text{H}_2; \text{Soln } X | \text{Soln } B, \text{RX;R},}_{E_2} \qquad (1.1)$$

where the double vertical line marks the glass membrane and the single
vertical line the liquid junction. If one ignores the asymmetry potential
across the glass membrane, the total emf of the cell is the sum of E_i and
E_2. The pH of solution X is, however, the quantity to be measured, so
it is E_2 which is of primary interest:

$$E_2 = E_{\text{cell}} - E_i = E^{0\prime} + E_j + k\text{pH}_X, \qquad (1.2)$$

where k is written for $(RT \ln 10)/F$. Thus one can write

$$\text{pH}_X = (E_{\text{cell}} - E_c^0)/k, \qquad (1.3)$$

where

$$E_c^0 \equiv E_i + E^{0\prime} + E_j. \qquad (1.4)$$

The quantity $E^{0\prime}$ is a constant at each particular temperature and is
characteristic of the nature of both the electrode R;RX and the composition of solution B [4]. Likewise, E_i is a function of the temperature,
of the nature of the electrode M;MX, and also of the composition of
solution i.

The inner solution should not attack the surface of the glass membrane,
and its buffer capacity should be high enough to offset changes of pH
produced by leaching of alkali from the special glass of the electrode,
which is somewhat soluble. Likewise, the concentration of the ion to
which the inner electrode is reversible should be in a range that will
insure stability of the inner reference electrode. These desirable properties
of inner solutions have been discussed by Dole in his monograph [5].

Some of the most useful combinations of inner electrode and inner solution are listed in Table 1.1, together with comments regarding their

Table 1.1   Characteristics of Some Inner Cells

Inner electrode	Inner solution	Comments
Ag;AgCl	H^+, Cl^- (HCl or buffered chloride)	Thermally stable, low hysteresis, $AgCl_2^-$ forms at high Cl^- concentrations
$Hg;Hg_2Cl_2$	H^+, Cl^-	Deterioration at elevated temperatures, marked hysteresis
Tl(Hg);TlCl	H^+, Cl^-	Stable over a wide range of temperatures
Pt;Quinhydrone	H^+ (strong acid)	Poor time stability
Ag	Ag^+, H^+ (usually perchlorates)	
Hg	Hg_2^{2+}, H^+ (usually perchlorates)	
Pt	Pt^{4+}, H^+ (usually chloroplatinic acid)	
Conducting films (Hg, Ag, Pb)		Variable asymmetry potential (?)
Fused salt films (e.g., silver halides)		

characteristics. By far the most common *external* reference electrode RX;R is the calomel electrode (H_2Cl_2;Hg), although silver-silver chloride electrodes (AgCl;Ag) and thallium amalgam-thallous chloride electrodes (TlCl;Tl-Hg) possess certain advantages in specialized situations. They are, for example, more suitable than the calomel electrode at elevated temperatures; a disproportionation of calomel causes deterioration of the electrode at temperatures above 80°C [6]. Concentrated solutions of potassium chloride are almost invariably used as solution *B*, in view of their equitransferent properties [7].

As shown by (1.4), the quantity $E_c{}^0$ is a combination of the standard potential $E^{0\prime}$ of the reference electrode, the liquid-junction potential E_j (which under the best of experimental conditions remains reasonably constant), and the emf E_i of the inner cell. It may be regarded as the standard emf of the total cell. For practical pH measurements, it would be most desirable that $E_c{}^0$ not vary with the temperature of the cell. If such were the case, the restriction of accurate pH measurements to the temperature at which the assembly was standardized would be removed. It would then be possible for example, to standardize the glass electrode at room temperature, and thereafter to make accurate measurements at blood temperature or in a process stream with temperature continually varying.

The standard emf of most cells is a quadratic function of the temperature. In view of the composite nature of $E_c{}^0$, it is possible to modify the temperature coefficient by proper choice of electrodes and inner solution, and even to nullify it completely over limited ranges of temperature. There also exists the possibility of altering the magnitude of $E_c{}^0$ and thus the absolute value of the emf which corresponds to a given pH value. This flexibility is of interest to the designers of pH instrumentation. The discussion of this chapter will deal primarily with the temperature range 10–40°C.

1.3 NONISOTHERMAL TEMPERATURE COEFFICIENTS OF ELECTRODE POTENTIALS

When the two electrodes of the pH cell are of widely different sizes and heat capacities, or when the two elecrodes are of necessity located in different thermal environments, difficulties may arise from the variation of the temperature of one electrode while that of the other remains constant. Although the true temperature coefficients of the potentials of single electrodes cannot be obtained by thermodynamic means, a useful indication of the changes in electrode potential can be derived from observation of the change of emf produced when nonisothermal conditions are deliberately created. For example, when a saturated calomel electrode is warmed from 25 to 26 or 27°C while the other electrode of the cell is kept at 25°C, the calomel electrode appears to become more positive by 0.22 mV per degree change of temperature[8]. A hydrogen electrode maintained in a solution of pH 7, on the other hand, appears to become more negative by 0.37 mV deg^{-1} when its temperature is increased in a like manner[9].

Some nonisothermal temperature coefficients for single electrodes obtained in this way are listed in Table 1.2. Although the values given lack thermodynamic rigor, they are of use in estimating the errors to be encountered when unavoidable thermal gradients exist in the pH cell.

Table 1.2   Nonisothermal Temperature Coefficients for the Potentials of Single Electrodes

Electrode	$\dfrac{dE}{dt}$ (mV deg^{-1})
Hg;Hg$_2$Cl$_2$, KCl (satd)	+0.22
Hg;Hg$_2$Cl$_2$, KCl (0.1 M)	+0.80
Ag;AgCl, HCl (0.1 M)	+0.55
Pt;H$_2$, H$^+$ (pH 7)	−0.37
Pt;H$_2$, H$^+$ (a = 1)	+0.92

1.4   TEMPERATURE VARIATION OF $E^{0\prime} + E_j$

The quantity $E^{0\prime} + E_j$ is often regarded as the standard potential of the external reference electrode. It is in actuality the standard emf of the cell

$$\text{Pt;H}_2(\text{g}), \text{H}^+(\text{a} = 1)|\text{Soln } B, \text{RX};\text{R}, \tag{1.5}$$

including the liquid-junction potential. The latter is, of course, a function of the composition of the "unknown" solution in the cell (X in Fig. 1.1). The most practical means of determining $E^{0\prime} + E_j$ is through measurement of the emf E of cell 1.5 with a series of standard buffer solutions. If the pH of these solutions lies between 2 and 12 and the solutions contain only simple ions in concentrations less than $0.2\,M$, good constancy of E_j is found. This value of $E^{0\prime} + E_j$ consequently relates specifically to solutions which are similarly constituted.

Values of $E^{0\prime} + E_j$ for four types of reference electrode from 10 to 40°C are listed in Table 1.3. The value of $d(E^{0\prime} + E_j)/dt$ at 25°C is given in the next to the last column, and the reference to the source of the data is entered in the last column. Unfortunately, no similar data for the thallium amalgam-thallous chloride electrode appear to be available.

The $3.5\,M$ calomel electrode, it should be noted, is nearly as effective as the saturated electrode in reducing the liquid-junction potentials to small constant values. Furthermore, the $3.5\,M$ reference electrode is

Table 1.3 Standard emf ($E^{0'} + E_j$) and Temperature Coefficient for Cells of the Type: Pt;H$_2$, H$^+$(a = 1) Soln B, RX;R

RX;R	Soln B	$E^{0'} + E_j$, in V, at (°C)							$\dfrac{d(E^{0'} + E_j)}{dt}$ at 25°C mV deg^{-1}	Ref.
		10	15	20	25	30	35	40		
Hg$_2$Cl$_2$;Hg	3.5 M KCl (at 25°C)	0.2556	0.2538	0.2520	0.2501	0.2481	0.2460	0.2439	−0.39	[9]
	Satd KCl	0.2542	0.2510	0.2478	0.2444	0.2411	0.2376	0.2341	−0.67	[9, 10]
AgCl;Ag	3.5 M KCl (at 25°C)	0.2152	0.2117	0.2082	0.2046	0.2009	0.1971	0.1933	−0.73	*
	Satd KCl	0.2138	0.2089	0.2040	0.1989	0.1939	0.1887	0.1835	−1.01	*

*Calculated from the data for the calomel electrode and the emf of the cell Ag;AgCl, Cl$^-$, Hg$_2$Cl$_2$;Hg, as given by Pouradier and Chateau[11].

free from the disturbances caused by crystallization of potassium chloride and interruption of the free flow of the bridge solution necessary for a stable junction potential. Hysteresis effects on change of temperature are smaller with the 3.5 M electrode than with the saturated electrode. For these reasons, the unsaturated electrode seems to be the better choice for routine work.

1.5   INNER CELL EMF (E_i) AND ITS TEMPERATURE COEFFICIENT

Metal-filled glass electrodes have never been widely employed, and it seems possible that their use might provoke an exaggerated and variable asymmetry potential when the temperature of the cell is changed. Some commercial electrodes employ metal-metal ion systems such as mercury immersed in mercurous perchlorate and perchloric acid. Nevertheless, by far the greatest number of glass electrodes are filled with solutions of hydrochloric acid or buffered chloride into which a silver-silver chloride electrode is immersed. This electrode is known for its high stability and reproducibility. One of its few faults stems from the enhanced solubility of silver chloride in concentrated chloride solutions. This reaction is troublesome when the silver-silver chloride electrode is employed as an external reference in contact with saturated or nearly saturated solutions of potassium chloride.

Data for the emf (in volts) of several types of inner cells are shown in Table 1.4. The emf given relates to cells of the type

$$Pt;H_2(g), Soln\ i, MX;M \qquad (1.6)$$

The values of E listed correspond, therefore, to $-E_i$ of the cell scheme given earlier (cell 1.1). The value of dE_i/dt at 25°C is given in the next to the last column, and references to the sources of the data are listed in the last column.

It is evident from an examination of Table 1.3 that the temperature coefficient of $E^{0\prime} + E_j$ at 25°C is negative for all four of the reference electrodes for which data were available. Inasmuch as the emf E given in Table 1.4 is $-E_i$, it is also evident that dE/dt should be negative for the inner cell to compensate the temperature effect on the reference electrode. It appears from the data given in the table that this result cannot be achieved with chloride inner solutions buffered with weak acids (uncharged or negatively charged) and their salts. Negative values of dE/dt appear in Table 1.4 only for solutions of strong acids and for buffers composed of positively charged weak acids and their neutral

Table 1.4   Emf (E) and its Temperature Coefficient for Inner Cells of the Type: Pt;H$_2$, Soln i, MX;M

Soln i (molality)	E, in V, at (°C)							$\dfrac{dE}{dt}$ at 25°C mV deg^{-1}	Ref.
	10	15	20	25	30	35	40		
	MX;M = AgCl:Ag								
HCl (0.1)	0.3544	0.3539	0.3532	0.3523	0.3513	0.3503	0.3489	−0.18	[12]
HCl (0.2)	0.3223	0.3212	0.3200	0.3187	0.3172	0.3156	0.3139	−0.28	[13]
HCl (0.5)	0.2780	0.2761	0.2743	0.2723	0.2702	0.2679	0.2655	−0.41	[13]
HCl (1.0)	0.2404	0.2381	0.2358	0.2333	0.2307	0.2279	0.2250	−0.51	[13]
HCl (2.0)	0.1951	0.1922	0.1893	0.1863	0.1832	0.1800	0.1765	−0.61	[13]
HCl (3.0)	0.1613	0.1582	0.1551	0.1518	0.1485	0.1451	0.1415	−0.65	[13]
HCl (4.0)	0.1321	0.1289	0.1255	0.1221	0.1187	0.1151	0·1114	−0.68	[13]
HCl (0.1), KCl (3.5)	0.3170	0.3161	0.3151	0.3141	0.3130	0.3118	0.3104	−0.21	[14]
Acetic acid (0.04922), NaAc (0.04737), NaCl (0.05042)	0.5714	0.5743	0.5771	0.5799	0.5827	0.5854	0.5882	+0.56	[15]
HAc (0.09056), NaAc (0.08716), NaCl (0.09276)	0.5568	0.5593	0.5619	0.5644	0.5669	0.5693	0.5717	+0.50	[15]
KH$_2$Citrate (0.1), KCl (0.01)	0.5611	0.5632	0.5651	0.5671	0.5690	0.5710	0.5728	+0.39	[16]
NaHSuccinate (0.1), NaCl (0.1)	0.5590	0.5612	0.5635	0.5656	0.5678	0.5701	0.5723	+0.44	[17]

Table 1.4 (Continued)

| Soln i (molality) | \multicolumn{7}{c}{E, in V, at (°C)} | $\dfrac{dE}{dt}$ at 25°C mV deg^{-1} | Ref. |
	10	15	20	25	30	35	40		
	\multicolumn{7}{c}{MX;M = AgCl;Ag}		
KH$_2$PO$_4$ (0.025), Na$_2$HP$_4$O (0.025), KCl (0.015)	0.7281	0.7326	0.7372	0.7421	0.7464	0.7514	0.7557	+0.94	[18]
NaHCO$_3$ (0.03254), Na$_2$CO$_3$ (0.03241), NaCl (0.03241)	0.8897	0.8952	0.9007	0.9062	0.9115	0.9173	0.9229	+1.10	[19]
NH$_4$Cl (0.1077), NH$_3$ (0.1077)	0.8457	0.8442	0.8426	0.8409	0.8390	0.8370	0.8348	−0.36	[20]
NH$_4$Cl (0.03017), NH$_3$ (0.03017)	0.8714	0.8704	0.8693	0.8680	0.8665	0.8649	0.8631	−0.28	[20]
Ethanolammonium chloride (0.04746), Ethanolamine (0.04717)	0.8760	0.8753	0.8745	0.8735	0.8723	0.8712	0.8679	−0.22	[21]
t-Butylammonium chloride (0.05037), t-Butylamine (0.02516)	0.7750	0.7743	0.7735	0.7726	0.7714	0.7701	—	−0.22	[22]

4-Aminopyridinium chloride (0.1007), 4-Aminopyridine (0.09914)	0.8378	0.8370	0.8361	0.8350	0.8338	0.8326	0.8311	−0.22	[23]
Tris(hydroxymethyl)-aminomethane·HCl (0.1015), THAM (0.1029)	0.7793	0.7773	0.7753	0.7734	0.7710	0.7688	0.7665	−0.42	[24]
Tris(hydroxymethyl)amino-methane·HCl (0.06046), THAM 0.06130	0.7896	0.7878	0.7859	0.7842	0.7820	0.7799	0.7778	−0.38	[24]
2-Ammonium-2-methylpropanediol chloride (0.09558), Ampd (0.09561)	0.8221	0.8206	0.8189	0.8171	0.8152	0.8132	0.8112	−0.37	[25]

$MX;M = AgBr;Ag$

HBr (0.1)	0.2003	0.2005	0.2006	0.2005	0.2002	0.1998	0.1993	−0.03	[26]
HBr (0.5)	0.1125	0.1215	0.1203	0.1190	0.1175	0.1158	0.1141	−0.28	[26]
HBr (1.0)	0.0834	0.0818	0.0800	0.0781	0.0761	0.0740	0.0717	−0.39	[26]
HBr (2.0)	0.0338	0.0314	0.0288	0.0263	0.0236	0.0208	0.0177	−0.52	[27]

$MX;M = Hg_2Cl_2;Hg$

HCl (0.01991)	0.4696	0.4720	0.4741	0.4762	0.4781	0.4798	0.4813	+0.41	[28]
KH₂PO₄ (0.02650), Na₂HPO₄ (0.01291), NaCl (0.04305)	0.7252	0.7307	0.7362	0.7418	0.7473	0.7529	0.7584	+1.11	[29]

conjugate bases; for example, mixtures of neutral ammonia bases and their hydrochloride salts. Furthermore, large negative values of dE/dt are not common. For this reason, it is easier to compensate successfully the temperature variation of the calomel electrode than that of the silver-silver chloride electrode by choice of a suitable inner cell and solution. Likewise, the Ag;AgBr inner electrode seems to have no advantages over the Ag;AgCl inner electrode.

Some of the data given in Tables 1.3 and 1.4 are plotted in Fig. 1.2. The value of the emf and of $E^{0'} + E_j$ have been normalized to 25°C. The solid lines represent the variation of $E^{0'} + E_j$ for three reference electrodes with temperature in the range from 10 to 40°C. The dashed lines

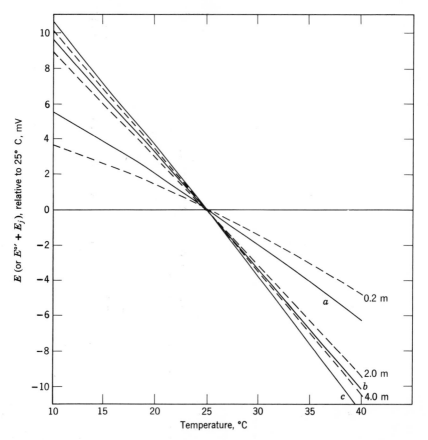

Fig. 1.2. Temperature effects on cell potentials (normalized to 25°C). Solid lines, $E^{0'} + E_j$ for reference electrodes; (*a*) 3.5 *M* calomel, (*b*) saturated calomel, (*c*) 3.5 *M* Ag;AgCl. Dashed lines, E for cells H_2;HCl(*m*), AgCl;Ag.

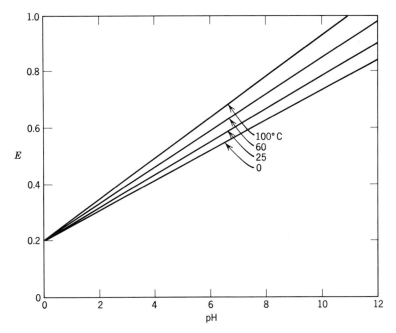

Fig. 1.6. Variation of emf (E) with pH when standard emf is temperature-invariant.

regardless of the temperature. The magnitude of the slope of each line is, of course, the value of k, that is $(RT \ln 10)/F$, at that particular temperature.

A pH cell with these characteristics can be standardized at any convenient temperature and used for measurements of pH at another temperature without appreciable loss of accuracy. It should be pointed out, however, that the successful application of this procedure depends on (1) absence of wide variations in the liquid-junction potential as the temperature is changed, and (2) absence of appreciable differences in the IR drop through the cell at the two temperatures. It is not unreasonable to expect that the first of these conditions will be fulfilled, for the mobilities of the ions at the liquid junction are, in general, similarly affected by temperature changes. Furthermore, the second condition should cause little concern when modern pH electrometers are employed for the measurement. Even though the resistance of the glass electrode is notoriously temperature-sensitive, the currents drawn by these modern pH instruments are so small that temperature-induced differences in the IR drop are unlikely to cause difficulty.

1.7 SYMMETRICAL pH CELLS

Symmetrical pH cells — in which the same type of electrode is found within the glass bulb and in the external reference half-cell, and in which solution B and solution i (see Fig. 1.1) are substantially the same in composition — possess certain special advantages. Consider, for example, the cell

$$\text{Hg;Hg}_2\text{Cl}_2, 3.5\,M\,\text{KCl, buffer}\|\text{Soln } X|3.5\,M\,\text{KCl, Hg}_2\text{Cl}_2;\text{Hg.} \quad (1.8)$$

The potentials of the two calomel electrodes are nearly identical. The only difference between them arises from the fact that the activity of chloride $(a_{\text{Cl}})_i$ in the inner (left) solution is slightly different from that $(a_{\text{Cl}})_0$ in the vicinity of the external reference electrode (right). This difference is a result of the addition of buffer substances to the 3.5 M potassium chloride solution within the glass electrode. The total emf of the cell can be written as follows:

$$E = \left[E_j - k \log \frac{(a_{\text{Cl}})_0}{(a_{\text{Cl}})_i} - k\text{pH}_i \right] + k\text{pH}_X \quad (1.9a)$$

$$= E_c^0 + k\text{pH}_X, \quad (1.9b)$$

where E_c^0 is written for the expression enclosed in square brackets. As before, E_c^0 may be regarded as the standard emf of the pH cell.

When the temperature of the cell changes under isothermal conditions, E_c^0 does not normally remain constant — although it would not be surprising if E_j and the ratio of chloride ion activities in the outer and inner solutions remained practically unchanged. However, alteration of the standard emf would be expected to occur because k and pH_i are both functions of the temperature. It may not be impossible to find buffer solutions whose pH is almost independent of temperature over a wide range [30]. Even under these circumstances, the quantity $k\text{pH}_i$ varies in a linear fashion with temperature because of the variation of k.

If it is desired to nullify completely the change of E_c^0 with temperature for a cell of the type shown in (1.8), the condition

$$\frac{d(k\text{pH}_i)}{dT} = 0, \quad (1.10)$$

where T is the temperature in $^\circ$K, must be met. Inasmuch as

$$\frac{dk}{dT} = \frac{R \ln 10}{F} = \frac{k}{T}, \quad (1.11)$$

$d\text{pH}_i/dT$ would have to be equal to $-\text{pH}_i/T$.

Once this condition is identified, it is possible to select those buffer systems most likely to be suitable. One way to achieve this aim is to examine data for the heats of ionization ($\Delta H°$) of weak acids available in the literature. As a first approximation

$$\frac{d\mathrm{pH}_i}{dT} \approx \frac{d\mathrm{pK}}{dT}, \tag{1.12}$$

where K is the acidic dissociation (ionization) constant of the buffer acid. This relationship can be expected to be most nearly fulfilled in dilute solutions and to hold only approximately in concentrated solutions of potassium chloride where the changes in the activity coefficients of the buffer species (and the corresponding relative partial molal enthalpies) may make an appreciable contribution[31].

In terms of the van't Hoff equation,

$$\frac{d\mathrm{pH}_i}{dT} \approx \frac{-\Delta H°}{RT^2 \ln 10}, \tag{1.13}$$

one then finds that $d\mathrm{pH}_i/dT$ should be given approximately by

$$\frac{d\mathrm{pH}_i}{dT} \approx -(2.46 \times 10^{-6})\Delta H° \tag{1.14}$$

when the temperature is 25°C and $\Delta H°$ is expressed in cal mole^{-1} (1 cal = 4.184 J).

Thus, if $d\mathrm{pH}_i/dT$ is to equal $-\mathrm{pH}_i/T$, it may be concluded that $\Delta H°$ should be about 1360 pH_i if the buffer solution is to be effective in nullifying changes in the standard emf $E_c{}^0$ of this symmetrical pH cell.

It is well known that plots of the pK of weak acids as a function of temperature are very close to parabolic in form; the pK passes through a minimum value (pK_m) at some temperature T_m. Moreover, when pK–pK_m is plotted as a function of T–T_m, the data for all weak acids fall very nearly on a single curve — such as that shown in Fig. 1.7 — regardless of the charge type of the weak acid. When the pK decreases with rising temperature (left branch of the curve), the heat of ionization $\Delta H°$ is positive; when pK rises with increasing temperature (right branch), $\Delta H°$ is negative.

In general, the temperature of the minimum for uncharged and negatively charged weak acids lies in the range 10–40°C or below. Hence the heat of ionization is either negative or has a small positive value for acids of such charge types. As a rule, only positively charged acids have pK

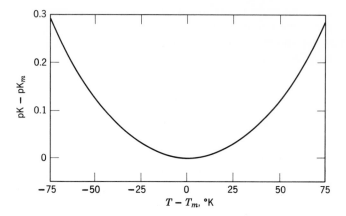

Fig. 1.7. Plot of pK–pK_m(*ordinate*) for weak acids as a function of T–T_m(*abscissa*).

values that fall on the left branch of the curve between 10 and 40°C and hence, show promise of fulfilling the condition set forth in 1.14.

The ratios $\Delta H°/pK$ for 11 protonated weak bases are summarized in Table 1.7. It is evident that tris(hydroxymethyl)aminomethane (THAM)

Table 1.7 $\Delta H°/pK$ for Weak Protonated Bases

Acidic species	$\Delta H°/pK$ (in cal mole^{-1})
Ammonium	1349
Methylammonium	1232
Dimethylammonium	1101
Trimethylammonium	899
Ethylammonium	1277
Ethanolammonium	1272
Triethanolammonium	1030
n-Butylammonium	1305
Piperidinium	1147
4-Aminopyridinium	1235
Protonated tris(hydroxymethyl)aminomethane (THAM·H$^+$)	1409

buffers again give promise in nullifying the temperature coefficient of the symmetrical pH cell, as do buffer solutions composed of ammonia and ammonium salts.

1.8   THE ISOPOTENTIAL pH

If the liquid-junction potential and the ratio of chloride ion activities are indeed unchanged upon a change of temperature, it is possible to write $(1.9a)$ in another way:

$$E = C + k(\text{pH}_X - \text{pH}_i). \tag{1.9c}$$

It has been found that an equation of this form may be valid even for some pH cells other than those of the symmetrical type. In such cases, the difference of standard potentials of the two dissimilar electrodes will be reflected in the value of the constant C. Furthermore, it is to be expected that temperature changes may have a larger effect on the activity (a_{Cl}) of chloride ion in the concentrated potassium chloride of the reference electrode than on $(a_{Cl})_i$ in the usually more dilute inner solution. Nevertheless, when $(1.9c)$ is followed, the straight lines representing the change of emf with pH_X at each temperature will intersect at a point (*Isothermenschnittpunkt*) the coordinates of which are $E = C$ and $\text{pH}_X = \text{pH}_i$. Under these conditions, the lines representing the change of cell emf as a function of pH at several different temperatures are disposed somewhat as shown in Fig. 1.8.

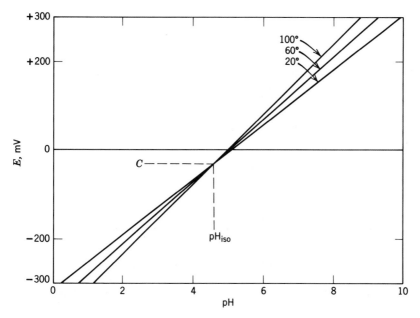

Fig. 1.8. Variation of emf (E) with pH for pH cells displaying isopotential pH.

As is quite evident, this intersection will be sharply defined only when pH_i does not vary with temperature. Although the variation of the pH of buffer solutions with temperature usually follows closely the change in pK for the buffer acid or base (see Fig. 1.7), it may be possible to compensate a large part of the pK change by the use of concentrated salt media. Fricke[30] has shown, for example, that the pH of a solution of acetic acid ($1\,M$) and sodium acetate ($1\,M$) in saturated potassium chloride changes but little over the temperature range 20–100°C. The minimum in the plot of the pK of acetic acid with respect to temperature occurs at 23°C, and the effect of the high concentration of salt appears to be that of considerably flattening the pH versus T curve in the vicinity of the minimum.

For many practical pH cells, the point of intersection of the emf/pH lines is not well defined. Nonetheless, it has been found possible, by empirical modification of the structure of the inner cell and the composition of the inner solution, to design pH cells with emf/pH relationships that obey an equation of the form of (1.9c) over considerable ranges of temperature. The intersection occurs at a pH characteristic of each type of assembly. This pH value has been designated the "isopotential pH"[32] or pH_{iso}. For these cells,

$$E = C + k(pH_X - pH_{iso}).$$

The relationship between E and pH under these circumstances is shown in Fig. 1.8. A particularly advantageous situation from the standpoint of the design of pH instrumentation occurs when the intersection of the curves is located sharply at $pH_{iso} = 7$ and $C = 0$. The uses of the isopotential pH have been elaborated by Mattock in his monograph[33]. The reader is also referred to two valuable contributions by A. E. Mans[34,35].

As may be seen in Tables 1.5 and 1.6 and in the figures presented in earlier sections of this chapter, the proper choice of inner solution can lead to very effective compensation of the temperature variation of the standard potential of a pH cell. However, it is also evident that the pH of zero emf obtained with those inner solutions most suitable for temperature compensation is not always in the most favorable range for the design of measuring equipment. In this respect the isopotential approach is superior to attempts to nullify completely the standard potential of the total cell, and under favorable conditions the same result can be achieved. In the usual case, however, it seems likely that the most effective compensation of temperature effects is accomplished by the approach outlined in earlier sections. Such would not be the case, however, if inner buffer solutions whose pH is truly invariant with temperature could be prepared.

1.9   INNER CELLS PROVIDING ISOPOTENTIAL pH OF 7

In an interesting and valuable study Wegmann and Simon[36] turned their attention to the selection of inner solutions which, used with the silver-silver chloride internal electrode and saturated external calomel or silver-silver chloride reference electrodes, form pH cells having an emf of 0 at an isopotential pH of 7.0. Their approach was empirical. The emf of the total cell and the change of emf with temperature were varied over wide ranges by altering the composition of the inner solution, and the most favorable compositions were selected. Inner solutions were prepared from a variety of organic acids and organic bases dissolved, together with hydrochloric acid, in aqueous glycerine. For each inner solution, the emf of the glass electrode pH cell was measured in standard phthalate buffer of known pH at 25 and 80°C. From the accumulated data, the following inner solutions were found to provide a cell with an emf near 0 at an iso-potential pH near 7.0, when the glass electrode was combined with the saturated calomel reference electrode:

soln 1	soln 2
$0.50\,M$ phenylphosphonic acid	$0.50\,M$ phosphoric acid
$0.82\,M$ morpholine	$0.623\,M$ diethanolamine
$0.10\,M$ hydrochloric acid	$0.188\,M$ triethanolamine
in 84% aqueous glycerine	$0.06\,M$ hydrochloric acid
	in 84% aqueous glycerine

Other solutions were selected for use with the saturated silver-silver chloride reference electrode.

REFERENCES

[1] R. N. Khuri, in *Glass Electrodes for Hydrogen and Other Cations* (G. Eisenman, ed.), Dekker, New York, 1967.
[2] M. Lavallée, *Circulation Res.*, **15**, 185 (1964).
[3] M. R. Thompson, *Bur. Std. J. Res.*, **9**, 833 (1932).
[4] R. G. Bates, *Chem. Rev.*, **42**, 1 (1948).
[5] M. Dole, *The Glass Electrode*, Wiley, New York, 1941.
[6] M. H. Lietzke, and J. V. Vaughen, *J. Am. Chem. Soc.*, **77**, 876 (1955).
[7] E. A. Guggenheim, *J. Am. Chem. Soc.*, **52**, 1315 (1930).
[8] N. Bjerrum, and A. Unmack, *Kgl, Danske Videnskab. Selskab, Mat. Fys. Skrifter*, **9**, 1 (1929).
[9] R. G. Bates, *Determination of pH*, Wiley, New York, 1964.
[10] G. J. Hills, and D. J. G. Ives, in *Reference Electrodes* (D. J. G. Ives and G. J. Janz, eds.) Academic, New York, 1961.

[11] J. Pouradier, and H. Chateau, *Compt. Rend. Acad. Sci.,* Paris, **237**, 711 (1953).
[12] R. G. Bates, and V. E. Bower, *J. Res. Nat. Bur. Std.,* **53**, 83 (1954).
[13] H. S. Harned, and R. W. Ehlers, *J. Am. Chem. Soc.,* **55**, 2179 (1933).
[14] H. S. Harned, and W. J. Hamer, *J. Am. Chem. Soc.,* **55**, 2194 (1933).
[15] H. S. Harned, and R. W. Ehlers, *J. Am. Chem. Soc.,* **54**, 1350 (1932).
[16] R. G. Bates, and G. D. Pinching, *J. Am. Chem. Soc.,* **71**, 1274 (1949).
[17] G. D. Pinching, and R. G. Bates, *J. Res. Nat. Bur. Std.,* **45**, 322 (1950).
[18] V. E. Bower, unpublished measurements made at the National Bureau of Standards.
[19] H. S. Harned, and S. R. Scholes, Jr., *J. Am. Chem. Soc.,* **63**, 1706 (1941).
[20] R. G. Bates, and G. D. Pinching, *J. Res. Nat. Bur. Std.,* **42**, 419 (1949)
[21] R. G. Bates, and G. D. Pinching, *J. Res. Nat. Bur. Std.,* **46**, 349 (1951).
[22] H. B. Hetzer, R. A. Robinson, and R. G. Bates, *J. Phys. Chem.* **66**, 2696 (1962).
[23] R. G. Bates, and H. B. Hetzer, *J. Res. Nat. Bur. Std.,* **64A**, 427 (1960).
[24] R. G. Bates, and H. B. Hetzer, *J. Phys. Chem.,* **65**, 667 (1961).
[25] H. B. Hetzer, and R. G. Bates, *J. Phys. Chem.,* **66** 308 (1962).
[26] H. S. Harned, A. S. Keston, and J. G. Donelson, *J. Am. Chem. Soc.,* **58**, 989 (1936).
[27] A. S. Keston, *thesis,* Yale University (1935).
[28] A. K. Grzybowski, *J. Phys. Chem.,* **62**, 550 (1958).
[29] A. K. Grzybowski, *J. Phys. Chem.,* **62**, 555 (1958).
[30] H. K. Fricke, in *Beitrage zur Angewandten Glasforschung,* (E. Schott, ed.), Wissenschaftliche Verlagsgesellschaft, Stuttgart, 1960.
[31] R. G. Bates, *Determination of pH*, Chapter 5, Wiley, New York, 1963.
[32] J. Jackson, *Chem. Ind. (London)*, 7 (1948).
[33] G. Mattock, *pH Measurement and Titration*, Macmillan., New York, 1961.
[34] A. E. Mans, *Chem. Weekblad*, **52**, 873 (1956).
[35] A. E. Mans, *Chem. Weekblad*, **54**, 418 (1958).
[36] D. Wegmann, and W. Simon, *Helv. Chim. Acta*, **47**, 1181 (1964).

CHAPTER 2

Properties of Microelectrode Glasses

NORMAND C. HEBERT

Research and Development Laboratories, Corning Glass Works

2.1 INTRODUCTION

A wide variety of glass compositions is presently being used in the fabrication of glass microelectrodes. These glasses differ in their physical, electrical, chemical, and electrochemical properties. Those properties that affect the behavior of microelectrodes are discussed in this chapter.

It is useful to know the typical values of the viscosity of each glass at various temperatures, as such information indicates at what temperature to melt, lampwork, seal, and anneal the specific glass. The changes in viscosity as a function of temperature for four microelectrode glasses are illustrated in Fig. 2.1. The NAS_{11-18} glass[1] is a sodium ion selective electrode glass. The NAS_{11-18} designation stands for the composition of the glass, that is, 11 mole % Na_2O, 18 mole % Al_2O_3, and the remainder silica. Code 7740 glass[2] is used in the fabrication of micropipette reference electrodes. The $NABS_{20-5-9}$ glass[3] is a cation-sensitive glass having a moderate selectivity for sodium ions in the presence of potassium ions. The $NCaS_{22-6}$ (Code 0150) glass is pH sensitive.

The ideal temperature for melting glass is that at which the viscosity of the melt is about 10^2 poises (P)[2]. Figure 2.1 indicates that a temperature of about 1800°C is desired for melting the NAS_{11-18} glass, whereas the $NCaS_{22-6}$ glass can be melted at approximately 1300°C.

The working temperature of glass is that temperature which yields a viscosity in the range of 10^4 to 10^5 P. It is this approximate temperature range which is needed for pulling microelectrodes from tubing. The four glasses described in Fig. 2.1 have fairly wide ranges of workability, but

I am indebted to Miss M. T. Splann for the electrical resistivity data, to Mr. P. B. Adams for the chemical durability data, and to Mr. E. H. Fontana for physical property measurements.

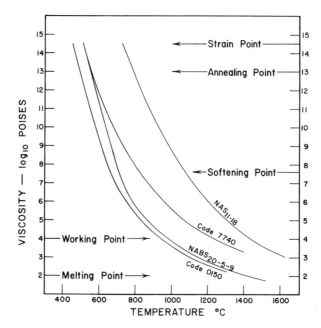

Fig. 2.1. Plot of viscosity (log Poises) versus temperature for four electrode glasses.

they differ appreciably in their actual working temperatures. The NAS_{11-18} glass has a working temperature of approximately 1450°C compared with 940°C for the $NCaS_{22-6}$ glass.

The softening point corresponds to a viscosity of $10^{7.6}P$[2]. At this temperature the glass deforms rapidly and starts to adhere to other bodies. In sealing two glasses together, it is important that they have similar thermal expansion coefficients and helpful if the softening points are in the same range.

The annealing point is the temperature at the upper end of the annealing range at which the internal stress is substantially relieved in a matter of minutes[2]. The annealing point corresponds to a viscosity of 10^{13} P.

The strain point is the temperature at the lower end of the annealing range at which the internal stress is substantially relieved in a matter of hours[2]. This point corresponds to a viscosity of $10^{14.5}$ P.

A characteristic that distinguishes glass from normal solids and liquids is its structure. Figure 2.2 demonstrates a plane representation of the structure of crystalline quartz and of fused silica[4]. Both materials have the same chemical composition (SiO_2), but they differ in that quartz has a regularly arranged crystalline structure whereas fused silica (being

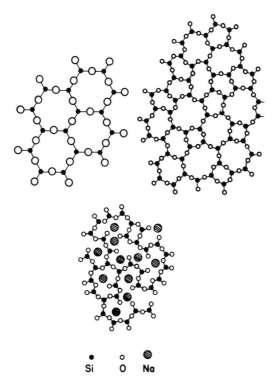

Fig. 2.2. Plane representation of structure of crystalline quartz, fused silica, and $Na_2O \cdot SiO_2$ binary glass. From G. W. Morey, *The Properties of Glass*, 2nd ed., Reinhold, New York, 1954.

a glassy material) possesses a random-type structure. In both structures the silicon is tetrahedrally coordinated with oxygen atoms.

In the formation of a typical glass, such as a sodium silicate glass composed of silica and sodium oxide, silicon-oxygen bonds are broken by ionic bonding of sodium ions to the oxygen sites. It is this site that is responsible for electrical conduction. The sodium ions of this glass are the current-carrying ions. The conductivity of the glass can be altered in two ways, that is, either by replacing the current-carrying ions by other monovalent or divalent ions, or by altering the field strength of the SiO^- sites. The latter can be effected by introducing network formers other than silica, such as Al_2O_3 or B_2O_3, into the glass. The Na^+ of this glass is known as a network modifier. Some elements, such as Al^{3+}, Be^{2+}, and Zn^{2+}, can occupy either position; they are referred to as intermediates.

2.2 ELECTRODE GLASSES FOR pH

Table 2.1 lists properties of two pH electrode glasses. The $NCaS_{22-6}$ glass has been used by numerous microelectrode workers in arriving at pH determinations. The $NCaS_{17-12.5}$ glass was formulated specifically for open-tip pH microelectrodes[5]. Its higher volume electrical resistivity reduced stray potentials believed to occur at the wall of the glass near the open tip of the electrode. Using the $NCaS_{17-12.5}$ glass as a closed-tip microelectrode yields an electrode having a high electrical resistance because the electrical resistance of a glass electrode is a function of the geometry and the specific volume resistivity of the glass. We have found that the lowest practical electrical resistance obtainable for a laboratory glass electrode is generally two orders of magnitude lower than the specific volume resistivity of the glass. The electrical resistance of fine microelectrodes, on the other hand, is generally of the same order or higher than the specific volume resistivity of the glass. A closed-tip microelectrode made from the $NCaS_{17-12.5}$ glass would be expected to have an electrical resistance in the range $10^{10.4}$ to $10^{12.4}\,\Omega$. The high electrical resistance would make it difficult to record potentials with conventional electrometers.

Table 2.1 Microelectrode pH Glasses

Glass	Log $\rho_{25°C}$ (ohm-cm)	Soft Point (°C)	Thermal Expansion (10^{-7} in./in./°C)	Chemical Durability
$NCaS_{22-6}$	10.8	655	110	0.42
$NCaS_{17-12.5}$	12.4	690	104	0.070

The chemical durabilities of these two glasses differ by a factor of about 10; that is, the $NCaS_{17-12.5}$ glass is more chemically durable than the $NCaS_{22-6}$ glass. Chemical durability is reported in Table 2.1 as percent alkali, represented in terms of percent Na_2O leached from the glass in a standard test. The test performed on all microelectrode glasses involves the digestion of 0.4 cc of powdered glass (40–50 mesh size) in a tube containing 5 ml of water for 4 hr at 90°C. After digestion, the liquid is decanted and analyzed for alkali by flame photometry. This test certainly does not duplicate conditions under which microelectrodes are used; however, the data do afford a comparison of the chemical durability of microelectrode glasses.

2.3   ELECTRODE GLASSES FOR Na$^+$

Table 2.2 gives some properties of microelectrode glasses possessing selectivity for sodium ions over potassium ions. The NAS$_{11-18}$ glass has high selectivity for sodium ions, low electrical resistivity, and is chemically quite durable. However, it requires a high temperature for melting and has a high softening point ($> 970°C$); this makes it difficult to use for precision tubing and electrode fabrication.

Table 2.2   Microelectrode pNa Glasses

Glass	Log $\rho_{25°C}$ (ohm-cm)	Soft Point (°C)	Thermal Expansion (10^{-7} in./in./°C)	Chemical Durability
NaS$_{11-18}$	9.45	> 970	53	0.0023
NAS$_{17.5-10}$	10.25	792	83	0.007
NABS$_{20-5-9}$	10.96	706	86	0.076
NAS$_{20-5}$	10.0	701	96	0.258

If a high sodium ion selectivity is not essential, then glasses such as NAS$_{17.5-10}$[6], NABS$_{20-5-9}$[3] or NAS$_{20-5}$ can be used. These glasses are more amenable to precision tube fabrication. The NAS$_{17.5-10}$ has a low specific volume resistivity, good chemical durability, and workability. The NABS$_{20-5-9}$ has good workability and chemical durability. Its high electrical resistivity, however, may preclude its usefulness as a closed-tip microelectrode for sodium. In comparing the NABS$_{20-5-9}$ with the NAS$_{20-5}$, one can clearly see that the substitution of 9 mole % boric oxide for 9 mole % silica increases the electrical resistivity of the glass and renders it more chemically durable.

2.4   ELECTRODE GLASSES FOR K$^+$

Table 2.3 presents properties of glasses possessing potassium ion selectivity over sodium ions. The NAS$_{27-4}$[7] and KAS$_{20-5}$[7] electrode glasses possess optimum selectivity for potassium ions ($k_{NaK} \approx 10$). The NAS$_{27-4}$ glass has a lower electrical resistivity and is less durable in water than the KAS$_{20-5}$ glass. Hollander[8] reports that microelectrodes made from the NAS$_{27-4}$ glass were found to lose their tips when filled by boiling in KCl solution. In view of this finding his electrodes were filled with methanol under reduced pressures and subsequently equilibrated in distilled water and KCl solution.

Table 2.3   Microelectrode pK$^+$ Glasses

Glass	Log $\rho_{25°C}$ (ohm-cm)	Soft Point (°C)	Thermal Expansion (10^{-7} in./in./°C)	Chemical Durability
NAS$_{27-4}$	9.00	647	123	1.79
KAS$_{20-5}$	11.18	796	102	1.04
KABS$_{20-5-9}$	12.5	799	89.5	0.016
KABS$_{20-9-5}$	12.0	802	98	0.013

The KABS$_{20-5-9}$ [6] and KABS$_{20-9-5}$ [9] glasses are more chemically durable K$^+$-sensing glasses, but they possess high electrical resistivities. The high electrical resistivity is favorable for the construction of open-tip electrodes, but not for those having closed tips.

2.5   ELECTRODE GLASSES FOR REFERENCE

Table 2.4 lists the properties of two glasses used as reference micro-electrodes. The Code 7740 glass is a borosilicate, whereas Code 1720 glass is an aluminosilicate. Both are glasses with high electrical resistivity and have excellent chemical durability in water.

Table 2.4   Microelectrode pREF Glasses

Glass	Log $\rho_{25°C}$ (ohm-cm)	Soft Point (°C)	Thermal Expansion (10^{-7} in./in./°C)	Chemical Durability
Code 7740	15	820	33	0.0015
Code 1720	> 17	915	42	0.00023

REFERENCES

[1] G. Eisenman, D. O. Rudin and J. U. Casby, *Science*, **126**, 831 (1957).
[2] "Properties of Selected Commercial Glasses," Bulletin B-83, Corning Glass Works, Corning, N.Y.
[3] A. A. Lev, *Nature*, **201**, 1132 (1964).
[4] G. W. Morey, *The Properties of Glass*, 2nd ed., Reinhold, New York, 1954.

[5] M. Lavallée, *Circulation Res.*, **15**, 185 (1964).
[6] A. A. Belyustin, and A. A. Lev, *Chemistry in Natural Sciences*, Leningrad State University, 32, 1965.
[7] G. Eisenman, *Biophys. J.*, **2**, 259 (1962).
[8] P. Hollander, paper presented at the Intracellular Glass Microelectrodes Conference, Montreal, May 1967.
[9] A. A. Lev, *Trans. Moscow Soc. Naturalists*, **9**, 30 (1964).

CHAPTER 3

The Ion-Exchange Characteristics of the Hydrated Surface of Na$^+$ Selective Glass Electrodes

GEORGE EISENMAN

Departments of Physiology and Biophysics, University of Chicago

3.1 INTRODUCTION

When exposed to aqueous solutions, or even to humid air, all glass electrodes hydrate to some extent. In certain glass compositions (typically, those which are selective electrodes for K$^+$), the hydration goes sufficiently deep that a hydrated layer thicker than 10,000 Å comes into existence in which the ionic mobilities and presumably the ion exchange affinities also are markedly different from those characteristics of the "dry" glass in the interior[1, 2]. In others (typified by Na$^+$ selective electrodes), the depth of hydration is much less extensive, being restricted (as the data to be presented here appear to indicate) to the immediate surface. For such glasses, diffusion studies are very difficult, but surface ion exchange equilibria can be characterized in a relatively straightforward manner.

This Chapter characterizes the ion exchange of the most superficial layers of Na$^+$ selective glass electrodes, as revealed by the competition of various cations for Cs^{134} — a species found to have sufficiently low mobility so that its diffusion into deeper regions of the glass can be neglected over normal experimental times. The properties of hydrated glass described here will be shown to be relevant to the understanding of a number of characteristics of glass microelectrodes. In particular, as hydration can increase the mobility of ions within the glass surface by as much as four orders of magnitude[1,2], such a hydrated layer can constitute an important electrical pathway parallel to the axis of an open-tipped microelectrode within the hydrated surface layers of the glass.

This work was supported in part by research grant GM-14404-01 from the U.S. Public Health Service.

32

Moreover, should the hydration extend throughout the micro-electrode wall, the material may no longer be the insulator it is usually assumed to be. The ion-exchange preference of K^+ over Na^+ found for the hydrated surface of Na^+ electrodes may also explain the K^+ response observed with incompletely sealed Na^+ selective glass electrodes[3–5].

On the other hand, since the hydrated surface of a typical Na^+ electrode appears to prefer K^+ to Na^+ as an ion exchanger, this may account for the transient K^+ response observed with such electrodes[6, 7].

3.2 METHODS

All experiments were performed at an ambient temperature of $22 \pm 2°C$. The experimental methods were generally the same as those described in detail on pp. 136–139 and 147–155 of *Glass Electrodes for Hydrogen and Other Cations*[1], the principal difference being a consequence of the much smaller uptakes of radioactivity by the glasses to be described here. This condition necessitated the use of radioactive isotopes of high specific activity and also careful controls to insure that no significant radioactivity remained in adherent fluid layers not removed by wiping.

Glasses whose uptakes were to be studied were made into closed surfaces — either "dumbbells" formed by sealing the ends of tubing of the desired compositions or "sealed bulbs" made by sealing the stems of bulb electrodes; these were of a size which would fit into the bottom 1.5 cm of the scintillation well detector within the region of maximum counting efficiency. Glasses so prepared were first aged by exposure to aqueous solutions for at least one week. They were then studied by immersion in radioactive solutions of known composition for varying periods of time. When their radioactivity was to be counted, they were removed from the solution, carefully wiped dry with absorbant tissue, and placed in a lusteroid tube for counting in the well. Following the wiping procedure, the level of radioactivity remaining in any adherent solution was estimated by measuring the radioactivity present when the uptake by the glass was suppressed by high competing concentrations of nonradioactive H^+. Such measurements showed that the residual corresponded to an aqueous film of no more than 200 Å thickness. For the present glasses, this meant that solutions more dilute than $0.01 N$ contributed only a negligible increment to the radioactivity actually taken up by the glass; all measurements discussed here will be restricted to radioactive solutions of no greater concentration. In addition, most experiments were performed with the concentration of radioactive species constant and varying the concentrations of the competing, nonradioactive species in the manner described in Fig. 3.3.

3.2.1 Preparation of Solutions

The preparation of K^{42}- and Na^{24}-labeled solutions followed procedures described elsewhere[1]. The solutions labeled with Cs^{134} were prepared from several aliquots of Cs^{134} (p) supplied by the Union Carbide Nuclear Co. (from the Oak Ridge National Laboratory) and specified to have greater than 98% radiochemical purity. Most experiments reported here were based on the use of solutions prepared from a 0.6-ml aliquot of CsCl containing 1.08 mg of Cs^{134} (p) in 0.019 N HCl solution and having a specific activity of 19.839 mc/g. The entire 0.6 ml was removed from the shipping vial with 100 ml of deionized distilled water. To the resulting solution, 10 ml of 0.01 N Cs_2CO_3 and 10 ml of 0.1 N CsCl (Penn-Rare Metals; purity greater than 99.9%) were added to make 120 ml of a mildly alkaline stock solution known to contain 0.0092 equivalents of Cs per liter, labeled with Cs^{134} to have a specific activity of 1.93×10^4 mc/mole.

To prepare solutions of varying Cs concentration at a constant pH of 7, an aliquot of the stock solution was serially diluted with 0.005 N tris (hydroxylmethyl) aminomethane buffer. Mixtures with various cations were prepared similarly. Figure 3.1 illustrates the accuracy of the dilutions and serves as a control that no Cs^{134} was inadvertently lost into the glassware in preparing the solutions. In this figure the measured radioactivity of 1 ml aliquots of a 1:100 dilution of the various solutions is plotted as a function of the calculated concentrations. From Fig. 3.1 the specific radioactivity can be calculated as 1.455×10^{13} counts/min/mole of labeled Cs. Although glassware was used in preparing them, the solutions were routinely stored in polyethylene containers to minimize losses by ion exchange. The pH of all solutions was measured before, during, and after the experiments with pHydrion paper and was always within ± 0.5 pH of the expected pH of 7.0 (unless specifically noted otherwise).

3.2.2 Counting Procedures

Radioactivity was measured by counting emission for a preset time interval (usually 1 min) with a γ-sensitive Tl-activated NaI crystal scintillation well detector in conjunction with a decade scaler. For Cs^{134}, 1400 V corresponded to the setting at which the count rate was found to be least dependent on voltage. The system was found to be completely linear for all the count rates measured.

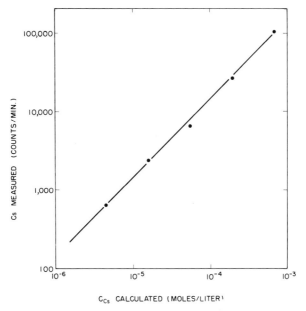

Fig. 3.1. Radioactivity in serial dilutions of Cs^{134}-labeled CsCl solutions. Cs^{134} activity in counts per minutes per 0.01 ml of solution plotted on ordinate as function of calculated Cs concentration in solution. Agreement with the line whose slope is 1 indicates accuracy of solution preparation.

3.3 RESULTS

3.3.1 The Time Course of Radioactive Tracer Uptake for Na^{24} and K^{42}

In contrast to the Na^{24} or K^{42} uptakes by typical K^+ electrodes — which are linear with the square root of time over several days[1, 2] — the time course for typical Na^+ electrode glasses is more complicated, as is illustrated in Fig. 3.2 for three typical Na^+ electrode compositions. Whereas the linearity of uptake for the typical K^+ electrode in A and E indicates that the diffusion coefficient is independent of depth, the nonlinearities for the Na^+ selective compositions signify that the diffusion coefficient varies with depth, decreasing markedly with increasing time (i.e., with increasing depth of penetration). However, even such complications as inflections in the time course (see F) are not uncommon.

These data indicate that both Na^{24} and K^{42} are taken up by Na^+ selective glasses, although at a considerably lower rate than by K^+ selective glasses. (Some idea of the relative rates of uptake can be obtained by comparing A–D, as all the solutions had the same specific activity and the surface

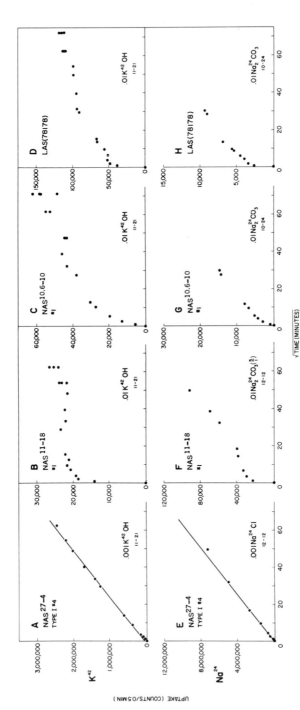

Fig. 3.2. Uptake of K^{42} and Na^{24} as a function of time. Data for typical K^+ selective glass electrode (NAS_{27-4}, type I, electrode no. 4) are presented in (a) and (e) for comparison with three typical Na^+ selective electrode compositions (NAS_{11-18}, $NAS_{10.6-10}$, and $LAS_{26.2-12.4}$—Beckman type 78178). Uptakes from indicated solutions were measured for glasses previously pre-equilibrated with nonradioactive K_2CO_3 or N_2CO_3. Surface areas and ion-exchange capacities (calculated in the manner indicated on page 170 of Ref. [1]. are as follows: NAS_{27-4} (Area 0.671 cm², C 3.2 × 10⁻³ moles cm⁻³), NAS_{11-18} (Area 1.7 cm², C 8.11 × 10⁻³ moles cm⁻³), $NAS_{10.6-10}$ (Area 3 cm², C 7.75 × 10⁻³ moles cm⁻³). $LAS_{26.2-12.4}$ (Area 3.42 cm², C 10.5 × 10⁻³ moles cm⁻³). From these and uptake data at earliest times, the following minimum estimates of self-diffusion coefficients of K^+ in the surface can be made for the four compositions, respectively: 1.1 × 10⁻¹¹, 3.6 × 10⁻¹⁴, 1.89 × 10⁻¹⁵, and 4.47 × 10⁻¹⁴ cm² sec⁻¹. These values are three to five orders of magnitude faster than those characteristic of later slopes. In the case of Na^{24} no estimates of self-diffusion coefficient are possible from these data, but the initial uptakes indicate surface diffusion coefficients some two to three orders of magnitude larger than in deeper layers.

36

areas are roughly comparable.) The quantitative interpretation of such data for the Na^+ selective glasses will not be attempted here; it has, however, been successfully accomplished for K^+ electrodes[1, 2]. It does seem reasonable to conclude from Fig. 3.2 that the initial rapid uptake of Na^{24} and K^{42} by Na^+ selective glasses corresponds to diffusion in superficial hydrated layers, whereas the slower uptakes at later times represent diffusion in the unhydrated interior of the glass. The reason for not trying to go further in the analysis of Na^{24} or K^{42} uptakes by Na^+ electrodes at present is because the effect of H^+ on these uptakes has not yet been eliminated, despite the use of high-pH solutions. As can be seen from the data in Table 3.1, which indicate that Cs is preferred 9.3 times as strongly as K^+ and 66 times as strongly as Na^+, the high H^+ affinity relative to Na^+ and K^+ makes H^+ concentrations important with these cations while its effects are entirely negligible in relation to Cs^+.

We can, however, examine the direct measurement of the ion exchange of the glass surface by using Cs^{134} — an ion which has the dual advantage of being one of the best competitors for H^+ among the alkali metal cations and also is the species which appears to penetrate most slowly into the present glasses. It is therefore the one for which it is easiest to separate the surface ion exchange from uptakes by deeper layers.

3.3.2 Studies of Surface Ion Exchange Using Cs^{134}

Figure 3.3 illustrates the reversible uptake of Cs^{134} by the typical Na^+ electrode glass, NAS_{11-18}, from solutions having a constant CsCl concentration of 0.00091 N in the presence of 0.005 N tris buffer and the indicated concentrations of KCl. The upper portion illustrates the time course as a function of solution conditions, while the lower portion plots Cs/K in the glass as a function of Cs/K in solution. (The concentration of K^+ in the glass was calculated as the difference between the maximum uptake of Cs^{134} in the absence of any competing K^+ and the uptakes observed in the presence of the indicated competing concentrations of K^+.) Notice that K^+ is capable of displacing essentially all of the Cs^{134}, a finding which serves as a control that there are no errors caused by residual radioactivity in any unremoved surface fluid.

For the ion exchange between cations I^+ and J^+:

$$J^+_{\text{aqueous}} + I^+ \text{ glass} \rightleftharpoons I^+_{\text{aqueous}} + J^+ \text{ glass}, \qquad (3.1)$$

the ideal law of mass action is:

$$K = \frac{a_i}{a_j} \cdot \frac{C_j(\text{glass})}{C_i(\text{glass})}, \qquad (3.2)$$

where a_i and a_j are the activities of ions I^+ and J^+ in solution and C_i and

C_j are their concentrations in the glass. Equation 3.2 can be written in logarithmic form as:

$$\log \frac{C_j(\text{glass})}{C_i(\text{glass})} = \log \frac{a_j}{a_i} + \log K, \qquad (3.2a)$$

from which it can be seen that if one plots the logarithm of C_j/C_i in glass as a function of the logarithm of a_j/a_i in solution (as has been done in Fig. 3.3), the data should fall on a straight line of slope 1. Such a theoretical line has been drawn in Fig. 3.3, indicating that the data presented are adequately represented by an ideal law of mass action for an ion exchanger having 7.7 times the affinity for Cs$^+$ as for K$^+$.

Not all ion exchange equilibria obey this simple form of mass law. Deviations from the ideal are common in the ion exchange of aluminosilicate minerals, but can be adequately described by the empirical equation:

$$K = \frac{a_i}{a_j} \left[\frac{C_j(\text{glass})}{C_i(\text{glass})} \right]^n, \qquad (3.3)$$

in which n is an empirical constant characteristic of the "n-type" nonideal behavior[8–12]. Equation 3.3 can also be written in logarithmic form as:

$$\log \frac{C_j(\text{glass})}{C_i(\text{glass})} = \frac{1}{n} \log \frac{a_j}{a_i} + \frac{1}{n} \log K, \qquad (3.3a)$$

from which it is clear that on a plot of $\log C_j/C_i$ in glass versus $\log a_j/a_i$ in solution, a straight line is also expected—but with a slope of $1/n$. In both (3.2a and 3.3a)], K is given by the ratio a_i/a_j, at which C_j glass $= C_j$ glass. Typical data for aluminosilicate ion exchangers plotted in this manner will be found in Fig. 3.9.

From experiments similar to that of Fig. 3.3, but in which the uptakes were measured after 15-min exposures at pH 7.0 in the presence of competing concentrations of Rb, K, Na, Li, and Ca^{2+}, ion-exchange isotherms characteristic of the glass surface were obtained. Representative examples are presented in Fig. 3.4 for NAS$_{11-18}$ glass and in Fig. 3.5 for NAS$_{10.6-10}$ glass, while the complete data are summarized in Table 3.1. Fifteen minutes of exposure was chosen in obtaining the data for Figs. 3.4 and 3.5 and Table 3.1 in the hope that this would give a better representation of the properties of the most superficial layers of the glasses than would an exposure of longer duration. (The justification for this step will be given later.)

It can be seen that the surface of these two typical sodium selective glasses is, without exception, characterized by an ion-exchange equilibria described by (3.3 or 3.3a) having the indicated values of n and K. The selectivities between various cations are large, with Cs being the

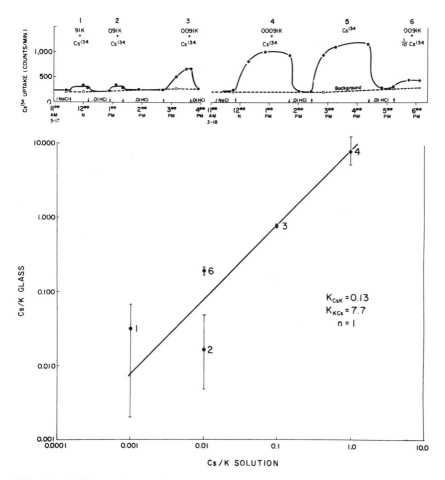

Fig. 3.3. Cs^{134}-K exchange in NAS_{11-18} glass. Upper portion of figure presents time course of uptake of Cs^{134} from 0.00091 N solutions of constant CsCl concentration in presence of .005 N tris buffer and indicated molar concentrations of KCl. Points are experimental uptakes at indicated times; solid curve is drawn to make visualization easier. Open circles and dashed curve give background radioactivity level. Notice break in abscissa between 4:00 P.M. on 3/17 and 11:00 A.M. on 3/18. Above abscissa are labeled the concentrations of nonradioactive solutions (e.g. 0.1 NaCl, 0.01 HCl) which were used to remove Cs^{134} between uptake experiments. Arrows pointing up and down indicate times of initial exposure to radioactive and nonradioactive solutions, respectively.

Lower portion of figure presents Cs/K ratio in glass as a function of Cs/K ratio in aqueous solution. Numbers by the points correspond to those in upper portion of figure: ranges indicate extreme values of uptakes for the indicated steady states. For comparison, solid straight line represents expectations of ideal ion-exchange equilibrium[3.2] having equilibrium constant of 7.7 in favor of Cs relative to K.

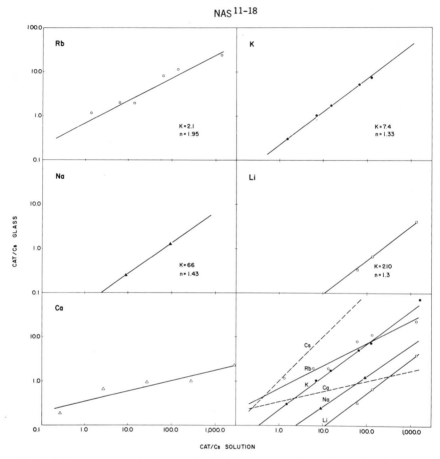

Fig. 3.4. Ion-exchange isotherms for NAS_{11-18} glass. Experimental points were obtained in manner of Fig. 3.3. Lines are theoretical expectations of (3.3) with the indicated values of K and n. Isotherms for individual cations relative to Cs^{134} are presented separately and are also combined in lower right-hand corner.

cation most preferred by NAS_{11-18} glass, and Rb the one most preferred by $NAS_{10.6-10}$ glass. It is noteworthy that both glasses markedly prefer Cs^+ to Na^+ by ion exchange in contrast to their opposite behavior as electrodes, as will be discussed in relation to Fig. 3.10. Not only is Cs^+ preferred to Na^+ but so is K^+, a finding the significance of which will be examined in the Discussion.

From Table 3.1 it can be seen that both glasses markedly prefer those cations which have the larger crystal radius (i.e., the smaller energies of hydration)—NAS_{11-18} being characterized by the sequence Cs > Rb >

NAS $^{10.6-10}$

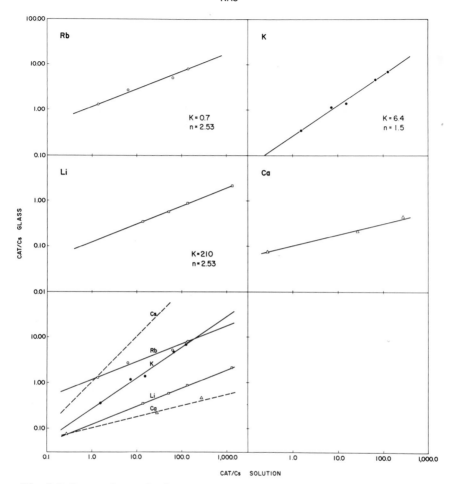

Fig. 3.5. Ion-exchange isotherms for $NAS_{10.6-10}$ glass. Data are plotted in the same manner as Fig. 3.4.

K > Na > Li (order $I^{1.9}$), whereas the sequence for $NAS_{10.6-10}$ corresponds to order II (Rb > Cs > K > Li). The ion exchange between K and Cs is nearly ideal, n having an average value of 1.23. That for other pairs is less so, the most nonideal behavior occurring with H^+, for which the average value of n was 3.7. It can also be seen that H^+ is markedly preferred to Cs^+, but the method of assessing the H^+-Cs^+ exchange should be described first, as it differs in some details from that used for the other cations.

Table 3.1   Summary of Surface Ion-Exchange Constants

	Li–Cs		Na–Cs		K–Cs		Rb–Cs		H–Cs	
	K	n	K	n	K	n	K	n	K	n
NAS 11–18										
f	230	1.3			6.8	1.33	1.5	1.48	10^{-4}–10^{-5}	3.17
g	580	1.81			15.0	1.38	2.1	1.95		3.26
l					7.7	1.0			10^{-5}	3.94
F					7.5	1.2				
G									1–6×10^{-4}	1.5
A			66	1.43					3×10^{-5}	3.9
12									2×10^{-5}	5.1
20										4.2
21									10^{-7}	4.2
Av.	405	1.6	66	1.43	9.3	1.23	1.8	1.72	10^{-5}	3.7
NAS 10.6–10										
2	210	2.53			6.6	1.52	0.7	2.53		
3	400	1.77			14.0	1.69			10^{-5}–10^{-6}	3.0
Av.	305	2.15			10.3	1.6	0.7	2.5	10^{-5}	3.0

3.3.3   The Ion Exchange between H$^+$ and Cs^{134}

The ion-exchange equilibrium between Cs^{134} and H$^+$ is more difficult to characterize completely than is that between Cs^{134} and the previous cations. Whereas we could restrict our previous studies to the properties of glass at neutral pH (where presumably the weak acid SiOH groups are undissociated), studies over a complete range of Cs/H ratios will involve exchange not only with the strong acid [Al–O–Si]$^-$ sites but also with the weak acid SiO$^-$ sites [1, 13].

The H–Cs^{134} exchange was studied both at constant Cs with variable H and at constant H with variable Cs. The results of these two types of experiment differed chiefly in that the values of n measured by the first type of experiment were considerably larger than those in the latter type, a finding that was expected from the observation that with highly alkaline pH Cs^{134} uptake *decreased*, which suggests that in this pH range glass as a whole dissolves faster than Cs^{134} is taken up. Because of this difficulty, only the results of studies at pH 7.00 will be presented here. These results are directly comparable to those for the alkali metal cations; typical data are illustrated in Fig. 3.6, where the upper portion presents the Cs^{134} uptake as a function of the Cs/H ratio in solution and the lower portion presents the calculated ratio of H to Cs in the glass as a function of the ratio of H to Cs in solution.

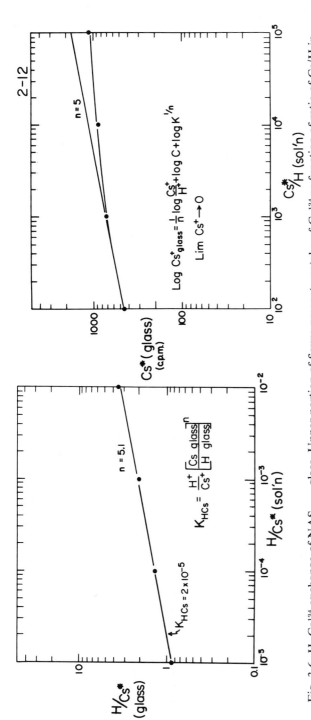

Fig. 3.6. H–Cs^{134} exchange of NAS$_{11-18}$ glass. Upper portion of figure presents uptake of Cs^{134} as function of ratio of Cs/H in solution. Concentration of H$^+$ was maintained by 0.005 N tris buffer, while the concentration of Cs^{134} labeled CsCl varied between 10^{-2} and 10^{-5}. Points are experimental uptakes, corrected for background. Diagonal straight line labeled $n = 5$ has slope of $\frac{1}{5}$ in the limit of low Cs/H values.

Lower portion represents ratio of H/Cs in glass, calculated from data of upper portion – assuming maximum ion-exchange capacity corresponding to 2,100 cpm (the H uptake is calculated as the difference between 2,100 and the observed Cs uptake at the given H/Cs ratio in solution). Straight line corresponds to (3.3a) with a value of $n = 5.1$ and $K = 2 \times 10^{-5}$. Note agreement between value of n from lower portion and limiting value in upper portion.

In H–Cs^{134} exchanges it is easy to examine the regions of low Cs^{134} uptake, but it is usually considerably more difficult to study the region of high Cs^{134} uptake where all ion exchange sites have Cs as their counterion – although for certain glasses a limiting maximum uptake of Cs^{134} is evident (cf. Fig. 3.8b and e). Nevertheless, a limiting behavior of (3.3a) makes it possible to assess n in a straightforward manner and then to estimate K even when this maximum uptake is not evident.

In the limit as C_j approaches zero (in which case C_i approaches the ion exchange capacity of the glass, C), equation (3.3a) simplifies to:

$$\log C_j \text{(glass)} = \frac{1}{n} \log \frac{a_j}{a_i} + \log \frac{K^{1/n}}{C} \tag{3.4}$$

$$\lim \frac{C_j \text{(glass)}}{C_i \text{(glass)}} \to 0,$$

which indicates that a plot of $\log C_j$ (glass) versus $\log(a_j/a_i)$ should be a straight line of slope $1/n$ in this limit, as illustrated in the upper portion of Fig. 3.6, where J^+ is Cs$^+$ and I^+ is H$^+$ (more extensive data of this type will also be presented in Fig. 3.8a, b, c). The limiting slope in Fig. 3.6 indicates a highly nonideal behavior for which n is 5.

Once the value of n is known, K can be calculated from (3.3) by successive approximations – assuming different tentative values for C and accepting that value for which the best agreement with (3.3) results. The lower portion of Fig. 3.6 has been calculated in this way, the plotted points corresponding to the H/Cs ratios in the glass, assuming C to correspond to 2,100 counts/min and calculating H as the difference between C and the Cs^{134} uptake for each point.

3.3.4   The Effect of the Duration of Exposure

All the ion exchange constants extracted above were calculated from Cs^{134} uptakes observed after brief exposures (usually 15 min) to the radioactive solutions. The data to be presented here show that the values of K and n so extracted are similar to those measured after 10 or 100 times longer exposures.

A complete description of the H–Cs^{134} exchange over the time range from 15 to 1,500 min is given in Figs. 3.7 and 3.8, as well as in Table 3.2. Table 3.2 summarizes the experimental data, the first four columns giving the conditions, while the remaining columns give the observations for two electrodes of lower Na$^+$ specificity than NAS$_{11-18}$ (cf. Fig. 3.10 for electrode potential data). The data for NAS$_{24.7-9.9}$, whose composition begins to approach that of the K$^+$ selective NAS$_{27-4}$ glass previously characterized[1, 2], provide a bridge between the present results of Na$^+$

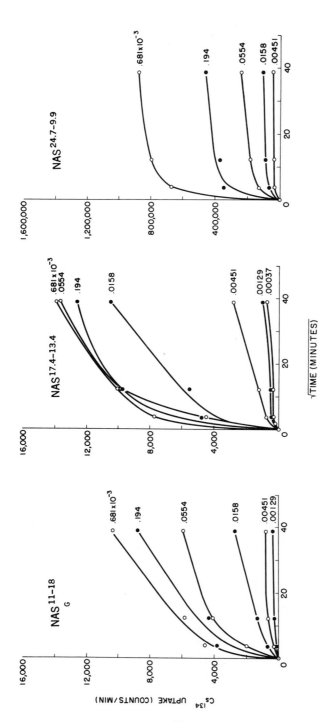

Fig. 3.7. Cs^{134} uptake as a function of time.

45

Table 3.2 Cs^{134} Uptake as a Function of Time

Conditions		Time in Cs^{134} Solution (min)	Background Radio-activity (cpm)	NAS$_{17.4-13.4}$		NAS$_{11-18G}$		NAS$_{24.7-9.9}$	
Time	Solution			Radio-activity in Glass (cpm)	Cs^{134} Uptake (counts-initial counts)	Radio-activity in Glass (cpm)	Cs^{134} Uptake (counts-initial counts)	Radio-activity in Glass (cpm)	Cs^{134} Uptake (counts-initial counts)
5/12; 11:15 A.M.	0.01 N HCl								
5/13; 9:42 A.M.	0.000681 N Cs^{134}, pH 7	0	434	429	0	2,970	0	1,620	0
10:03 A.M.	0.000681 N Cs^{134}, pH 7	15	434	8,105	7,721	7,503	4,533	671,174	669,554
12:18 P.M.	0.000681 N Cs^{134}, pH 7	156	563	10,784	10,355	8,765	5,795	796,846	795,226
5/14; 11:18 A.M.	0.000681 N Cs^{134}, pH 7	1536	554	14,307	13,878	13,319	10,349	876,696	875,076
11:30	0.01 N HCl								
5/15; 10:24	0.000194 N Cs^{134}, pH 7	0	553	700	0	3,110	0	74,278	0
10:39	0.000194 N Cs^{134}, pH 7	15		6,924	6,224	6,926	3,816	418,212	343,934
12:55 P.M.	0.000194 N Cs^{134}, pH 7	145	222	10,458	9,758	7,397	4,287	443,368	369,090
5/16; 11:26 A.M.	0.000194 N Cs^{134}, pH 7	1502	582	13,257	12,557	11,217	8,774	532,170	457,892
11:40	0.01 N HCl								
5/19; 12:58 P.M.	0.0000554 N Cs^{134}, pH 7	0	230	310	0	2,443	0	70,934	0
1:18	0.0000554 N Cs^{134}, pH 7	20	264	4,773	4,463	4,461	2,018	199,290	128,356
3:43	0.0000554 N Cs^{134}, pH 7	165	601	10,038	9,728	6,503	4,060	251,242	180,308
5/20; 2:17 P.M.	0.0000554 N Cs^{134}, pH 7	1519	493	13,971	13,661	8,346	5,903	312,234	241,300
2:30	0.01 N HCl								

5/23; 1:00 P.M. 0.000000369 N Cs¹³⁴, pH 7	0	490	641	0	3,149	0	18,449	0
1:21 0.000000369 N Cs¹³⁴, pH 7	21	502	971	330	3,299	150	25,446	6,997
3:41 0.000000369 N Cs¹³⁴, pH 7	161	489	1022	381	3,262	113	32,772	14,323
5/24; 2:00 P.M. 0.000000369 N Cs¹³⁴, pH 7	1500	453	1376	735	3,456	307	36,133	17,684
2:10 0.01 N HCl								
5/26; 10:36 A.M. 0.00000129 N Cs¹³⁴, pH 7	0	492	590	0	3,098	0	19,905	0
10:51 0.00000129 N Cs¹³⁴, pH 7	15	—	983	393	3,242	144	31,355	11,450
1:21 P.M. 0.00000129 N Cs¹³⁴, pH 7	165	509	1,064	474	3,322	224	35,835	15,920
5/27; 11:53 A.M. 0.00000129 N Cs¹³⁴, pH 7	1517	452	1,587	997	3,482	384	38,938	19,033
12:05 P.M. 0.01 N HCl								
5/28; 1:00 P.M. 0.00000451 N Cs¹³⁴, pH 7	0	464	568	0	2,984	0	21,166	0
1:30 0.00000451 N Cs¹³⁴, pH 7	30	—	1,498	930	3,333	349	51,518	30,352
4:00 0.00000451 N Cs¹³⁴, pH 7	180	458	1,672	1,124	3,619	635	56,117	34,951
5/29; 1:40 P.M. 0.00000451 N Cs¹³⁴, pH 7	1440	447	3,369	2,801	3,769	785	66,785	45,619
1:55 0.01 N HCl								
6/2; 11:06 A.M. 0.0000158 N Cs¹³⁴, pH 7	0	550	674	0	2,990	0	25,288	0
11:21 0.0000158 N Cs¹³⁴, pH 7	15	—	5,410	4,736	3,645	655	87,561	62,273
1:36 P.M. 0.0000158 N Cs¹³⁴, pH 7	150	564	6,176	5,502	4,316	1,326	109,469	84,181
1:34 P.M. 0.0000158 N Cs¹³⁴, pH 7	1588	516	11,116	10,442	5,728	2,738	131,439	106,151

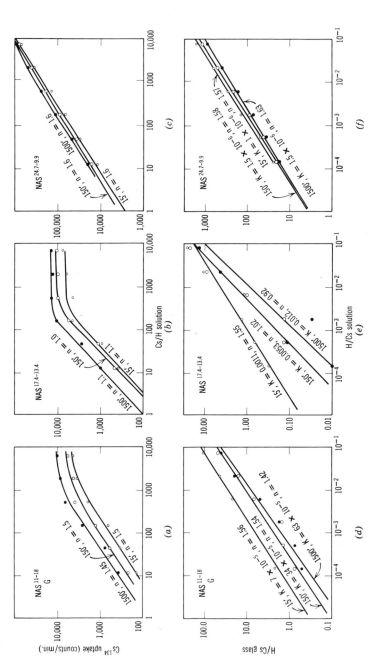

Fig. 3.8. Approximate time invariance of n and K for H–Cs exchange. Data of upper portion are plotted in same manner as Fig. 3.6 upper: those in lower portion in same manner as Fig. 3.6 lower. (Further details in text.) Time and calculated values of K and n are indicated on each curve. In calculating H/Cs ratios in glass in the *lower* portion of Fig. 3.8, following values for ion-exchange capacity (in counts per min) were used: for NAS_{11-18}: 14,000, 10,000, and 15,000; for $NAS_{17.4-13.4}$: 10,000, 11,000, and 14,000; for $NAS_{24.7-9.9}$: 16,000,000, 16,000,000, and 16,000,000 at 15, 150 and 1,500 min, respectively. For $NAS_{17.4-13.4}$, values were read directly from B as the limiting values of Cs^{134} uptake for large values of Cs/H in solution. For $NAS_{24.7-9.9}$, value was estimated by solving for (H/Cs) in glass, trying successive values of ion-exchange capacity until the data in F gave slopes corresponding to values of n in good agreement with those measured directly in C. For NAS_{11-18} both procedures were used.

selective electrodes and those for K^+ selective electrodes. (From Fig. 3.10 it will be seen that $NAS_{24.7-9.9}$ is nearly equally responsive as an electrode to K^+ and Na^+ at pH 7.)

After an initial exposure for 22.5 hr to 0.01 N HCl to remove any superficial residual radioactivity remaining from previous studies, it was found that $NAS_{17.4-13.4}$ had an apparent radioactivity of 429 cpm (i.e., identical to background; see row "5/13; 9,42 A.M.", which corresponds to the radioactivity measured just before immersion in the "0.000681 N Cs^{134}, pH 7" solution). The NAS_{11-18} had a residual radioactivity of 2,970 cpm (i.e., some 2,500 cpm greater than background) and $NAS_{24.7-9.9}$ had a residual activity of 1,186 (i.e., 1,620–434) cpm. The glasses were then immersed in 0.000681 N Cs^{134}, at pH 7 (0.005 tris buffer) and their radioactivity counted at the indicated times. The results are tabulated in the columns labeled "Radioactivity in Glass" (the specific activity of Cs^{134} of all solutions uncorrected for background being 1.455×10^{13} cpm/mole of labeled Cs).

The reversibility of these uptakes is seen by the decrease in radioactivity on elution in 0.01 N HCl. For $NAS_{17.4-13.4}$ glass, the radioactivity is reduced by this procedure to the background level; for NAS_{11-18} the counts are decreased to the initial level of 2,500 counts/min above background (which represents a deep residual radioactivity from previous experiments with Na^{22} and Cs^{134}). Even the large Cs^{134} uptake by the $NAS_{24.7-9.9}$ composition is reversible to a considerable extent, although a significant diffusion into deeper layers is evident from the uneluted residual radioactivity. From these data it is possible to calculate the Cs^{134} uptake on exposure to the various radioactive solutions by subtracting the initial radioactivity from each set of measurements, with the results given in the columns labeled "Cs^{134} Uptake" in Table 3.2.

These data are plotted in Fig. 3.7 as a function of the square root of time; they show that the rate of uptake is not linear with the square root of time, but decreases markedly with increasing time. For each of the times of Fig. 3.7, the ion exchange isotherms of Fig. 3.8 have been constructed in the manner of Fig. 3.6 and illustrate the extent to which the values of n and K are independent of the time at which the uptakes were measured.

The upper portion of Fig. 3.8 shows that the value of n is independent of the time at which the measurements are made. The lower portion of Fig. 3.8 indicates that, although the values of K are less independent of time, the estimated values are still of the same magnitude over the time range of 15–1,500 min.

The information of Fig. 3.8 covers a wide range of types of H–Cs exchange equilibria. The data in Fig. 3.8b for the composition $NAS_{17.4-13.4}$

are characteristic of relatively strong acid sites in that the uptake of Cs[134] becomes independent of the Cs/H ratio toward the right—indicating a complete displacement of H$^+$ by Cs[134] for those exchange sites available at pH 7. This exchange presumably represents one almost purely for the strong acid [AlOSi]$^-$ sites of the glass (notice from e that the value of K is of the order of 10^{-2}–10^{-3}); it is of some interest to note that the ion exchange is close to ideal in this case. On the other hand, the data for NAS$_{24.7-9.9}$ in Fig. 3.8c show not only that 100 times as many Cs ions are taken up by this glass within the experimental range but also that we are still very far from displacing all the H$^+$ ions. Such chemical behavior indicates a participation by a large number of weak acid (presumably SiO$^-$) sites.

These details bear some relationship, although not a very tight one, to the chemical composition of the glasses. The Na$_2$O/Al$_2$O$_3$ ratio is closest to unity in NAS$_{17.4-13.4}$ glass; thus, from the theory described in a previous publication[1,9], almost all the sites are expected to be of the strong acid [AlOSi]$^-$ type; whereas for NAS$_{24.7-9.9}$ there are 14.8 weak acid SiO$^-$ sites for every 9.9 strong acid [AlOSi]$^-$ sites and NAS$_{11-18}$ is complicated by the presence of 9 Al atoms in an "alumina-like" state for every 11 [AlOSi]$^-$ sites.*

3.4 DISCUSSION

3.4.1 The Ion-Exchange Capacity of the Hydrated Glass Surface

It is of interest to compare the ion exchange capacity measured in Fig. 3.8 with that calculated from the chemical composition of the glass. Such a comparison provides an estimate of the depth to which the ion exchange takes place over the experimental times we used. It also provides a measure of counterion concentration at the surface.

For NAS$_{17.4-13.4}$ glass, the ion exchange capacity C was found to be equivalent to 10,000 cpm at 15 min, 11,000 at 150 and 14,000 at 1,500. As the surface area of this sample was about 2 cm^2 and the specific activity of the Cs solution was 1.455×10^{13} cpm/mole, these values of C correspond to a range of uptake between 0.35×10^{-9} moles/cm^2 at 15 min and 0.48×10^{-9} moles/cm^2 at 1500 min—or 2.1×10^{14} atoms/cm^2, respectively. For these atoms to exist in a monolayer, they would have to be spaced from 5.9 to 6.9 Å apart. The ion exchange capacity for this glass (calculated for the strong acid [Al–O–Si]$^-$ sites in the manner

---

*The ion exchange properties of hydrated alumina have been recently reported upon by Churms[14].

described on page 170 in *Glass Electrodes for Hydrogen and Other Cations*[1]) is 9×10^{-3} moles/cm^3, corresponding to 5.4×10^{21} atoms/cm^3 — or 1 atom every 5.7 Å. It therefore appears that the Cs^{134} uptake by this composition could easily be accounted for by a monolayer of exchangeable glass.

The markedly low rate of change of the value of C between 15 and 1,500 min suggests that the exchange of Cs^{134} with the immediate surface layers of the glass is much more rapid than is further diffusion into the bulk glass; this supports the assumption that the ion exchanges here studied are chiefly those of the glass surface, uncomplicated by further diffusion into the interior.

Although the estimates of C are less accurate for NAS$_{11-18}$, the ion exchange capacity could be measured for a specimen of 2 cm^2 surface area to be between 0.35×10^{-9} and 0.52×10^{-9} moles/cm^2, corresponding to sites spaced 6–7 Å apart. Comparison with the calculated ion exchange capacity of 8×10^{-3} moles/cm^3 (corresponding to 1 atom every 6 Å) again indicates that the uptake could have occurred into a monolayer of sites spaced at the distances expected from the ion exchange capacity of the glass.

In contrast to the above-noted findings with Na$^+$ electrode glasses is the situation for the NAS$_{24.7-9.9}$ composition. Here C is estimated to be 16,000,000 cpm for 2 cm^2 of glass surface, indicating that 1,000 times as many atoms are capable of entering into this glass as into the others. As the calculated ion exchange capacity of 7.5×10^{-3} moles/cm^3 is nearly the same as that of the others, we can infer that Cs^{134} atoms can penetrate this glass to a depth of 1,000 monolayers, or about 7,000 Å.

3.4.2 Comparison with the Ion-Exchange Behavior of Aluminosilicate Minerals

The general similarity between the ionic selectivity observed in the electrode potentials of aluminosilicate glasses and the equilibrium ion exchange selectivity of aluminosilicate minerals has already been noted (cf. Fig. 8a and 8b of Ref.[9]). It has been suggested that this similarity exists because the ion exchange equilibrium constant K contributes to the selectivities for both phenomena, although the mobility ratios also enter into the electrode potentials[1,2,15]. Moreover, it has been shown for K$^+$ selective composition that the calculated values of K are in good agreement with those directly measured[1,2]. Comparison of the ion exchange properties of the surfaces of the Na$^+$ electrode glasses in Figs. 3.4 and 3.5 with those of two typical aluminosilicate ion exchangers[8,16] illustrated in Fig. 3.9 indicates a close similarity, both qualitatively and quantitatively. This similarity provides further support for

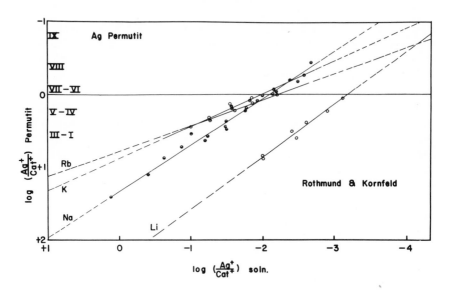

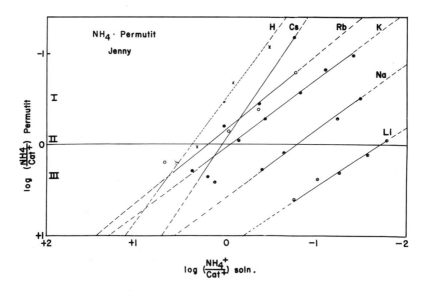

Fig. 3.9. Ion-exchange selectivity of typical aluminosilicate ion exchangers, the Permutits. Data for exchange equilibria of indicated cations for NH_4 and Ag in Permutit, are plotted in the manner of Figs. 6.4 and 6.5, after References [8] and [16].

the notion that the aluminosilicate ion exchangers are excellent protypes for aluminosilicate glasses.

3.4.3 Comparison with Electrode Potentials

The steady-state electrode potential selectivities of electrodes representative of those whose ion-exchange properties have been presented in this chapter are illustrated in Fig. 3.10, where the values of n and K^{pot}, the potential selectivity constant, are given; K^{pot} — the superscript has been omitted in the figure — corresponds to the permeability ratio of the ions in the glass, and it is not to be equated with the ion exchange equilibrium constant because it also contains a contribution from the mobility ratio of the ions[2, 10, 17]. The most obvious conclusion that can be drawn by comparison of the potential selectivity data for NAS_{11-18} with its ion exchange selectivity data as presented in Table 3.1 is that there is no correlation whatsoever between the surface ion-exchange properties and the (steady-state) electrode potential selectivities, except for similar values of n for H–Cs mixtures. This, in itself, offers no theoretical difficulties because it is conceivable that mobility differences entirely offset the ion exchange selectivities to produce the resultant potential selectivity properties, a phenomenon already demonstrated for K^+ selective glasses[2]. However, the values of K and K^{pot} would require a mobility ratio of 7,000 : 1 higher for Na^+ than K^+ for such a complete offset because K^{pot} indicates a 1000 : 1 Na^+ to K^+ potential selectivity while K implies a reverse preference of K^+ over Na^+ of 7 : 1. Since the mobility ratio of $Na^+ : K^+$ in the hydrated glass surface is unlikely to be as high as the 7000 : 1 necessary to offset completely the ion exchange preference for K^+, I would prefer to suggest that the hydrated glass surface is less Na^+ selective as an electrode than the interior (even perhaps being more selective to K^+ than to Na^+). If the $Na^+ : K^+$ permeability ratio of the glass surface differed from that of the interior, this should appear in the *transient* behavior of the electrodes, as will be discussed in the concluding section of the Discussion. Such a transient response to K^+ ions by NAS_{11-18} glass has been observed by Friedman[6] and is shown in Fig. 3.11. I have recently confirmed this finding and have measured a transient Na : K selectivity of 1 : 1. More recently, such behavior has been confirmed by Rechnitz[7], who has also found a transient response to Ca^{2+}. This transient response is important theoretically and may also be of practical usefulness in extending the range of the glass electrode to the measurement of poorly mobile ions to which it is not usefully responsive in the steady state (e.g., Ca^{2+}, quaternary amines, amino acids).

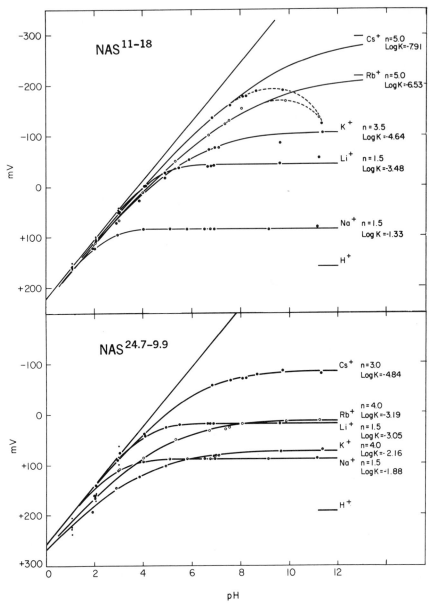

Fig. 3.10. Electrode potentials of four glasses typical of those for which surface ion-exchange properties have been examined. Observed electrode potentials of labeled glass compositions in 0.1 N solutions of indicated cations as a function of pH. To the right of each curve is given the value of n and log K^{pot} (superscript omitted) characteristic of the electrode potential selectivities. (Data for

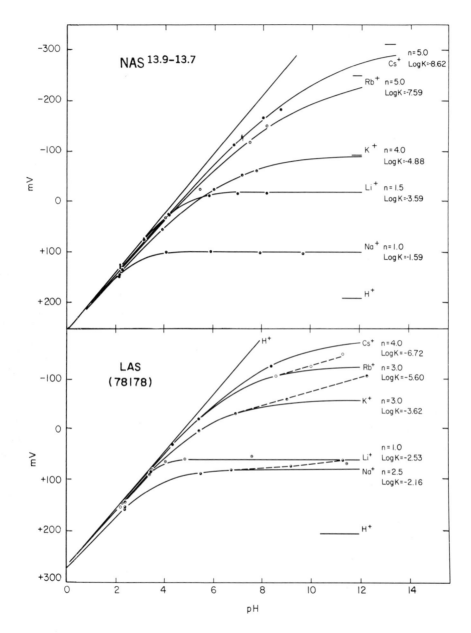

NAS$_{13.9-13.7}$ glass can be taken to represent both the NAS$_{10.6-10}$ and the NAS$_{17.4-13.4}$ compositions.) Notice for NAS$_{11-18}$ that the value of n observed by potential for H–Cs mixture is of the same order of magnitude as that characteristic of H–Cs ion exchange.

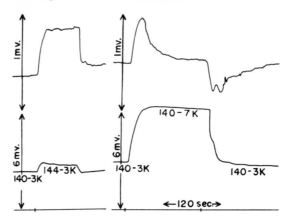

Fig. 3.11. Transient responses of electrode potential of NAS_{11-18} glass. Simultaneous observations for Na⁺ selective NAS_{11-18} and K⁺ selective NAS_{27-6} glass electrodes (upper and lower, respectively). Traces at the left show step response characteristic of both for a small step change in Na⁺ concentration (from 140 to 144 mEq/L) in presence of a constant 3 mEq/L of K⁺. Right-hand traces give potential changes observed for step change of K⁺ (from 3 to 7 mEq/L) or in presence of a constant 140 mEq/L of Na⁺. Notice that whereas there is step response of potential for step change in K⁺, the Na⁺ electrode exhibits transient K⁺ response despite negligible effects of K⁺ in the steady state.

(Reproduced after Fig. 4 of Friedman[3], with the permission of John Wiley and Sons.)

3.4.4   Previous Tracer Studies on the Surface of Glass

It should be noted that in addition to studies of Na²⁴ and K⁴² diffusion and ion exchange in thickly hydrated surface of K⁺ glass electrodes[1, 2] fundamental studies have been carried out by Doremus[11, 12] and by Garfinkel[18] on the same phenomena for dry glass. Recently it has also been found possible to study the current-voltage characteristics of a glass membrane which is hydrated throughout its entire thickness[19].

For pH-responsive glasses exposed to aqueous solutions, Schwabe and his colleagues[20, 21] have measured the uptake and loss of alkali cations and concluded that the hydrated glass surface undergoes cation exchange, greater quantities of Na²⁴ being taken up than required for monolayer coverage (an observation confirmed by the present author for Corning Code 0150 glass). Nevertheless, for Na⁺ selective glasses, the present results indicate that the Cs¹³⁴ uptake at ordinary times is essentially restricted to a monolayer or at most a few layers. Schwabe has attempted to estimate the values of K from his studies, and more recently, Lengyel and his co-workers[22] have assessed K values for pH-responsive glasses from measurements of ion uptake together with electrode potentials. However, the values calculated by these authors have been validly criticized by Doremus[12] as being "questionable because the authors have assumed that there are no diffusion potentials in the glasses." An anion exchange for the hydroxyl group of the Si–OH groups of glass has also been demonstrated[23].

Of prime importance has been Doremus' demonstration[11] of the validity of the Nernst-Planck equations and of ideal ion exchange behavior (i.e., $n = 1 = dlna/dlnc$) for Na^+–Ag^+ exchange in a typical dry soda-lime glass, for which he also verified the Nernst-Einstein relationship between diffusion constant and electrical conductivity. Doremus found D_{Na}/D_{Ag} to be 10 in this glass. He also measured diffusion constants and equilibrium constant for the K^+–Na^+ exchange, finding D_{Na}/D_K to be 500 and K to be 200 to 1 in favor of Na^+ over K^+. (Note that K is referred to a molten nitrate phase and is therefore not directly comparable to its value referred to an aqueous solution phase.) Despite the ideal behavior of Ag^+ and Na^+ ions, Doremus finds that K^+ is partially excluded from all but the immediate surface layers of soda-lime glass[12]. Garfinkel [18] has obtained results on dry NAS_{27-3} and NAS_{11-18} glass which are in general accord with those of Doremus. However, as his detailed results have not yet been published, they cannot be commented on further here.

Doremus also observed an anomalous behavior for a dry borosilicate glass (commercial Pyrex brand), for which the values of K for the Na^+–Ag^+ exchange were neither constant (as expected from equation 3.2) nor describable by the n-type nonideal behavior of (3.3). He found that this anomaly was caused by a separation of pyrex glass into two phases, one phase being almost pure silica, the other being a sodium borosilicate containing 95% of the total Na^+ (although it constituted only 14% of the volume composition). The finding of such phenomena must be anticipated in hydrated glasses, and Doremus' studies on non-hydrated glasses will be useful prototypes for such analyses.

3.4.5 The Role of the Hydrated Glass Surface in the Transient K^+ Response of Na^+ Electrodes and in the K^+ Selectivity of Incompletely Sealed Na^+ Electrodes

From the present results it is clear that glass electrodes exposed to aqueous solutions should, in general, not be regarded as having uniform chemical properties over their entire thickness, but rather to consist of an unaltered "dry" interior bounded by hydrated layers of significantly different properties. This can be shown diagrammatically as follows:

	─────GLASS MEMBRANE─────			
aqueous	HYDRATED :	"DRY"	:HYDRATED	aqueous
solution	LAYER :	INTERIOR:	LAYER	solution

The thickness of the hydrated layer appears to be a function of the glass composition and varies, on the basis of the present results, from no more than a few monolayers (i.e., about 10 Å) for Na$^+$ selective glasses to as much as 100,000 Å in the case of the K$^+$ selective glasses described previously[1, 2]. Moreover, the ionic mobilities (and, by inference, the ion-exchange affinities) appear to be different within the hydrated surface from those of the "dry" interior. These findings bear on two properties of glass electrodes: (a) the transient K$^+$ response of Na$^+$ electrodes[6, 7]; and (b) the cation response of "incompletely sealed" microelectrodes[3] and the related surface conductivity of open-tip microelectrodes[4].

In the first case, in which the electrode potential is measured across the thickness of the glass, the transient K$^+$ response as illustrated in Fig. 3.11 represents the properties of the hydrated layers seen in a direction *normal* to the glass surface. This transient behavior can be explained as an initial response to K$^+$ by the K$^+$-preferring surface, succeeded by the steady-state Na$^+$ selectivity characteristic of the deeper layers of "dry" glass.*

In the second case, where the glass forms a tube, the properties of the hydrated layers manifest themselves in the direction *parallel* to the glass surface by providing additional electrically conductive pathways both within the hydrated glass as well as through the conductance of (highly concentrated) counterions in the adjoining aqueous solutions[3, 4]. Indeed, depending upon the extent to which the glass hydrates, the conductivity of the glass may even be the most important electrical pathway because of the large increase in ionic mobility within the hydrated layers. The K$^+$ over Na$^+$ selectivity of the surface of Na$^+$ selective glasses would in this way explain Lev's observation that "incompletely sealed microelectrodes of Na$^+$ selective glasses show a K$^+$ selectivity"[3, 25].

For some compositions (e.g., NAS$_{27-4}$) the entire thickness of the microelectrode wall must be hydrated (cf. the complete hydration of 1μ thick films of this glass[19] so that the wall resistance is so much lower than that expected from the resistivity of dry glass that the existence of the open tip can be completely negligible. This property of glass would explain the fact that geometrically identical open-tip microelectrodes have been reported to have a much lower resistance when made of NAS$_{27-4}$ glass than when made of NAS$_{11-18}$ glass[25]. (Recall that the resistivities of dry NAS$_{11-18}$ and NAS$_{27-4}$ glasses are nearly the same[26].)

---

*It has been shown elsewhere[17] that a simple step response of potential to a step change in solution composition is expected when diffusion of ions is restricted to a chemically homogeneous region of glass or, somewhat more generally, to regions of glass in which the permeability ratio (i.e., the *product* of ion exchange equilibrium constant and mobility ratio) is constant[24]. This has been discussed by Doremus in some detail on pp. 111–112 of our recent book[11].

It is unfortunate that so little is known at present about the hydration and electrical resistance of the glass compositions from which the usual micropipette reference electrodes have been made[5]. However, the elegant analysis by Lavallée and Szabo of the "tip potentials" of high and low resistivity glasses[4] is particularly pertinent and should stimulate the improved design of reference microelectrodes.

3.5 CONCLUSIONS

Despite the difficulties inherent in trying rigorously to characterize the ionic mobilities and ion-exchange equilibrium constants for systems which are not chemically uniform (such as the Na^+ selective glasses studied here, wherein surface chemical properties appear to differ from those at a depth), an attempt has been made to characterize the ion-exchange behavior of the surface of such glasses. The following conclusions have been reached.

1. The surface of typical Na^+ selective glass electrodes (in particular NAS_{11-18} glass) exhibits classical ion-exchange properties with a preference for cations in the decreasing sequence H, Cs, Rb, K, Na, Li. The entire Cs^{134} exchange appears to occur over ordinary times in a monolayer of sites spaced 5–6 Å apart, or, at most, a few layers of more widely spaced sites.

2. The ion exchange equilibrium is described by a law of mass action obeying n-type nonideal behavior (corresponding to $dlna/dlnc = n$). The values of n observed by equilibrium ion exchange are of the same magnitude as those observed in the electrode potentials.

3. The properties of the surface differ from those of the bulk, most notably in that the diffusion constant is much larger in the surface than in the bulk of glass. This suggests the likelihood of important surface conductance effects along the interior of an open-tipped microelectrode as well as the possibility of a much higher conductivity of the supposedly insulating electrode wall than had previously been assumed.

4. The existence of a thin surface layer of properties different from those of the bulk interior implies the possibility of a transient response to ions (e.g., K^+ and Ca^{2+}) to which glass electrodes are not highly responsive in their steady-state. As the selectivity of the glass surface favors K^+ over Na^+ by ion exchange, even for such highly selective Na^+ electrodes as NAS_{11-18}, and it seems unlikely that this equilibrium selectivity is entirely offset by a sufficiently higher mobility in the immediate surface of Na^+ than K^+, the immediate surface of NAS_{11-18} glass should respond to K^+ better than the interior will. This constitutes an explana-

tion of both the K$^+$ transient observed for NAS$_{11-18}$ glass by Friedman and by Rechnitz and also of the K$^+$ selectivity of incompletely sealed Na$^+$ electrodes described by Lev.

REFERENCES

[1] G. Eisenman, in *Glass Electrodes for Hydrogen and Other Cations* (G. Eisenman, ed.), Dekker, New York, 1967.

[2] G. Eisenman, *Ann. N.Y. Acad. Sci.* **148**, 5 (1968).

[3] A. Lev, in *Glass Microelectrodes* (M. Lavallée, O. Schanne, and N. Hébert, eds.), Wiley, New York, 1969.

[4] M. Lavallée, and G. Szabo, in *Glass Microelectrodes* (M. Lavallée, O. Schanne, and N. Hébert, eds.), Wiley, New York, 1969.

[5] D. P. Agin, in *Glass Microelectrodes*, (M. Lavallée, O. Schanne and N. Hébert, eds.), Wiley, New York, 1969.

[6] S. M. Friedman, in *The Glass Electrode* (G. Eisenman, R. Bates, G. Mattock, and S. M. Friedman, eds.), Interscience New York, 1966.

[7] G. A. Rechnitz, and G. C. Kugler, *Anal. Chem.*, in press.

[8] V. Rothmund, and G. Kornfeld, *Z. Anorg. Allgem. Chem.*, **103**, 129 (1918).

[9] G. Eisenman, *Biophys. J.*, **2**, 259 (1962).

[10] G. Karreman, and G. Eisenman, *Bull. Math. Biophys.*, **24**, 413 (1962).

[11] R. H. Doremus, in *Glass Electrodes for Hydrogen and Other Cations* (G. Eisenman, ed.), Dekker, New York, 1967.

[12] R. H. Doremus, in *Ion Exchange*, Vol. 2 (J. A. Marinsky, ed.), Dekker, New York. In press.

[13] B. P. Nicolsky, M. M. Shultz, A. A. Belijustin, and A. A. Lev, in *Glass Electrodes for Hydrogen and other Cations* (G. Eisenman, ed.), Dekker, New York, 1967.

[14] S. C. Churms, *J. S. African Chem. Inst.*, **19**, 98 (1966).

[15] G. Eisenman, in *Proc. International Union Physiolog. Sci.*, **4**, 489 (1965); International Congress Series No. 87, Excerpta Medica Foundation, Amsterdam and New York (1965).

[16] H. Jenny, *Kolloid Chem. Beiheft*, **23**, 428; *J. Phys. Chem.*, **36**, 2217 (1927, 1932).

[17] F. Conti, and G. Eisenman, *Biophys. J.*, **5**, 247 (1965).

[18] H. M. Garfinkel, private communications, also in *Membranes: A Series of Advances* (G. Eisenman, ed.), Vol. 1, Dekker, New York, 1969, in press.

[19] G. Eisenman, J. P. Sandblom, and J. L. Walker, Jr., *Science*, **155**, 965 (1967).

[20] K. Schwabe, and H. Dahms, *Z. Elektrochem.*, **65**, 518 (1961).

[21] K. Schwabe, and H. D. Suschke, *Angew. Chem. Intern. Ed. Engl.*, **3**, 36 (1964).

[22] B. Lengyel, D. Csakvari, and J. Toperczer, *Acta Chem. Hung.*, **45**, 177 (1965); **50**, 119 (1966).

[23] N. A. Izmailov, and A. G. Vasil'yev. *Dokl. Akad. Nauk SSSR*, **95**, 579 (1954).

[24] J. P. Sandblom, and G. Eisenman, *Biophys. J.*, **7**, 217 (1967).
[25] T. B. Hollander, and J. C. Rock, paper presented to the Intracellular Glass Microelectrode Conference, May 23–25, 1967, Montreal, Canada.
[26] N. C. Hébert, in *Glass Microelectrodes*, (M. Lavallée, O. Schanne, and N. Hébert, eds.), Wiley, New York, 1969.

CHAPTER 4

Electrochemical Properties of Glass Microelectrodes

DANIEL P. AGIN

Department of Physiology, University of Chicago

4.1 INTRODUCTION

During the past several decades only a dozen or so published works have dealt with the electrochemical properties of open-tip microelectrodes [1–13]. First used by Ling and Gerard[1] to measure resting potentials in muscle fibers, these electrodes are still the principal means of recording resting and action potentials in biological material; thus an understanding of their electrochemical behavior is of importance. My purpose is to call attention to some problems which are of immediate practical concern, and at the same time to provide a framework for future research.

4.2 SOME PROPERTIES OF GLASS

The general properties of microelectrode glasses have been discussed in Chapter 2 by Hébert, and we need here emphasize only those properties that should be of importance for electrochemical behavior. These fall into two categories: (a) electrical conduction mechanisms in glass; (b) interactions of glass with water, ions, and organic molecules.

As has been pointed out, the introduction of sodium oxide into an ordinary glass results in the existence of nonbridging oxygens and sodium

This work was supported by a research grant from the U.S. Public Health Service and by the Wallace C. and Clare A. Abbott Memorial Fund of the University of Chicago.

ions in the interstices. The sodium ions may drift from one interstice to another under the influence of an electric field or concentration gradient. This is the basic view of the mechanism for almost all electrical conductivity in glass. The conduction process is considered to be an activated one, so that an alkali cation, for example, can jump from one interstice to another only when its vibrational energy has attained a certain threshold value determined by the height of the intervening energy barrier. One would therefore expect the conductivity to be related to temperature in accordance with the Boltzmann relation, and this is what is observed experimentally [14]. There is also in glass experimental evidence of a deviation from Ohm's law, a phenomenon which is explained in terms of the theory of the electrolytic conductivity of solids [15, 16].

The next question concerns what happens when an ordinary glass is exposed to aqueous solutions. It is generally agreed that the effect of exposing a glass surface to water is to produce a silicate hydrolysis and the formation of a porous silica film [17, 18]. The thickness of this film will depend on the type of glass, the nature of the attacking solution, and the duration of exposure. If the glass contains alkali cations, an ion-exchange interaction will also occur; protons will migrate into the glass and alkali cations out of the glass. If the aqueous solution contains cations other than hydrogen, the situation is more complicated and the various types of ion selectivity to be discussed elsewhere in this book become possible. It should, however, be noted that the pH response of Pyrex glass (Code 7740) is approximately 18 mV per pH unit [19,20].

4.3   THE GLASS-WATER INTERFACE

The brief description above of the properties of glass is certainly a very primitive picture, but it allows us to make some general conclusions which will be useful later on. A glass-water interface must be considered as a three-phase system (glass-hydrated layer-water) which seeks to equilibrate chemically, mechanically, and electrostatically. The attainment of chemical equilibrium is associated with ion-exchange processes in a cation-permeable silica gel layer. If this gel layer becomes sufficiently hydrated to incorporate bulk water, the polymer matrix will be subjected to a swelling pressure, and a mechanical term must be introduced into the thermodynamic equation of state for the system. Similarly, as ions are involved, the system will have an electrical structure as defined by potential profiles in all three phases; thus an electrostatic term must be included in a thermodynamic analysis.

4.4   THE ELECTRICAL STRUCTURE OF THE GLASS MICROELECTRODE

All surfaces are associated with electric double layers. In some cases this is because the material of the surface is ionized; in others, because of adsorbed ions[28]. At the aqueous interface the glass gel layer will be associated with a negative surface-charge density, and this must be compensated for if macroscopic electroneutrality is to be maintained. Part of the compensation is provided by the presence of alkali cations in the surface. The remainder is taken care of by a cloud of cations in the aqueous solution, the density of which rapidly decays with distance. This is the famous "double layer." There is evidence to suggest that part of the cloud is held rigidly associated with the surface. The exact potential profile from the surface has been the subject of intensive theoretical work for many years[21, 22]. It should be similar to what appears in Fig. 4.1, which depicts the older Gouy-Chapman model[23, 24] without the Stern modification[25] to account for the rigid layer. The potential profiles are obtained by a one-dimensional integration of the Poisson equation; they differ according to the concentration of electrolyte [21]. In any system of this type there should be, according to hydrodynamic analysis, a region adjacent to the solid surface which is immobile. A slipping plane or plane of shear will therefore exist and is

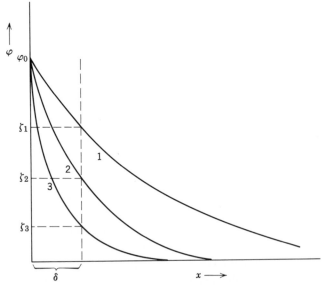

Fig. 4.1. Profile of electric potential (ϕ) at interface (Gouy–Chapman model).

show as δ in Fig. 4.1. The intersection of this plane with the potential profile determines the electrokinetic potential, zeta, at least according to current theory. Hydrodynamic movement of the liquid with respect to the solid involves the movement of charges with respect to each other; a streaming current is thus created and with it a streaming IR drop or streaming potential. Similarly, the application of an electric field to this system can cause the liquid to move with respect to the solid, which is the phenomenon of electro-osmosis. The two other electrokinetic phenomena, sedimentation potentials and electrophoresis, can be described in a similar manner.

Figure 4.2a shows the Stern modification of the Gouy-Chapman model to include the presence of a rigidly adsorbed layer of counterions; 2b illustrates a so-called reversal-of-charge phenomenon. This is essentially an overcompensation of the surface-charge density by the rigid counterion layer, and a reversal in sign as well as change in magnitude of the zeta potential.

The zeta potential cannot be directly measured, but must be inferred from the findings of electrokinetic experiments. In Fig. 4.3 we see some results of the classical experiments of Rutgers and deSmet[26] on the zeta potential, deduced from streaming potentials measured in Jena glass (16 III) capillaries. This material is a borosilicate thermometer glass. A strong dependence of the zeta potential on electrolyte properties is quite

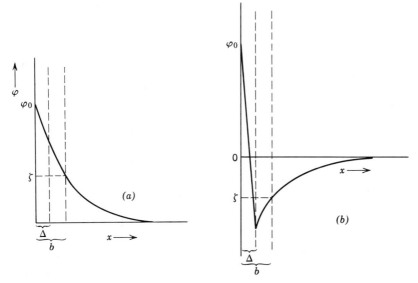

Fig. 4.2. Stern modification of Gouy–Chapman model. Rigid Helmholtz layer is indicated by $\triangle$. (a), ordinary profile; (b), reversal of charge phenomenon.

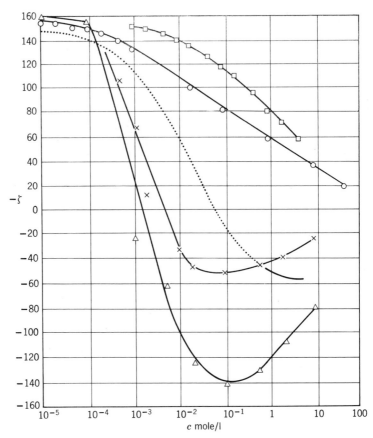

Fig. 4.3. Dependence of zeta potential on electrolyte concentration. □, KCl; ○, Ca(NO₃)₂; x, Al(NO₃)₃; △, Th(NO₃)₄; ○ crystal violet. (*Rutgers and De Smet*[26]. *Reprinted by permission of the Faraday Society.*)

evident, with a rather startling effect for thorium ions. Rutgers and deSmet found that the zeta-potential–log-concentration curves for 1 : 1 electrolytes are the same, with a slope of 42 mV between 0.2 and 10 mM (e.g., KCl in the figure). Similarly, for 2 : 1 electrolytes the slope was 26 mV. However, for 3 : 1 and 4 : 1 electrolytes, linearity of the zeta-potential–log-concentration curves is not present. Similar results are obtained with surface-active electrolytes [22, 27].

The anomalous behavior of heavy metal cations has been known for many decades, but has never been completely explained. The strong adsorption of thorium on silica gel is also well known [29]. Similar behavior is observed in connection with charge-reversal phenomena in colloidal systems[22]. In some cases the classical Schulze–Hardy rule is obeyed and the important parameter seems to be the valence of the cation raised to the sixth power. However, this is an empirical rule and it is probably more useful to consider the fact that cations such as thorium may have rather unusual properties — particularly the ability to engage in large complexes in aqueous solutions. There is good experimental evidence for hydrolysis, and the existence of entities carrying a positive charge of 8 or more has been suggested[29]. Presumably these highly charged entities are very strongly adsorbed to the glass surface, resulting in an overcompensation of the surface-charge density and a reversal in sign. Whatever the mechanism, however, the experimental fact is that extremely small quantities of thorium present in the aqueous solution can completely alter the electrokinetic behavior of glass and other surfaces. It is usually observed in ion-exchange membranes, for example, that the presence of thorium ions will produce negative electroosmosis[30].

Let us now consider the glass microcapillary filled with an aqueous salt solution. One question that immediately arises concerns the relation between the diffuse double layer and the lumen of the capillary. Table 4.1

Table 4.1   Extension of Double Layer for Different Concentrations of 1:1 Electrolyte

c (moles/liter)	Debye length (cm)
10^{-5}	10^{-5}
10^{-3}	10^{-6}
10^{-1}	10^{-7}

shows the calculated Debye length for various electrolyte concentrations. The Debye length is given by

$$L_D = \left[\frac{10^3 \epsilon k T}{4\pi e^2 N \sum_i c_i z_i^2} \right]^{1/2}, \tag{4.1}$$

where ϵ is the absolute dielectric constant of the solvent, k is the Boltzmann constant, T is the absolute temperature, e is the electronic charge, N is Avogadro's number, c_i is the concentration (mole per liter) of the ith ionic species, and z_i is the valence of ith ionic species.

The dimensions of the diffuse layer are thus quite small. It is important to note that (4.1) results from the use of the Boltzmann relation in the

Poisson equation and then the application of the Debye–Huckel approximation; therefore it does *not* depend upon the coordinate geometry assumed for the Poisson equation, and is identical for both a cylindrical capillary and a plane. In Fig. 4.4, two possible electric potential profiles in the lumen are illustrated. In one case the lumen is wide enough to allow for complete decay to zero; in the other, this is not possible and there is a nonzero charge density throughout. This problem has been studied in some detail in connection with plane parallel double layers[21] and cylindrical capillaries[31], and is of some importance in membrane biophysics [32]. Considering the dimensions of microelectrodes and the concentrations of electrolytes usually involved, overlapping of double layers should not occur.

4.5 TIP POTENTIALS OF GLASS MICROELECTRODES

Glass microelectrodes are ordinarily used as salt bridges for establishing direct electrical contact with the interior of cells. Many of the associated technical problems have been dealt with elsewhere[6,8] and will not be

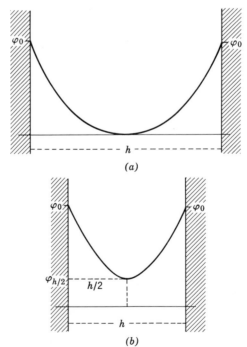

Fig. 4.4. Electric potential profile in (*a*) thick and (*b*) thin layers.

treated here. In connection with use of these devices in measuring resting potentials the existence of an important artifact has been recognized for some time[3]. This is the so-called *microelectrode tip potential*. Its existence precludes accurate determination of cell membrane resting potentials – particularly with extracellular solutions of low ionic strength, in which the error may be as large as 50 mV.

It is probably best to operationally define the tip potential as simply the difference between a total-circuit emf before and after breaking the tip of the microelectrode, with the sign of the tip potential taken as that of the inside of the electrode relative to that of the outside solution. The situation with respect to the resting potential is illustrated in Fig. 4.5.

Adrian[3], using a somewhat different definition of the tip potential, in 1956 reported that the tip potential of electrodes filled with 3 M KCl was logarithmically related to the external concentration of salt (in an inverse way) and correlated with electrode resistance. These observations were confirmed by Holtzman and myself[12] in an extensive series of experiments on nearly two-thousand microelectrodes. We found that trace concentrations of thorium chloride in the outside solution can eliminate the tip potential and that physiological concentrations of calcium chloride significantly reduce it. The observations are summarized in Table 4.2.

Thorium has recently been used to eliminate tip potentials in studies concerning the behavior of the resting potential of striated muscle fibers in solutions of low ionic strength[34].

It is important to note that at certain concentrations of thorium chloride, the sign of the tip potential may be reversed – an observation which has been confirmed by Szabo[13]. If electrodes are filled with 3 M KCl containing 10 μm $ThCl_4$ and $ThCl_4$ is *not* present in the outside solution, the

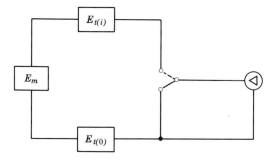

Fig. 4.5. Schematic diagram of tip potential artifact. $E_m{}^*$ is the recorded membrane potential; E_m is the actual membrane potential. $E_{t(i)}$ and $E_{t(0)}$ are tip potentials of electrode in intracellular and extracellular solutions, respectively. (Reference electrode potentials are not shown.) $E_m{}^* = E_m + E_{t(i)} - E_{t(0)}$.

Table 4.2   Mean and Standard Error of Microelectrode Tip Potentials (in mV) with Outside Solutions of Various Electrolyte Compositions

Composition	Sodium or Potassium Concentration of Outside Solution			
	1 mM	10 mM	100 mM	1 M
NaCl	-33.9 ± 3.3	-22.2 ± 2.0	-11.9 ± 0.9	-5.5 ± 0.5
NaCl + 2 mM $CaCl_2$	-12.6 ± 1.2	-13.8 ± 1.1	-10.0 ± 0.9	-6.0 ± 0.5
NaCl + 10 μM $ThCl_4$	$+5.0 \pm 0.9$	$+2.3 \pm 1.0$	$+0.8 \pm 0.9$	-5.4 ± 0.4
KCl	-30.0 ± 2.0	-20.3 ± 1.8	-10.5 ± 0.7	-5.0 ± 0.3
KCl + 2 mM $CaCl_2$	-15.3 ± 1.1	-15.2 ± 1.4	-10.3 ± 0.8	-5.1 ± 0.3
KCl + 10 μM $ThCl_4$	$+9.1 \pm 1.0$	$+6.4 \pm 0.9$	-0.3 ± 0.9	-4.9 ± 0.2
NaCl + 1 μM $ThCl_4$		-8.6 ± 0.9		
NaCl + 50 μM $ThCl_4$		$+1.2 \pm 0.9$		
NaCl + 100 μM $ThCl_4$		-3.8 ± 0.6		

tip potentials are *not* affected (as Szabo also confirmed). In addition, Szabo observed that if the internal thorium chloride concentration is increased to about 50 mM, a reduction in tip potential is produced. This reduction may possibly be a result of leakage of thorium.

More recently, in an attempt to quantitatively define the behavior of these electrodes, I made a number of experiments with KCl solutions of which conductance was measured. The results for an electrode filled with 1 M KCl are shown in Fig. 4.6. This electrode has unusually large tip potentials, with a slope of 53 mV per log unit. The abscissa represents the ratio of the conductance of the external solution to the conductance of 1 M KCl; KCl solutions at ten different concentrations are thus represented. The electrode is seen to be behaving as a nearly ideal cation-exchange membrane. (Some of the possible implications of this will be discussed later.)

If we are to understand the behavior of these electrodes, we must know something about the structure of the microelectrode tip in aqueous salt solutions. If Pyrex glass membranes approximately 0.1 μ thick (as evidenced by the presence of interference colors) are exposed to water, they rapidly change their shape and become soft. This is probably an indication of extensive hydration and swelling. It is known that Pyrex can absorb 1.8 mg of water per cubic centimeter of glass[33], and that probably the electrical resistivity of such thin membranes falls by 5 to 10 orders of magnitude on exposure to water.

Another observation of some interest is that if electrodes are filled with a dilute salt solution, for example, 10 mM, and exposed to an identical solution, substantial tip potentials (e.g., 10 mV), either positive or

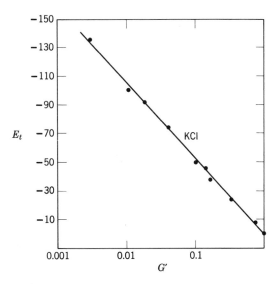

Fig. 4.6. Tip potentials of microelectrode filled with 1 M KCl. G' is ratio of measured conductance of external KCl solution to conductance of 1 M KCl. E_t is the measured tip potential.

negative, can be measured. The high electrode resistances, 1,000 MΩ or more, prevent a definitive quantitative analysis. (This result will be discussed below.)

I am sure there are other possible interpretations, but at present the following ones seem to me to be the simplest means of explaining tip-potential observations.

The first possibility is that noted by Holtzman and myself[12] and briefly suggested by Bingley[9]. The tip potential is considered to originate from one or more interfacial potentials between glass and electrolyte solutions, and the system is approximately described by the equivalent circuit shown in Fig. 4.7.

It is assumed that the phase-boundary potentials are greatly reduced by high cation concentrations in the ambient solutions. If an electrode contains 3 M KCl, the internal interfacial potential should be practically nonexistent. The outside interfacial potential, however, can be substantial and will depend on the ionic strength and species of ions in the outside solution. With *symmetrical* solutions, differences between the inner and outer surfaces will cause a net potential difference in a manner similar to that presumed for asymmetry potentials in glass electrodes[35]. It is important to note that there is no implied violation of any thermodynamic principle. It must be remembered that these glass-water systems are *not*

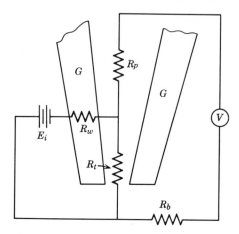

Fig. 4.7. Equivalent circuit illustrating possible origin of tip potential. E_i is inter-facial potential; G is glass wall of the microelectrode; R_p is resistance of bulk electrolyte in the pipette; R_t is resistance of the tip; R_w is resistance of the wall. R_b is resistance of the bath electrolyte; V is the voltage-measuring device.

equilibrated. There is an exchange reaction occurring at each surface. If the surfaces differ, there will be two different rate processes occurring and the existence of a potential difference between identical solutions is to be expected.

Near the tip of the electrode the glass wall should be about $0.1\ \mu$ thick. The critical point in this interpretation is the magnitude of change in specific resistivity when glass this thin is exposed to water. The observations on membranes showing interference colors suggest that extensive hydration and swelling occur. The specific resistivity of Pyrex 7740 glass is 10^{14}–$10^{15}\ \Omega$-cm. When the glass is $0.1\ \mu$ thick, the areal resistance is 10^9–$10^{10}\ \Omega/cm^2$. If the wall is thin over an area of $25\ \mu^2$, the dry resistance of the wall near the tip would be of the order of 10^{16}–$10^{17}\ \Omega$. If hydration, swelling, and perhaps dissolution cause the resistance to decrease by 10 orders of magnitude, the amplitude of the tip potential would be related to a voltage divider network as shown in Fig. 4.7, and the observed correlation of tip potential with electrode resistance would follow.

In summary, this first interpretation (or hypothesis) considers the system to be essentially that of a punctured glass membrane.

The second interpretation is closely related to the first: Hydration and swelling of the wall near the tip completely close the lumen, and the system is that of a complete glass membrane. Such an electrode would behave as a nearly perfect cation-exchange membrane and would explain the results observed in Fig. 4.6.

Some people have suggested that the microelectrode tip potential is a streaming potential. Experimental results, however, do not support this view. Against a 10-mM bath solution, electrodes filled with 100 mM KCl almost always have lower tip potentials than those filled with 1 M KCl. At these concentrations, however, streaming potentials are expected to increase when the capillary electrolyte concentration is lowered.

4.6 CURRENT-VOLTAGE CHARACTERISTICS

Glass microelectrodes are used not only to record membrane potentials, but also to pass electric current. The two important associated problems concern the rectifying properties of the electrode and the dielectric breakdown of glass at high field strengths.

In general the tip of a microcapillary containing a salt solution differing from that in the surrounding bath should be considered a region the composition of which may change owing to electroosmosis produced by the passage of electric current. The integral resistance of this region, by the simplest approach, will be determined by

$$\frac{1}{F} \int_a^b \frac{dx}{\sum_i \mu_i z_i^2 c_i}, \tag{4.2}$$

where a and b are the ends of the tip region, F is the Faraday constant, μ_i is the mobility of the ith species, z_i is its valence, c_i is its concentration, and x is the axial coordinate. As the concentration profiles will change with an applied emf, the value of this integral (i.e., the resistance of the tip) is voltage dependent, and the tip region exhibits a nonlinear current-voltage characteristic. The behavior of the microcapillary is therefore very similar to that of the porous membranes discussed in detail by Teorell[36] and later by Kobatake[37]. If the capillary is fine enough, or the electrolyte dilute enough (so that the total conductance is dominated by the surface conductance), the system is more complicated and some treatment similar to that discussed by Lavallée[7] must be used. In any case the rectification behavior of glass microelectrodes is expected, and we need not postulate "clogging" of the tip in order to explain what is observed.

It might also be noted that the change in concentration profiles with an applied emf is not instantaneous. A "redistribution time" exists, and the current-voltage characteristic will be dependent on time.

Some aspects of the current-voltage characteristics of glass microelectrodes, including the effects of hydrostatic pressure, have been reported by Rubio and Zubieta[5].

The intrinsic breakdown field strength of glasses is generally of the order of 10^7 V/cm, and it is believed that the breakdown process is probably an "electron avalanche" process[14]. It has been found, however, that for glass membranes less than 5 μ thick, the breakdown strength begins to increase[16,38]. Similar behavior has been observed in crystals [39]. A further complication is that the glass wall near the tip of the electrode (as has been mentioned) is probably extensively hydrated. The field strength necessary for dielectric breakdown, therefore, will vary from electrode to electrode.

4.7 INTRACELLULAR INJECTION OF IONS

Glass microelectrodes have been given an important use as means of injecting ions into the interior of cells, particularly nerve cells[2,10]. In order to avoid cellular damage, these electrodes must have fine tips, preferably not more than 0.5 μ in diameter. As has been indicated, such electrodes often behave as excellent cation-exchange membranes — suggesting that the transport numbers of *anions* through such electrodes must be very small, if not zero[30]. This fact has unfortunately not been considered in calculations concerning "ionophoretic injection" of anions into nerve cells. The problem exists only with intracellular injections; for extracellular injections, electrodes with tip diameters of 2–5 μ are normally used. Experimental determination of transport numbers of anions through fine electrodes must be made before anion injection experiments can be fully evaluated.

REFERENCES

[1]   G. Ling, and R. W. Gerard, *J. Cellular Comp. Physiol.*, **34**, 383 (1949).
[2]   J. S. Coombs, et al., *J. Physiol. (London)*, **130**, 326 (1955).
[3]   R. H. Adrian, *J. Physiol. (London)*, **133**, 631 (1956).
[4]   J. H. Emck, *Phys. Med. Biol.*, **3**, 339 (1959).
[5]   R. Rubio, and G. Zubieta, *Acta Physiol. latinoam.*, **11**, 91 (1961).
[6]   P. Fatt, in *Methods in Medical Research* (J. H. Quastel, ed.), Vol. 9, Year Book, Chicago, 1961.
[7]   M. Lavallée, "On the Zeta Potentials in Glass Microelectrodes," Ph.D. thesis, University of Southern California, (1963).
[8]   K. Frank, and M. C. Becker, in *Physical Techniques in Biological Research*, (W. L. Nastuk, ed.), Academic, New York, 1964.
[9]   M. S. Bingley, *Nature (London)*, **202**, 1218 (1964).
[10] D. R. Curtis, in *Physical Techniques in Biological Research* (W. L. Nastuk, ed.), Academic, New York, 1964.

[11] R. F. Bils, and M. Lavallée, *Experientia*, **20**, 231 (1964).

[12] D. Agin, and D. Holtzman, *Nature (London)*, **211**, 1194 (1966).

[13] G. Szabo, "Etude des Phénomènes d'Amplification Ionique dans les Membranes Artificielles, M.S. thesis, University of Montreal (1966).

[14] J. M. Stevels, *Handbuch der Physik*, **20**, 350 (1957).

[15] R. J. Maurer, *J. Chem. Phys.*, **9**, 579 (1941).

[16] J. Vermeer, *Physica*, **22**, 1257 (1956).

[17] G. W. Morey, *The Properties of Glass*, Reinhold, New York, 1954.

[18] L. Holland, *The Properties of Glass Surfaces*, Chapman and Hall, London, 1964.

[19] R. G. Bates, in *Reference Electrodes* (D. J. G. Ives and G. J. Janz, eds.), Academic, New York, 1961.

[20] R. G. Bates, *Determination of pH*, Wiley, New York, 1964.

[21] E. J. W. Verwey and J. T. G. Overbeek, *Theory of the Stability of Lyophobic Colloids*, Elsevier, Amsterdam, 1948.

[22] H. R. Kruyt (ed), *Colloid Science*, Elsevier, Amsterdam, 1952.

[23] M. Gouy, *J. Phys. Radium*, **9**, 457 (1910).

[24] D. L. Chapman, *Philo. Mag.*, **25**, 475 (1913).

[25] O. Stern, *Z. Elektrochem.*, **30**, 508 (1924).

[26] A. J. Rutgers, and M. DeSmet, *Trans. Faraday Soc.*, **41**, 758 (1945).

[27] J. T. Davis, and E. K. Rideal, *Interfacial Phenomena*, Academic, New York, 1963.

[28] G. Halsey, and A. R. Burkin, *Nature (London)*, **193**, 1177 (1962).

[29] P. Albert et al., *Nouveau Traité de Chimie Minérale*, Vol. 9, Masson, Paris, 1963.

[30] F. Helfferich, *Ion Exchanges*, McGraw–Hill, New York, 1962.

[31] E. L. Rice, and R. Whitehead, *J. Phys. Chem.*, **69**, 4017 (1965).

[32] D. Agin, *Proc. Nat. Acad. Sci.*, **57**, 1232 (1967).

[33] D. Hubbard, *J. Res. Nat. Bur. Std.*, **36**, 511 (1946).

[34] D. Holtzman, *J. Gen. Physiol.*, **50**, 1485 (1967).

[35] M. Dole, *The Glass Electrode*, Wiley, New York, 1941.

[36] T. Teorell, *Progr. Biophys. Biophys. Chem.*, **3**, 305 (1953).

[37] Y. Kobatake, and H. Fujita, *J. Chem. Phys.*, **40**, 2212, 2219 (1964).

[38] P. M. Sutton, *Progr. Dielectrics*, **2**, 115 (1960).

[39] A. F. Jaffe, *The Physics of Crystals*, McGraw–Hill, New York, 1928.

CHAPTER 5

Electrochemical Properties of "Incompletely Sealed" Cation-Sensitive Microelectrodes

A. A. LEV

Laboratory of Cell Physiology, Institute of Cytology, Academy of Sciences of the U.S.S.R. (Leningrad)

5.1 INTRODUCTION

In the course of our work with potassium- and sodium-selective micro-electrodes[1–4], two unexpected facts were established: First, the resistance of a large number of potassium-selective microelectrodes made from $NAS_{27.5-4.8}$ glass was almost the same as that of the conventional micropipette electrodes with tip diameter less than 0.4μ. Second, potassium rather than sodium specificity was found in some microelectrodes made from NAS_{20-10} glass, a typical sodium-specific electrode glass.

Fine-tipped cation-sensitive microelectrodes are necessary[1,3–5] for measurement of cation activity in the protoplasm of striated muscle fibers and other cells. We fabricated microelectrodes of Hinke's design[6] with lengths of the noninsulated "working" part about 3–5 μ and tips less than 1 μ in diameter. The diameter of the tip of the insulating capillary made from Pyrex was about 3 μ. The procedure of sealing micropipettes of cation-sensitive glass was most delicate. A platinum loop microforge was mounted on a micromanipulator of the Fonbrune type. This loop heated to yellow incandescence was used for sealing micropipettes under microscopic control of 200–400 magnification. Changes in light refraction and small changes in the shape of the micropipette tip served as an indication of sufficient sealing. No water penetration into the microelectrodes from the outside by capillary forces could be noticed after this procedure. After sealing, microelectrodes were filled with $1.0 M$ KCl (except for a

I wish to thank Drs. Raya Selkova and E. P. Buzhinsky for their help in this work.

series in which 1.0 M NaCl was used) and covered with Pyrex insulation by the usual technique [5].

Resistance measurement is a most simple method of checking for open channels in the sealed tip. If one assumes that the thickness of the capillary wall at the tip region is about 0.2 μ, it is easy to calculate the resistance of the microelectrode using the above dimensions of the microelectrode working tip and the specific electrical resistivity of the glass [7,8]. Such calculations for microelectrodes made from potassium-specific $NAS_{27.5-4.8}$ and sodium-specific NAS_{20-10} glass yield 3.10^{11} and $6.10^{11}\,\Omega$, respectively. For some of the microelectrodes made from NAS_{20-10} glass which showed Na^+ specificity, the mean value of the measured resistance was $4.76 \pm 1.33 \cdot 10^{11}\,\Omega$. But in cases where microelectrodes made from the same glass manifested unexpected potassium specificity the resistance was significantly lower. For series of 19 microelectrodes made from sodium specific glass, but behaving as potassium specific, the resistance was $1.1 \pm 0.4 \cdot 10^{11}\,\Omega$. Especially low resistance was found for microelectrodes made from $NAS_{27.5-4.8}$ glass; the mean resistance value for a series of 36 microelectrodes was $1.68 \pm 0.94 \cdot 10^9\,\Omega$. A difference between resistance values of microelectrodes made from NAS_{20-10} and $NAS_{27.5-4.8}$ glasses cannot be explained by the small difference in the specific resistivity of these glasses. Thus we concluded that in a considerable number of cation-sensitive microelectrodes prepared by the usual technique, there were narrow pores in the sealed part of the capillary tip. Such microelectrodes are "incompletely sealed."

Aside from their practical importance for biological microelectrode measurements, electrochemical properties of pores formed in such highly negatively charged material as glass are of interest for the theory of cation specificity.

5.2 RESULTS

The following properties of incompletely sealed microelectrodes were studied: (a) slope of emf dependence on the concentration of KCl or NaCl in outside solution, and cation transference numbers determined with the aid of these emf measurements; (b) specificity constants for cations of Ia group; (c) dependence of microelectrode potentials on pH of outside solutions; (d) kinetics of the establishment of steady emf level; and (e) resistance of microelectrodes immersed in different salt solutions and changes of the resistance in the course of transitionary processes.

The characteristic dependence of the emf of incompletely sealed microelectrodes on $\log_{10}$ of activity of outside KCl solutions is shown in Fig.

5.1*a*. A deviation from the straight line was found only at about 0.005 *M* and in more dilute solutions. Slope of the straight line was found to be near the thermodynamically predicted value for a perfect permselective electrode system (see the mean values given in Table 5.1). Cation transference numbers calculated from such a slope by the method of Oda and Yawetaya[9] should be of equal unity.

The specificity constants K_{ij} for any pair of monovalent cations may be derived from the data on the corresponding biionic potentials:

$$K_{ij} = \exp\left(\frac{\Psi_{ij}}{RT/F}\right), \tag{5.1}$$

where Ψ_{ij} is a biionic potential in the case of cations *i* and *j*. In this work biionic potentials were not measured directly, but were obtained as a difference of emf $(E_i - E_j = \Psi_{ij})$ when microelectrodes were subsequently immersed in chloride solutions having equal activities of cations *i* and *j*.*

*Activities of single ions may be determined by measurement with the aid of galvanic cells with appropriate reversible electrodes if some basic assumptions (e.g., the equality of activity of potassium and chloride ions in diluted KCl solutions) have been made.

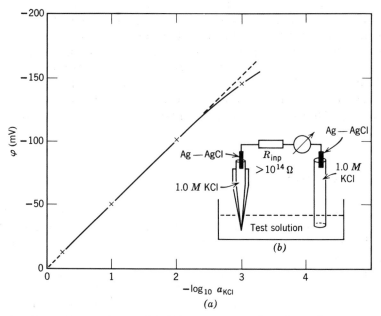

Fig. 5.1. Dependence of potential difference (*a*) of microelectrode, measured in circuit shown in (*b*), when KCl of different concentration was used as test solution.

Potassium in respect to sodium specificity found for some incompletely sealed microelectrodes was near the highest known for inorganic potassium-specific membranes, for example, the best potassium-specific glasses[10,11]. The minimal values of K_{KNa} equal to 0.11–0.13 appeared to be the same as for a series of incompletely sealed microelectrodes made from potassium-specific $NAS_{27.5-4.8}$ and from sodium-specific NAS_{20-10} glasses. The mean value of K_{KNa} for a series consisting of 36 microelectrodes from $NAS_{27.5-4.8}$ was 0.29. For microelectrodes from NAS_{20-10} glass, it was 0.38. No correlation was found between the magnitude of K–Na biionic potentials and the resistance of microelectrodes made from $NAS_{27.5-4.8}$ glass (correlation coefficient $r < 0.1$). For the series of microelectrodes from NAS_{20-10} glass, a very weak correlation between the mentioned parameters was found ($r = -0.33$).

The emf dependence on pH for imcompletely sealed $NAS_{27.5-4.8}$ glass microelectrodes immersed in 0.1 M KCl and NaCl solutions is shown in Fig. 5.2. From these data the hydrogen-potassium specificity constant K_{HK} was found equal to 0.11. In an outside solution of pH 1, incompletely sealed microelectrodes immediately lose their permselectivity and cation specificity. This electrode poisoning is transient, as initial microelectrode properties were restored after aging the microelectrode in KCl solutions or in air.

For the series consisting of six incompletely sealed microelectrodes from $NAS_{27.5-4.8}$ glass, a special determination of specificity values for Li$^+$, Na$^+$, Rb$^+$, and Cs$^+$ was performed. The data are presented in Fig. 5.3a as values of biionic potentials measured in 0.1 M chlorides in respect to 0.1 M KCl. The mean values of specificity constants for this series of microelectrodes was found to be as follows:

$$K_{KLi} = 0.32; K_{KNa} = 0.36; K_{KCs} = 0.94; K_{KRb} = 1.11.$$

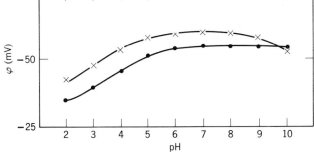

Fig. 5.2. Dependence of microelectrode potential difference on pH of 0.1 M KCl solution (dots) and 0.1 M NaCl solution (crosses).

Table 5.1 Biionic Potentials, Slopes of Potential-Concentration Dependence, and Resistances of Incompletely Sealed Microelectrodes (Mean Values and S.E. of Mean)

Inside solution	Number of electrodes	K/Na Specificity (in 0.01 M Cl solutions)		Slope of Potential/Concentration Dependence (mV/log$_{10}a_{K}$)	Resistance ($\Omega \cdot 10^9$)	Correlation Coefficients (r)	
		K – Na Biionic Potential (mV)	K_{KNa}			Biionic Potential/ Resistance	Slope/ Resistance
				NAS$_{27.5-4.8}$			
1.0 M KCl	36	30.3 ± 2.3	0.29	55.6 ± 0.7	1.68 ± 0.24	< 0.1	0.25
				NAS$_{20-10}$			
1.0 M KCl	19	24.0 ± 3.1 (40.2)*	0.38 (0.20)	57.6 ± 1.3	110 ± 42 (145)	−0.33 ± 0.2	< 0.1
1.0 M NaCl	12	19.3 ± 3.3	0.46	57.7 ± 1.3	56†	—	—
1.0 M KCl	8	−29.2 ± 5.0	3.24	57.4 ± 1.8	476 ± 133	−0.15	−0.56 ± 0.24

*Figures in parentheses represent values for the pore channel (see text).
†The resistances in this case were measured for three microelectrodes of the series.

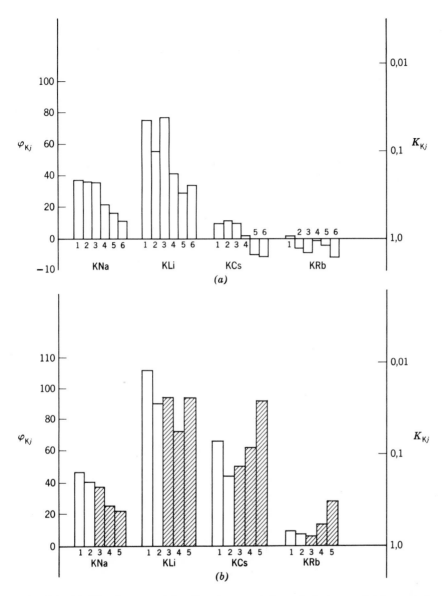

Fig. 5.3. (a) Biionic potentials of series consisting of six incompletely sealed microelectrodes from NAS$_{27.5-4.8}$ glass measured in 0.1 M chloride solutions. (b) Biionic potentials of five bulb macroelectrodes from glasses of slightly different content: (1) NAS$_{27.0-4.0}$; (2) NAS$_{27.0-4.5}$; (3, 4 and 5) NAS$_{27.5-4.8}$. Biionic potentials were measured in same conditions as in (a). Macroelectrodes 3–5 were prepared from same sample of glass used for microelectrodes, with biionic potentials presented in (a). Electrode 5 had been stored in 1.0 M KCl solution for a year.

Accordingly, a sequence of specificities for the cations of $1a$ group for these microelectrodes was:

$$H > Rb \geqslant K \geqslant Cs > Na > Li.$$

It was of interest to compare cations specificity of incompletely sealed microelectrodes with the specificity of usual bulb macroelectrodes made from the same glass. Taking into consideration possible slight changes of the glass composition in the course of pulling and sealing microcapillaries, in addition to the $NAS_{27.5-4.8}$ glass, we studied two electrodes with small variations of the glass composition. These data are shown in Fig. 5.3b also as values of biionic potentials. Specificity constants for three electrodes of the same $NAS_{27.5-4.8}$ glass were found to be slightly different. The mean values of constants for these electrodes were:

$$K_{KLi} = 0.07; K_{KNa} = 0.34; K_{KCs} = 0.09; K_{KRb} = 0.59.$$

The sequence of specificities accordingly was:

$$H > K > Rb > Cs > Na > Li.$$

This sequence and the values of specificity constants for the macroelectrodes from $NAS_{27.5-4.8}$, as well as from glasses with small variations of the content, significantly differed from those for incompletely sealed microelectrodes.

Another important difference between the conventional potassium-specific glass macroelectrodes and those of porous type was found in the kinetic properties. Time required for the establishment of a steady level of emf after changing one outside solution for another with different concentration was much shorter in the case of incompletely sealed microelectrodes when compared with ordinary bulb glass electrodes having the similar electrical time constant.

A specific behavior of incompletely sealed microelectrodes was revealed when some of these electrodes were immersed in 0.01–0.1 M NaCl and LiCl or sometimes in dilute (0.01–0.001 M) KCl solutions. In Fig. 5.4 typical changes of potential differences of this kind are shown. As seen from the given curves, the potential differences are quite stable when microelectrodes are in contact with KCl, RbCl, and CsCl solutions. After these solutions have been substituted by NaCl or LiCl of the same molarity, a new level of potential difference persists for several minutes. Then the potential falls to another level which seems to be a second stable state. It was shown that changes in the potential differences of this kind are accompanied by a simultaneous rise of microelectrode resistance by one order or greater. A quick descending phase of the transition process takes about 0.5 to 1 min, but usually one may notice a more or less

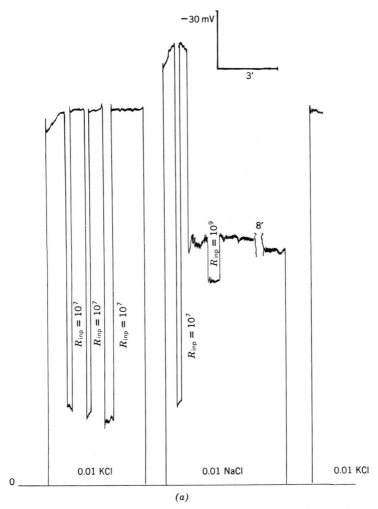

Fig. 5.4. Changes of potential differences of incompletely sealed microelectrodes: (a) Microelectrode in 0.01 M KCl and NaCl solutions. Changes of input resistance of electrometer from $R_{inp} > 10^{14}\Omega$ to $R_{inp} = 10^7$ or $10^9\Omega$ indicated on curves help in estimating rise of electrode resistance after fall of potential difference when microelectrode is in NaCl solution.

prominent foregoing decrease of the potential difference and, correspondingly, a slow rise of resistance. It can be easily shown that the fall of potential difference is not a result of some electric breakdown, because much higher stable potentials may be obtained if the same microelectrodes are in more dilute KCl solutions. These alterations of the electrode properties are completely reversible. After the fall of the microelectrode potential difference in NaCl or LiCl solutions it is enough to immerse the microelectrode for several seconds in KCl solution and restore the initial potential value then to return to previous solutions of NaCl or LiCl (see Fig. 5.4b). The restituted high potential difference persists nearly as before and then there is a new fall of potential. These processes may be repeated many times. No correlation between the latency of transition and the time of previous microelectrode soaking in KCl solution or on the concentration of that solution was found.

Occasionally the microelectrode behavior in NaCl or LiCl solutions has a more complicated character. In Fig. 5.5a examples of spontaneous slow changes in microelectrode potential difference and resistance are shown. In this case the emf of the microelectrode immersed in 0.01 M NaCl has several damped relaxation oscillations before coming to a new stable value. Small relaxation-type oscillations are a usual phenomenon for incompletely sealed microelectrodes immersed in diluted NaCl or LiCl solutions. These emf oscillations may be of different frequencies and regularity. Figures 5.5b and c show two examples of fast relaxation oscillations. For microelectrodes immersed in KCl, RbCl, or CsCl solutions, such oscillations were not found — except when microelectrodes were in dilute KCl solutions.

5.3 DISCUSSION

Permselectivity and cation specificity of narrow glass capillaries have been described by several authors [12–14], but the physicochemical mechanism of these phenomena is still not fully understood.

It is obvious that in the case of incompletely sealed microelectrodes there are at least two parallel ionic permeability channels determining the resulting potential difference: The first is a passway through a glass wall and the second goes through a pore (see Fig. 5.6a). The resulting potential difference may be determined from the simple equation:

$$V = \frac{E_1 + E_2 \, (R_1/R_2)}{1 + R_1/R_2},$$ (5.2)

where E_1, E_2 and R_1, R_2 are emf generated in these channels and corresponding inner resistances. It is clear that the first passway may be neglected in the case of incompletely sealed microelectrodes made from $NAS_{27.5-4.8}$ glass, in which resulting resistance was found to be two orders less than would reflect possible resistance of the glass wall at the working tip. However, in the case of microelectrodes from NAS_{20-10} in which the

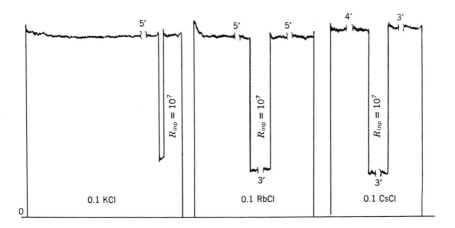

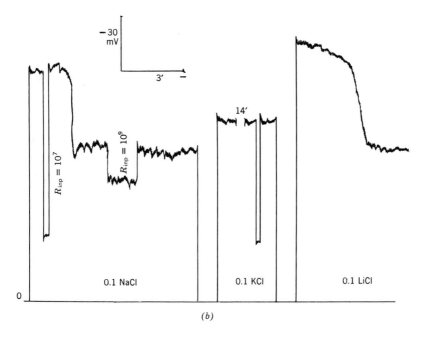

Fig. 5.4. (b) Very stable level of potential difference of microelectrode in 0.1 M KCl, RbCl, and CsCl solutions. Uniform changes of potential difference in 0.1 M NaCl and LiCl solutions. After NaCl is replaced by KCl of some concentration, microelectrode properties are immediately restored.

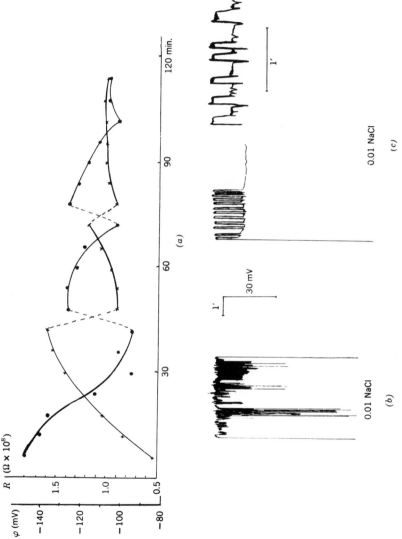

Fig. 5.5. (a) Slow reciprocal changes of microelectrode potential difference (crosses) and resistance (dots). Damped relaxation oscillations of microelectrode parameters in course of establishment of new stable state. (b) Irregular fast potential oscillations when microelectrode is immersed in 0.01 M NaCl. (c) Regular potential oscillations of another microelectrode in same condition. Spontaneous disappearance of oscillations is shown. Second part of curve was recorded at higher speed.

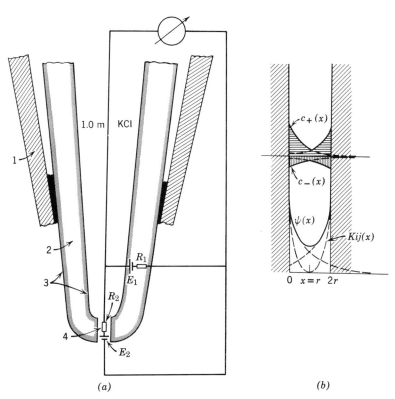

Fig. 5.6. (*a*) Schematic representation of tip of incompletely sealed microelectrode: (1) isolating Pyrex glass capillary; (2) cation-sensitive glass capillary; (3) hydrated glass layer; (4) pore in glass; E_1 and E_2, emf's generated in tip wall and in pore; R_1 and R_2, inner resistances of the corresponding generators. (*b*) The profiles of cation (c_+) and anion (c_-) concentrations and potential (ψ) as a function of distance (x) from the wall of negatively charged material (1,1-electrolyte case). Dotted line indicates suggested profile of cation specificity (K_{ij}).

ratio of measured resulting resistance and that predicted for the tip wall is less than six, the influence of the first channel has to be taken into consideration. If selectivity and specificity properties of the channels mentioned are different (e.g., one channel is specific for potassium and another for sodium), the parameters experimentally determined for the microelectrode may be somewhere in between the two levels. Electrochemical properties of the glass wall at the tip should be within the limits of variations usual for bulb macroelectrodes made of the same glass; this was shown by the similarity of properties in high-resistance sodium-specific microelectrodes and ordinary bulb macroelectrodes of the same sodium-specific glass.

One of the most probable explanations for the mechanism of selectivity and cation specificity of the pore channel is based on the possible changes of the ion concentration and ion mobilities in the pore formed in highly charged material (glass).

For diffusion potential (E_d) of the media with steady flows of charged particles, the following general expression may be used[15]:

$$E_d = \frac{RT}{F} \sum_i \int_{a_p}^{a_{i^*}} (t_i/z_i)\, d\log a_i, \qquad (5.3)$$

where $t_i = u_i \cdot c_i / \sum_{i=1}^{n} u_i \cdot c_i$ is a transference number of i-type ion and u_i, z_i, c_i, and a_i denote mobility, valence, concentration, and activities of this ion, respectively. It is easy to show that except for the dependence of the galvanic cell's emf on the cation activities in solutions separated by the porous glass, there should also be dependence of emf on concentrations and mobilities of the ions inside the pore. In narrow pores the electric potential profile (Fig. 5.6b) should depend on the wall charge, the magnitude of charge interaction (expressed in dielectric constant of the water solution in the pore, which may differ from the constant in a free solution), and the pore radii.* In the case of a highly negatively charged pore wall, the concentration of positive counterions should be much larger than that of coions. If the surface charge of the pore wall and pore dimensions are known, the bulk concentration of counterions and coions inside the pore may be estimated. In the case of equal mobilities of counterions and coions the concentration ratio should be equal to the ratio of corresponding transference numbers. This is a probable explanation for permselectivity of narrow pore systems. Unfortunately, the predominance in the accumulation of different counterions as well as the mobilities of ions inside the narrow pore cannot be determined without special investigation.†

From Eisenman's work[11] it became known that a hydrated surface layer (at least one of potassium-specific glasses of NAS_{27-4} type) has ion-exchange specificity for potassium in comparison with sodium in the order of 80:1. This specificity, much exceeding the levels indicated

---

*The theory for flat double layer interaction was developed by Verwey and Overbeek[16].

†Using the method developed by Conti and Eisenman[17,18], it is possible to estimate the role of two different parameters determining cation specificity: the ion-exchange constant and the ratio of cation mobilities.

by electrode-potential measurements, depends on the properties of the hydrated glass layer and the probable steep decrease in specific cation accumulation with a distance in plane normal to the inner glass surface of the pore (see Fig. 5.6b). If it is true that the ratio of cation concentrations is a determining factor for specificity, the width of the pore should be a more critical parameter for cation specificity than for permselectivity. Indeed, many microelectrodes with comparatively low resistance and perfect permselectivity lacked cation specificity, but none with significant specificity lacked permselectivity.

Mobility of counterions in the dense Helmholtz part of the double layer is probably small, but the mobility in the diffuse Gouy-Chapman layer should be sufficient to provide specific conductivity of about $2 \cdot 10^{-4} \, \Omega^{-1}$ cm^{-1}.* Equivalent ion conductivities for the narrow capillary system are not known, but if the equivalent conductivity of K$^+$ is of the same order as the limiting equivalent conductivity (conductivity of Cl$^-$ present is of no interest because $t_- = 0$), the bulk concentration of K$^+$ in the pore should be about several milliequivalents per liter. Such a conclusion is doubtful, however, both because of diffusion of $1.0 \, M$ KCl solution from the inside of the microelectrode and the high density of negative charges of the glass surface, which should be compensated by cations of the solution. If the concentration of counterions in the pore is higher, the mobilities should be proportionally less.

For *very narrow* glass pores it is not reasonable to divide the space of the lumen into two spaces, one occupied by a diffuse part of a double layer and the other free from it. As is seen in Fig. 5.6b, however, the resulting specificity (and also permselectivity for wider capillaries) should depend on the pore diameter. For a precise comparison of specificity properties of pores in different glasses and for establishing the dependence of specificity on the glass composition, the equality of pore diameters should be maintained. For potassium-specific incompletely sealed microelectrodes from NAS$_{20-10}$ glass, the resulting potential difference can also be influenced by the emf generated through the capillary wall. Fortunately we can estimate the emf and resistance of this channel, as data on sodium-specific (completely sealed) microelectrodes made of the same glass and having the same dimensions of the working tip are available. Mean values for K-Na biionic potentials and resistance of sodium-specific microelectrodes were -29.24 mV and $4.76 \cdot 10^{11} \, \Omega$, respectively. Setting up parallel connections of two channels in incompletely sealed microelectrodes (Fig. 5.6a) and using for the wall the resistance and potential differences (biionic potentials) of completely sealed micro-

*For the calculation, pore diameter of $0.2 \, \mu$ and length of $1.0 \, \mu$ were assumed.

electrodes, we found the resistance of pores in incompletely sealed microelectrodes equal to $1.45 \cdot 10^{11} \, \Omega$ and biionic potential (5.2) to be 40.2 mV. The cation specificity constant K_{KNa} of the pore channel thus calculated was 0.2. This constant is less than the mean values found for microelectrodes from $NAS_{27.5-4.8}$ glass, which indicates better potassium specificity in the pores of microelectrodes made from sodium-specific glass.

If the conductivity of the pore is proportional to the area of the pore section, the diameter of the pore in microelectrodes from NAS_{20-10} glass should be about 10 times less than in microelectrodes made from $NAS_{27.5-4.8}$ glass. The corresponding difference in potassium-sodium specificity between microelectrodes of these two glasses is comparatively small even if the above correction for the influence of the wall channel is made. If our estimation of the ratio of pore area in two series of micro-electrodes is correct, a small difference in the specificity in these cases may be an indication of a special role played by the innermost part of the double layer for the pore cation specificities.

Except for the pore dimensions, a decrease of pore conductivity in microelectrodes made from NAS_{20-10} glass, if compared with those of $NAS_{27.5-4.8}$, may depend on a decrease of cation mobility and/or cation concentration in the pore. A smaller cation concentration in the pores of NAS_{20-10} glass is doubtful, because this glass has a greater anionic field strength than $NAS_{27.5-4.8}$ does. Of course, there may be a large decrease of the cationic mobilities in the pores of the first glass both for K^+ and Na^+, but for unknown reasons. Thus the concept of a difference in pore diameters for these two cases seems to be a more probable explanation.

The preparation technique used for microelectrodes from $NAS_{27.5-4.8}$ and NAS_{20-10} glass was the same, and it is difficult to explain a smaller pore width in NAS_{20-10} glass with higher softening point. It was also strange to obtain a comparatively small variation in the resistance in series of handmade microelectrodes (see S.E. of mean values given in Table 5.1). As an explanation we suggest that the thickness of the hydrated layer, different for different glasses, is extremely important in determining the dimension of pores in microelectrodes.

The process of dealkalization of the glass surface layer when a glass is kept in water solutions or in moist atmosphere is well known. The glasses with large amounts of alkaline oxides (such as $NAS_{27.5-4.8}$) are especially subject to this process, and the deteriorated layer in these glasses is much thicker than in glasses with low alkaline oxides content [11]. The process of dealkalization leads to the formation of a network of $SiO_2 - Al_2O_3$ with low content of alkaline oxides that constitute the surface layer, with a higher melting point compared with the remaining glass phase. In the process of delicate sealing of the micropipette tip with a platinum loop this layer of deteriorated glass may remain unmelted—thus preventing complete sealing of the tip. A pore channel of sponge structure connecting the outside to the inside of the microelectrode may be formed in this way. A visible change in the shape of the tip does not necessarily indicate the absence of a pore inside the core of the glass (Fig. 5.6a).

The pore channel in microelectrodes prepared from NAS_{20-10} glass having less alkaline oxide content and, consequently, a thinner hydrated layer, should be of smaller diameter than microelectrodes of extremely hydrated $NAS_{27.5-4.8}$ glass. If such is the case, the pore diameter should depend predominantly on the thickness of the hydrated layer. This is a possible explanation of the similarity of resistance of microelectrodes in series and the difference of resistance for microelectrodes of different glasses.

If ion-exchange potassium specificity of the surface layer of both glasses used is high enough for the "charging" of the pore surface with K^+, the ratio of K^+ concentration to the concentration of such ions as Na^+ or Li^+ in the outside solution in some limits is not important. Even when we prevented this "charging" process from the inside solution of the microelectrode by filling it with $1.0\ M$ NaCl, we had the same potassium specificity after a short period of contact of this microelectrode with the solutions containing K^+ ions. In Table 5.1 the data for the series of 12 microelectrodes filled with $1.0\ M$ NaCl are presented. The small difference in biionic potentials in this series compared with the series of microelectrodes filled with $1.0\ M$ KCl is not statistically significant.

For incompletely sealed microelectrodes with resistances about $10^8\ \Omega$ and higher, no dependence of microelectrode resistance on the content of outside solutions was found within a 10% error limit. Microelectrodes with comparatively low resistances were subjected to various kinds of cation in the outside solution. In Table 5.2 mean values of the resistance of a microelectrode immersed in $0.1\ M$ chlorides of different alkali metals are given. In this case microelectrode resistance was extremely low, but the slope of potential-concentration dependence measured in the range of 0.1 to $1.0\ M$ KCl was about $57\ mV/\log_{10} a_K$ and the specificity constant K_{KNa} was equal to 0.48.

One of the most distinguishing features of the behavior of incompletely sealed microelectrodes was the sharp change in resistance after the electrodes were immersed in some of the outside solutions. Further investigations are needed to develop a theory to explain this interesting phenomenon, but the following qualitative analysis offers one possibility.

Two facts ought to be taken into consideration when the nature of these

Table 5.2   Mean Resistance Values ($\Omega \cdot 10^7$) of Incompletely sealed Microelectrode Immersed in 0.1 M Chloride Solutions

CsCl	RbCl	KCl	NaCl*	LiCl
3.2	2.0	1.6	3.1	4.2

*After the transition to the second stable state, the resistance in this solution was 5.0 $\Omega \cdot 10^8$.

resistance changes is discussed. First, the rise of the resistance appears predominantly when microelectrodes are immersed in dilute salt solutions (0.01–0.001 M). Second, these changes arise when NaCl or LiCl—but not RbCl or CsCl—are used as outside solutions. In only a few cases did such changes occur in dilute KCl solutions. These facts indicate that both factors, the osmotic pressure of the solution and the nature of the cation in the outside solutions, are important for these transition processes. According to Chapman[19], the osmotic pressure in the double layer is in equilibrium with electric forces. On the other hand, the density of the double layer should depend on the radii of hydrated ions. Significant changes in one of these factors should be followed by the establishment of a new equilibrium state. Possibly the latter process is associated with the substitution of K^+ in the innermost parts of the double layer by less mobile Na^+. This substitution is manifested in the rise of microelectrode resistance, the fall of pore potassium specificity and, related to the latter, a decrease of potential difference in the biionic system. Cations of smaller hydrated radii (Rb^+ and Cs^+) cannot produce so large a change in the properties of the doublelayer and thus in the electrochemical behavior of the pore.

A similar explanation may be given for relaxation oscillations of potential difference such as are found for microelectrodes immersed in NaCl or LiCl solutions. In such cases, when the rise of microelectrode potential exceeds some limits, the change appears to disturb the equilibrium followed by the initiation of the electro-osmotic inward flow of diluted NaCl or LiCl outside solutions. This temporary intense increase of inward flow may be represented as a positive anomalous osmosis[20, 21].

Our investigations of electrochemical properties of microelectrodes were carried out with an electrometer having an input resistance greater than $10^{14}\,\Omega$. In this case, for electro-osmotic processes, the existence of local currents in the microelectrode tip region should be suggested. The influence of current flow on the discussed processes was shown by the shortening of the latency of transition from one state to another when the outer circuit resistance was reduced.

Possibly a change in the ion content of the double layer (the increase of Na^+ or Li^+ concentration) followed by a decrease of cation specificity corresponds to the decreasing phase of the potential oscillation curve. The new state is not stable because the decrease of potential difference leads to the cessation of electro-osmotic flow. Then, owing to the comparatively slow diffusion process, K^+ ions from inside the microelectrode enter the pore and substitute Na^+ ions because of the potassium ion-exchange specificity of the glass surface. This process, followed by a rise of cation specificity of the pore, corresponds to a rising phase of potential oscillations.

The described processes are similar to those in the Teorell membrane oscillatory model [22, 23]. The difference is in the presence of external emf in Teorell's model. In the case we are describing, the pore itself generates emf when a biionic system is involved. Another difference is in the cation specificity of our system and the changes of this parameter in the course of relaxation oscillations.

Electrochemical properties of pores in incompletely sealed microelectrodes in many aspects resemble those of bimolecular phospholipide membranes with cyclic dodecadepsipeptide (valinomycin)[24–26]. For the latter we established a low specificity if Cs^+, Rb^+, and K^+ ions are compared, but a very high specificity for the cations of this group in respect to Na^+ and Li^+. The sequence for these membranes for cations of Ia group was similar to that for incompletely sealed microelectrodes, except that there was no discrimination between Na^+ and Li^+ ions. The dependence of cation specificity on the ionic strength of the solution separated by the membrane, and the destructive effect of low pH and instability (relaxation oscillations) when the membrane separates KCl and NaCl (or LiCl) solutions were found in this case, too. In valinomycin-charged phospholipide bimolecular membranes, there probably exists an analog of porous system in which pores are formed by valinomycin molecules that penetrate in the membrane or move in it. Much higher potassium specificity of these valinomycin-induced pores may be associated with an extremely small pore diameter.

The question arises whether these porous site systems with prominent potassium specificity are relevant to the mechanism of potassium specificity in the passive ion permeability channel in the membrane of the living cell.

REFERENCES

[1] A. A. Lev and E. P. Buzhinsky, *Cytology (USSR)*, **3**, 614 (1961).

[2] A. A. Lev, *Trans. Moscow Soc. Naturalists*, **9**, 30 (1964).

[3] A. A. Lev, *Biophysica (USSR)*, **9**, 684 (1964).

[4] A. A. Lev, *Nature (London)*, **201**, 1132 (1964).

[5] J. A. M. Hinke, in *Glass Electrodes for Hydrogen and Other Cations* (G. Eisenman, ed.), Dekker, New York, 1967.

[6] J. A. M. Hinke, *Nature (London)*, **184**, 1257 (1959).

[7] V. A. Ioffe, G. I. Khvostenko, and I. S. Yanchevskaya, *The Glassy State*, Moscow-Leningrad, 1960.

[8] V. A. Ioffe, and G. I. Khvostenko, *Solid State Phy. (USSR)*, **2**, 509 (1960).

[9] Y. Oda, and T. Yawetaya, *Bull. Chem. Soc. Japan*, **29**, 673 (1956).

[10] G. Eisenman, D. O. Rudin, and J. U. Casby, *Science*, **126**, 831 (1957).

[11] G. Eisenman, in *Advances in Analytical Chemistry and Instrumentation*, Wiley, New York, 1965.

[12] R. H. Adrian, *J. Physiol.*, **133**, 631 (1956).

[13] P. G. Kostyuk, *Physiol. J. (USSR)*, **50**, 373 (1964).

[14] M. Lavallée, and G. Szabo, *Abstr. Second Intern. Biophys. Congress*, *Vienna*, (1966).

[15] B. P. Nicolsky, M. M. Shultz, A. A. Belijustin, and A. A. Lev, in *Glass Electrodes for Hydrogen and Other Cations* (G. Eisenman, ed.), Dekker, New York, 1967.

[16] E. J. W. Verwey, and J. T. G. Overbeek, *Theory of Stability of Lyophobic Colloids*, Elsevier, New York, 1948.

[17] F. Conti, and G. Eisenman, *Biophys. J.*, **5**, 511 (1965).

[18] F. Conti, and G. Eisenman, *Biophys. J.*, **6**, 227 (1966).

[19] D. L. Chapman, *Phil. Mag.*, **25**, 475 (1913).

[20] F. Helfferich, *Ionenaustauscher.*, Verlag Chemie, Weinheim, 1959.

[21] R. Schlogl, *Z. Phys. Chem.*, **3**, 73 (1955).

[22] T. Teorell, *J. Gen. Physiol*, **42**, 831 (1959).

[23] T. Teorell, *J. Gen. Physiol.*, **42**, 897 (1959).

[24] A. A. Lev, and E. P. Buzhinsky, *Cytology (USSR)*, **9**, 102 (1967).

[25] P. Mueller, and D. O. Rudin, *Biochem. Biophys. Res. Commun.*, **26**, 398 (1967).

[26] Valentina A. Gotlib, E. P. Buzhinsky, and A. A. Lev, *Biofizica* (U.S.S.R.), in press.

CHAPTER 6

The Effect of Glass Surface Conductivity Phenomena
on the Tip Potential of Micropipette Electrodes

MARC LAVALLÉE AND GABOR SZABO

Département de Biophysique. Faculté de Médecine. Université de Sherbrooke

6.1 INTRODUCTION

The noting of progressive reduction in size of electrolyte-filled micro-capillaries used as intracellular electrodes [1–3], was not accompanied by any modification of the theory concerning the junction between the aqueous solution of the electrolyte filling the electrode and the medium surrounding this electrode.

Using a system of half cells with reversible junctions and witnessing that every liquid junction between these two half cells was balanced by a junction similar but of the opposite direction, most workers assumed that such a system could operate without introducing any junction-potential error. In other words, the junction (Fig. 6.1a) between the medium filling the microelectrode and the test solution should, except for the direction, be equivalent to the junction between the test solution and the 3 M KCl medium filling the agar bridge.

It was later observed, however, that the two junctions are not completely equivalent, but show a difference of several millivolts [4–6], which disappears when the tip breaks [5, 6] (Fig. 6.1b); this mysterious potential was thereafter called the *tip potential*.

6.2 THEORY

When a glass capillary is filled with water, the silicic acid groups within the lumen dissociate and the fixed anionic groups and free hydrogen ions

---

This work was supported by a grant of the Medical Research Council of Canada. Professor Lavallée is an Associate of the Medical Research Council of Canada.

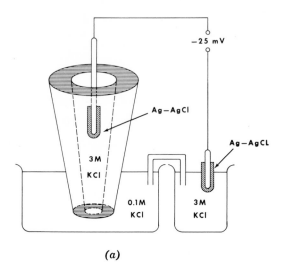

(a)

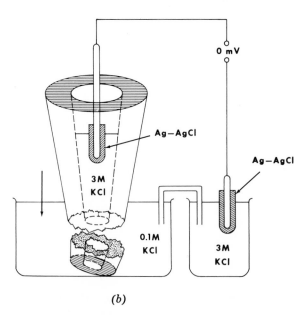

(b)

Fig. 6.1(a) Despite symmetry of system, potential can be recorded in Ling and Gerard[2] glass micropipette electrode. (b) When tip of micropipette breaks. potential is abolished.

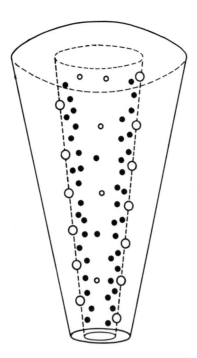

Fig. 6.2. Formation of double layer along glass wall of glass micropipette; $\bigcirc$, fixed negative charges; $\bigcirc$, cations; $\bullet$, anions.

form a double layer along the inner glass wall (Fig. 6.2). We propose that the formation of this ionic double layer accounts for the existence of the tip potential of glass micropipette electrodes.

6.2.1 The Diffuse Region of the Double Layer

The classical treatment of the diffuse double layer rests on the Poisson-Boltzman equation,

$$\frac{d^2\psi}{dx^2} = -\frac{4\pi}{D} \sum_i z_i \epsilon c_i(\infty) \exp\left(-\frac{z_i \epsilon \psi(x)}{kT}\right), \tag{6.1}$$

where ψ is the electrostatic potential, x, the distance from the glass water interface, $c_i(\infty)$ is the number of ions at zero potential, D is the dielectric constant of water, z is the valency, k is the Boltzman constant, and T is the absolute temperature.

Therefore the following assumptions can be made:

1. The mobile ions cannot penetrate into the glass wall.
2. The charge density of the surface is uniformally spread and varies only in a direction normal to the surface of the glass wall.

3. The dielectric constant is independent of the potential and the charge density. The small decrease of the dielectric constant in this area (caused by the intense electrical field) is compensated by the polarization of water around the counterion.
4. The counterions are point charges. In reality, the nonzero size of the counterion decreases the ionic concentration of the surface, but this effect in dilute solution ($10^{-3} M$) is almost cancelled by the increase of the ionic concentration brought about at the surface by the coulombic energy of interaction between different counterions.
5. The radius of the electrode is large in comparison to the thickness of the double layer. In our experiment the microelectrodes were filled with $0.5 M$ KCl solution; therefore the inner surface of the electrode can be considered flat. The relatively large dimension of the cross section of the micropipette in comparison with the thickness of the double layer validates the use of the Boltzman equation when there is a high concentration of electrolyte[7]. The fact remains, however, that at $0.5 M$ a possible important correction may have to be added to ψ_0 if the value of the electrostatic potential is larger than $100 \, mV$; the error brought by the ionic volume cannot be compensated.

The integration of the Poisson-Boltzman equation was carried out by Gouy[8] and Chapman[9]. The Gouy-Chapman equation,

$$\frac{d\psi}{dx} = -\left(\frac{32\pi kT c_i(\infty)}{D}\right)^{1/2} \sinh \frac{\epsilon\psi}{2kT}, \tag{6.2}$$

accounts for the distribution of the potential from the glass wall of the micropipette (Fig. 6.3, curve 1). The potential ψ, which appears in (6.2), is not equal to the wall potential ψ_0, but is the potential of the closest approach of the counterions to the wall. Because of the relatively large size of the micropipette, the potential value will be zero at the center of the lumen of the pipette.

The combination of the Gouy-Chapman equation and the Poisson equation for the charge density $\rho(x)$,

$$\rho(x) = -\frac{D}{4\pi} \frac{d^2\psi(x)}{dx^2}, \tag{6.3}$$

leads to the expression for σ_D, the charge density of the surface of the glass

$$\sigma_D = \left(\frac{2DkT c_i(\infty)}{\pi}\right) \sinh \frac{\epsilon\psi_0}{2kT}. \tag{6.4}$$

Assuming electroneutrality, we can state that the charge density of the Gouy layer is equal to the surface charge density of the glass wall. The surface conductivity, σ_ω (in Ω^{-1}), is directly related to σ_D as in the following equation[10]:

$$\sigma_\omega = \frac{\sigma_D l^+}{F}, \tag{6.5}$$

where l^+ is the ionic conductivity ($\Omega^{-1}\,cm^2$) of the counterion (increased by a small correction factor to account for electro-osmosis) and F is the Faraday constant.

We will see later that surface conductivity accounts for a large part of the total conductance of certain glass micropipettes filled with dilute solutions or made of a highly charged glass.

6.2.2 The Nondiffuse Region of the Double Layer

Stern[11] developed an equation to account for the effect of specific adsorption of counterions by the glass surface, that is, the formation of the so-called "Stern layer" along the glass-water interface. Of interest to us here is the specific adsorption of thorium by the glass of the micropipette. Depending on the relative concentration of thorium and other ions in the filling medium and also on the charge density of the glass surface, the existence of a Stern layer (Fig. 6.3) effectively displaces the locus of ψ_0 (as far as the nonspecifically adsorbed counterions are concerned) toward the center of the specifically adsorbed counterions (the inner Helmholtz plane). This displacement also reduces the algebraic value of the effective surface potential ψ_{01} to the lower values of the potential of the inner Helmholtz plane (ψ_{02}, ψ_{03}, ψ_{04}). According to (6.4), the surface charge density will decrease to zero (at $\psi = \psi_{03}$) and will even become negative (at $\psi = \psi_{04}$). Therefore the addition of thorium to the filling solution may reduce (curve 2), cancel (curve 3), or even reverse (curve 4) the effect of the wall charges of the glass. When thorium is added in sufficiently high concentration, anions may replace the cations within the double layer (curve 4).

6.2.3 The Importance of the Surface Conductivity in Glass Microcapillaries

It was found that the Helmholtz-Smoluchowski[12] equation did not hold when the streaming potential was measured on capillaries filled with an aqueous solution containing very small amounts of electrolyte[13]. Rutgers[14] has proposed a correction that accounts for the surface conductivity. Therefore σ, the specific electric conductivity of the Helmholtz-Smoluchowski equation, was replaced by the product $\sigma_0(1 +$

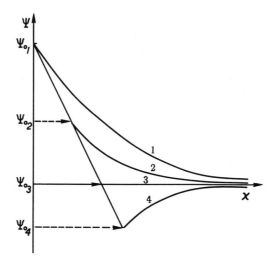

Fig. 6.3. Effect of thorium on potential distribution along wall of glass micropipette. Curve 1, distribution of potential in absence of specifically adsorbed ions; Ψ_{0_1} is assumed to be nearly equal to Ψ_0. Curve 2, potential distribution in presence of traces of thorium; Ψ_{0_2} is much smaller than Ψ_0. Curve 3, larger concentration of thorium cancels effect of surface charge; there is no diffuse region. Curve 4, in presence of still higher concentration of thorium, effects of fixed charges can be reversed, and diffuse region contains anions instead of cations.

$2\sigma_\omega/r_1\sigma_0$), in which σ_0 is the specific electric conductivity of the bulk solution, σ_ω is the specific electric surface conductivity, and r_1 is the radius of the microcapillary.

The corrected Helmholtz-Smoluchowski equation becomes

$$\zeta = \frac{4\pi\eta}{D}\,\sigma_0\!\left(1 + \frac{2\sigma_\omega}{r_1\sigma_0}\right)\!\frac{E}{P},\qquad(6.6)$$

where ζ is the true value of the zeta potential, η is the viscosity, E is the streaming potential, and P is the hydrostatic pressure. From (6.6) and the non-corrected Helmholtz-Smoluchowski equation we can write:

$$\zeta_1 = \frac{\zeta}{1 + 2\sigma_\omega/r_1\sigma_0},\qquad(6.7)$$

where ζ_1 is the apparent (or measured) zeta potential. The apparent zeta potential value will be reduced by one-half when $2\sigma_\omega/r_1\sigma_0$ equals unity, that is, when the diameter of the capillary equals $4\sigma_\omega/\sigma_0$. A curve of this critical diameter value plotted as a function of the concentration of the solution filling the microcapillary illustrates (Fig. 6.4). the importance of surface conductivity — even in the presence of relatively concentrated solutions. We see that the curve does not fall abruptly to zero around 40 μmole as expected by extrapolation of the values found in the presence of very dilute solutions. This change of slope is brought by an increase of σ_ω, the surface conductivity, as the concentration of the filling solution increases.

We did not extrapolate from the curve of Fig. 6.4 at higher concentration because of the imprecision of the last experimental points. We can see, however, that it is perfectly possible that at higher concentration (but with capillaries much smaller than those used by Rutgers) surface conductivity could account for a significant part of the total conductance of a capillary.

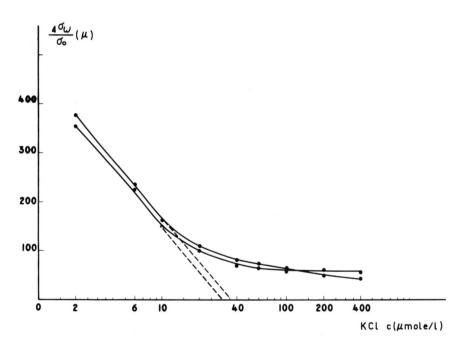

Fig. 6.4. Diameter of Jena Glass 16^{III} capillary at which apparent zeta-potential value is only one-half of true value of zeta potential as a function of concentration of solution filling capillary. Points of curve were calculated from experimental data of Rutgers[14]. Extrapolating only from data obtained in very dilute solution (smaller than 10 μM) one would conclude (dotted lines) that surface conductivity is no longer important at concentration larger than 40 μM. However, experiments made at higher concentrations (solid curves) invalidate this conclusion.

6.3 HIGH-RESISTIVITY GLASS MICROPIPETTE ELECTRODES

Micropipettes were made of two glasses of high resistivity: Pyrex Code 7740 ($\log \rho = 16$) and $NaCaS_{17-12.5}$ ($\log \rho = 12.4$). We propose that the equivalent circuit of Fig. 6.5 accounts for the resistance and tip potential of a micropipette made from glass of high resistivity. The resistance across the wall is so large that it can be neglected; therefore, the equivalent circuit amounts to a resistance R_ω along the surface of the glass (surface resistance) in parallel with another resistance, R_0, the resistance of the bulk solution. Our working hypothesis is that the pathway R_ω acts as a membrane permeable for cations only, as it is constituted by the diffuse region of the double layer.

When the micropipette is filled with a KCl solution of a given concentration and immersed in a KCl solution of a different concentration, a potential E_K will appear in series with R_ω. Therefore, under such conditions, the micropipette amounts to a potassium electrode (the Gouy layer) of a resistance R_ω shunted by a parallel resistance R_0, the latter constituted by the filling solution that is not in contact with the glass wall. The so-called "tip potential" results from this potential E_K in series with R_ω and in parallel with R_0. The maximum value that the tip potential can reach

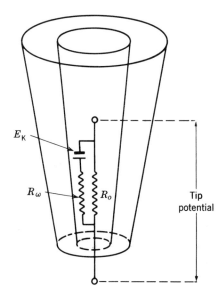

Fig. 6.5. Equivalent circuit of a high-resistivity glass micropipette electrode filled with KCl solution more concentrated than test solutions. (Symbols are defined in text.)

when R_0 is much larger than R_ω is E_K; that is, the Nernst potential depending on the concentration of the cations inside and outside the micropipette. It is also possible to speculate that the tip potential may be abolished by reducing the value of E_K to zero (with the same concentration outside and inside the pipette) or by bringing the ratio R_0/R_ω to zero (with a large micropipette diameter, a highly concentrated solution inside and outside, or a very low glass-surface charge). Accordingly, in high-resistivity glass micropipettes, the change of the tip potential and resistance can be looked upon as resulting from relative change in the value of three parameters E_K, R_ω, and R_0.

In order to enhance the effect of surface conductivity, we filled the micropipette with a more dilute KCl solution (0.5 M) than that (3.0 M) used conventionally by electrophysiologists. Code 7740 glass micropipettes filled with 0.5 M KCl were tested in KCl solutions of various concentrations, but all were maintained at pH 7. At the more dilute concentrations, the electrode had a response of 43 mV per log unit of concentration; at higher concentrations, the slope of this response decreased, presumably because of the decrease of R_0, which, at the very end of the pipette, depends also on the concentration of the solution outside. The closer the response to the thermodynamic limit of 58 mV, the more important the surface conductivity relative to the bulk electrolyte conductivity.

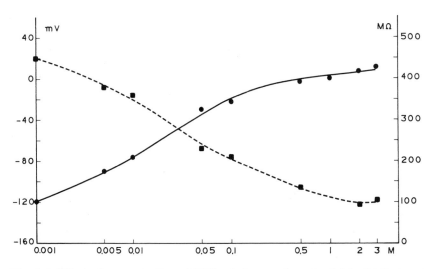

Fig. 6.6. Effect of concentration of KCl solution on tip potential (solid line) and resistance (dotted line) of Pyrex Code 7740 glass microelectrode filled with 0.5 M KCl solution.

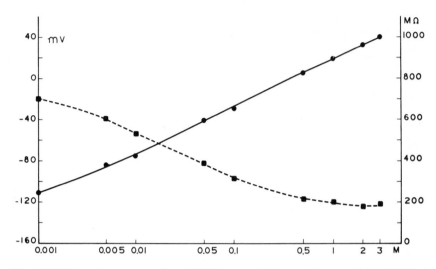

Fig. 6.7. Effect of concentration of KCl test solution on tip potential (solid line) and resistance (dotted line) of a $NaCaS_{17-12.5}$ glass microelectrode filled with 0.5 M KCl solution.

This experiment was repeated (Fig. 6.7) with micropipettes made of $NaCaS_{17-12.5}$ glass. According to findings of our experiments, the $NaCaS_{17-12.5}$ seems to carry more surface charges than Code 7740 glass; the $NaCaS_{17-12.5}$ glass micropipette usually had a larger response to variations of concentration of the test solution than did Code 7740 glass micropipettes of about the same resistance. It is not possible to compare the micropipette of Figs. 6.6 and 6.7 as they do not have the same resistance and presumably not the same terminal diameter. However, we can see that at higher concentrations the response to potassium does not diminish, which shows that, even in this range surface conductivity remains predominant over the bulk electrolyte conductivity. When the concentrations outside and inside are equal, the value of the tip potential is close to zero but not always exactly nil. We shall see later why there is sometimes a small departure of a few millivolts from zero in these conditions.

The change of resistance of the micropipettes during these experiments was brought about by an effect of a change in both R_0 and R_ω. In order to maintain R_0 constant while changing the value of R_ω, we tested the micropipettes in KCl solutions of the same concentration and conductivity but at various pH levels. The resistivity of the solutions of different pH but of the same concentration was carefully checked and not allowed to vary within $\pm 1\%$. The tip potential and resistance of very small Code

7740 glass micropipettes, when filled with dilute solutions, are sensitive to changes of pH (Table 6.1). According to our observations, it is reasonably safe to state that Pyrex glass micropipettes of a resistance smaller than 100 MΩ do not respond to pH. It can be seen that the tip potential of the Pyrex micropipette has a larger response to a change of concentration at high pH, surface conductivity then having a greater importance.

Table 6.1   The Effect of pH Value (Left) and Concentration (Top) of Test Solution on the Tip Potential and Resistance of a Code 7740 (Pyrex) Glass Microelectrode Filled with 0.5 M KCl at pH 7

	Potential (mV)			Resistance (MΩ)		
	0.01	0.1	1.0	0.01	0.1	1.0
4	−25	−12	−4.0	770	360	140
5	−36	−14	−4.2	630	342	140
6	−48	−18	−2.2	520	330	140
7	−55	−22	−2.2	450	300	140
9	−69	−25	−1.4	360	250	140

However, the resistance of the micropipette shows larger variations with a change of concentration at low pH, as surface conductivity is then reduced and can no longer mask the variation of the bulk electrolyte conductivity as it does at high values of pH.

The same experiments were repeated again with $NaCaS_{17-12.5}$ glass micropipettes (Table 6.2). Two main differences appeared in the behavior of these pipettes, presumably because of the higher surface charge. First,

Table 6.2   The Effect of pH Value (Left) and Concentration (Top) of Test Solution on the Tip Potential and Resistance of a $NaCaS_{17-12.5}$ Glass Microelectrode Filled with 0.5 M KCl at pH 7.

	Potential (mV)			Resistance (MΩ)		
	0.01	0.1	1.0	0.01	0.1	1.0
4	−53	−12	41	140	120	80
5	−62	−13	39	130	110	73
6	−66	−17	37	90	90	62
7	−67	−18	36	90	86	62
9	−72	−21	34	80	73	57

the micropipettes of low resistance could still respond to change of pH; the low value of their total resistance seemed to result from a very high surface conductivity (sensitive to pH) rather than from a low value of R_0. Second, as the tip potential already has quite large variations even at low pH, no further increase was observed at higher pH; therefore the tip potential was not very sensitive to pH.

Another way to express this effect is to state that a drop of pH from 9 to 6 did not reduce the surface-charge density (and thus the surface conductivity) very significantly in comparison with the bulk electrolyte conductivity. Consequently such micropipettes can be used as pH indicators. The tip potential of the Code 7740 glass[15] and the resistance of the $NaCaS_{17-12.5}$ glass[16] micropipettes are both sensitive to pH. The pH measurement of the Code 7740 glass (Pyrex) micropipette must be corrected for the value of the membrane resting potential while that of the $NaCaS_{17-12.5}$ glass micropipette has to account for the variation of the cytoplasmic resistivity[17], which itself influences the resistance of the electrode.

6.4 LOW-RESISTIVITY GLASS MICROPIPETTE ELECTRODES

Micropipettes were made of low-resistivity $NaCaS_{22-6}$ glass ($\log \rho = 10.8$) and filled with a $0.5\,M$ KCl solution at pH 7. The leak resistance through the wall of the micropipette is then no longer negligible, and the equivalent circuit (Fig. 6.8) has to account for it. Moreover, because the $NaCaS_{22-6}$ glass is pH sensitive, a voltage E_H will appear across the wall when the pH inside and outside the electrode are different. This voltage is in series with E_A, the asymmetry potential resulting from the dissimilarity of the two faces of glass wall. It should be noted that this asymmetry potential also exists in high-resistivity glass micropipettes, but in series with a very high wall resistance so that it is shunted by the comparatively low parallel resistances R_0 and R_ω. However, E_A is not entirely cancelled by the short-circuiting effect of R_0 and R_ω. When the solutions outside and inside the micropipette are identical, a residual voltage of a few millivolts can still be measured; this voltage originates from the glass asymmetry potential reduced by the shunting effect of the low resistances across the open end of the pipette.

$NaCaS_{22-6}$ glass micropipettes filled with $0.5\,M$ KCl at pH 7 were tested in KCl solutions of different pH and concentrations (Table 6.3) under the same conditions as in the experiments recorded in Tables 6.1 and 6.2. The tip potential results from the algebraic sum of E_K, E_H, and E_A, the contribution of each one depending on the relative values of the series resistance R_g, R_0, and R_ω.

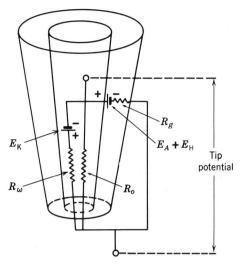

Fig. 6.8. Equivalent circuit of low resistivity glass micropipette electrode filled with KCl solution more concentrated than KCl test solution. R_g, being of the same order as that of R_0 and R_ω is no longer negligible and the equivalent circuit has to account for it. (Symbols are defined in the text.)

Table 6.3   The Effect of pH Value (Left) and Concentration (Top) of Test Solution on the Tip Potential and Resistance of a NaCaS$_{22-6}$ Glass Microelectrode Filled with 0.5 M KCl at pH 7

	Tip Potential (mV)			Resistance (MΩ)		
	0.01	0.1	1.0	0.01	0.1	1.0
4	122	130	147	180	150	130
5	80	88	106	170	140	110
6	35	50	88	150	120	100
7	18	42	74	150	120	90
9	−28	5	46	130	110	90

6.5.   IONIC AMPLIFICATION

It is possible with the equivalent circuit shown in Fig. 6.8 to predict that — if the potential and resistance values are properly chosen — one can obtain a larger voltage per pH unit than the thermodynamic limit of 58 mV at room temperature. For instance, we filled a NaCaS$_{22-6}$ glass micropipette with 1 M KCl solution at pH 9 and tested it in 0.1 M KCl solution

(Table 6.4) between pH 5 and 4. In these conditions E_H and E_K vary in the same direction and summate; the resistances of the glass and of the open tip are matched in order not to cancel E_H entirely.

Table 6.4   The Effect of pH Value of a 0.01 M KCl Solution on the Tip Potential of a NaCaS$_{22-6}$ Glass Microelectrode Filled with 1.0 M KCl at pH 9

pH	4	5	6	7	9
Potential (mV)	340	260	210	160	64
dV/dpH(mV/pH unit)		80	50	50	48

There is, of course, no break of the law of thermodynamics: the hydrogen ions of the solution exert on the charges of the wall an effect similar to that of the grid voltage in a triode. The potassium gradient through the open-tip plays the role of a plate voltage, so that a small change in pH of the solution affects the potassium potential through its action on the glass-wall charge density. The larger the difference of concentration of the KCl solutions inside and outside the pipette, the larger the variations of the micropipette tip potential with changes in pH of these solutions. Therefore, the NaCaS$_{22-6}$ glass micropipette was able to respond with a voltage of 80 mV to a pH change of 1 unit between 4 and 5 of a 0.1 M KCl test solution.

The resistivity of the test solutions did not change significantly between pH 4 and 9. However, the resistance of the electrode showed quite significant changes caused by the action of the hydrogen ions on the surface-charge density, that is, on the surface conductivity. When a battery is in series with the micropipette, the current that passes through the open end depends on the value of the pipette's resistance, which itself is a function of the pH of the solutions. This variable resistance in series with the source voltage provides all the conditions necessary for current amplification, as shown in previously reported experiments[18].

6.6   THE EFFECT OF THORIUM ON THE TIP POTENTIAL OF GLASS MICROPIPETTE ELECTRODES

Finally, we tested the effect of thorium added to the test solution. We observed that at a concentration of 0.05 M, thorium can abolish (Fig. 6.9) or even reverse (Fig. 6.10) the sign of the tip potential. In the latter case the Gouy layer presumably becomes a chloride electrode.

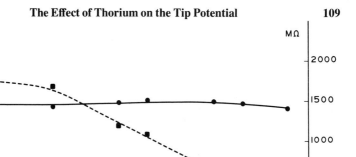

Fig. 6.9. Effect of concentration of KCl test solution on tip potential (solid line) and resistance (dotted line) of a Pyrex Code 7740 glass microelectrode filled with 0.5 M KCl and 0.05 M ThCl$_4$.

We also tested the effect of adding small amounts (10, 50, 100 and 500 μM, as well as 1 mM) of thorium chloride to the solutions outside and inside Pyrex glass micropipettes filled with 0.5 M KCl. No significant effect could be detected at 10, 50 and 100 μM of thorium chloride added to the test solutions or to the filling medium. A slight reduction of the tip potential at 500 μM and a reverse of the sign at 1 mM were observed in the acid range (pH 4–6).

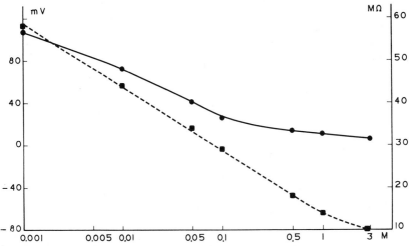

Fig. 6.10. Effect of concentration of KCl test solution on tip potential (solid line) and resistance (dotted line) of NaCaS$_{17-12.5}$ glass microelectrode containing 0.5 M KCl and 0.05 M ThCl$_4$.

Agin[19] found that concentrations of thorium chloride much lower than those used by us suffice to abolish the tip potential; however, his study dealt with micropipettes filled with $3\,M$ KCl, so that the tip potential was already much lower than that of our experiments. It is therefore perfectly conceivable that pipettes filled with more dilute solutions, in which the surface conductivity phenomena play a larger role, call for higher concentration of thorium to cancel the effect of surface charges.

In any case the selection of the exact concentration of thorium chloride to cancel the tip potential is still done on rather an empirical basis.

We have also attempted to neutralize the fixed charges of the wall by exposing the micropipette to vapor of dichlorodimethylsilane before filling. However, the wall of the electrode is made hydrophobic by this procedure, and it is virtually impossible to fill it with an aqueous electrolyte solution afterwards. A perfect method to eliminate the tip-potential artifact from the glass-micropipette electrode measurements remains to be found.

REFERENCES

[1] J. Graham, and R. W. Gerard, *J. Cellular Comp. Physiol.*, **23**, 99 (1946).

[2] G. Ling, and R. W. Gerard, *J. Cellular Comp. Physiol.*, **34**, 383 (1949).

[3] W. L. Nastuk, and A. L. Hodgkin, *J. Cellular Comp. Physiol.*, **35**, 39 (1950).

[4] W. L. Nastuk *J. Cellular Comp. Physiol.*, **42**, 249 (1953).

[5] J. del Castillo, and B. Katz, *J. Physiol. (London)*, **128**, 396 (1955).

[6] R. H. Adrian, *J. Physiol.*, **133**, 631 (1956).

[7] H. B. G. Casimir, in *Tweedie Symposium over sterke Electrolyten en de electrische Dubbelaag*, Sectie voor Kolloid-chemie, Med. Chem. Ver., Utrecht, 1944.

[8] M. Gouy, *J. Physique*, **9**, 457 (1910).

[9] D. L. Chapman, *Phil. Mag.*, **25**, 475 (1913).

[10] J. T. Davies, and E. K. Rideal, in *Interfacial Phenomena*, Academic, New York, 1963.

[11] O. Stern, *Z. Elektrochem.*, **30**, 508 (1924).

[12] M. von Smoluchowski, *Z. Physik.*, **6**, 529 (1905).

[13] H. Freundlich, and G. Ettisch, *Z. Physik. Chem.*, **116**, 401 (1925).

[14] A. J. Rutgers, *Trans. Faraday Soc.*, **36**, 69 (1940).

[15] M. Lavallée, *Circulation Res.*, **15**, 185 (1964).

[16] M. Lavallée, and G. Szabo, *Digest 6th Intern. Conf. Med. Electron. Biol. Eng.*, Tokyo (1965).

[17] O. Schanne, H. Kawata, B. Schäfer, and M. Lavallée, *J. Gen. Physiol.*, **49**, 897 (1966).

[18] M. Lavallée, and G. Szabo, *2nd Intern. Biophys. Congr. International Organization for Pure and Applied Biophysics*, Vienna (1966).

[19] D. Agin, in *Glass Microelectrodes* (M. Lavallée, O. Schanne, and N. Hébert, eds.), Wiley, New York, 1969.

CHAPTER 7

Some Electrical Properties of Fine-Tipped Pipette Microelectrodes

FRED M. SNELL

Department of Biophysics, State University of New York at Buffalo

7.1 INTRODUCTION

Many physiological processes at the cellular and tissue levels of biological organization are reflected in terms of measurable electrical events. The description of these electrical events now occupies a vast literature; further, since the introduction of the micropipette glass electrode by Ling and Gerard[1], much information has been gained that pertains to the more detailed localization of the electrical events. Biologists have gradually come to the realization that cells are heterogeneous and that the variables associated with biological processes must be measured in truly *micro* domains to gain full appreciation of the mechanisms that may be involved. In our laboratory, for instance, we have shown that the electrical potential associated with the transepithelial transport of sodium in frog skin and toad bladder is a monotonic function across the entire cell[2]. The total potential difference does not appear to be localized in the limiting membrane as had been suggested by the work of others[3, 4]. In our experiments we employed micropipette electrodes with tip diameters considerably smaller than those used by others, in order to obviate any tissue or membrane distortion as the electrode was being advanced through the cellular structures.

We have directed little attention to the properties of our microelectrodes per se, except to establish the fact that the results are reproducible and that the electrode is, in fact, sampling the tissue potential in a truly microdomain near the open-ended tip[5]*. In the course of our experiments we have noted that if the electrodes are filled with concentrated

---

*Penetration of isolated frog sartorius muscle shows a sharp localization of the potential difference at the limiting muscle membrane.

111

KCl, they generally show large tip potentials of the order of 20–50 mV. We attributed this phenomenon to an ion-exchange-like behavior at the tip, for the potential was considerably reduced when the fillant solution was NaCl.

In using microelectrodes we have been as delinquent as many other investigators in not carrying out the experiments necessary to gain a full understanding of electrode properties. I should like therefore to present some elementary theory that may be helpful to guide experimentalists in ascertaining the functional characteristics of micropipette glass electrodes. In particular, I shall direct the development toward the ultra-fine-tipped pipette electrodes, for it is in this category that we might expect properties to depart significantly from those of ordinary liquid junctions. It must be remembered that in the case of irreversible liquid junctions, thermodynamic principles do not localize the origin of any electrical potential. To do so we must invoke a model and resort to an analysis of it. The resultant theory must be interpreted with caution, because any theoretical development must of necessity involve assumptions and approximations that should be subjected to experimental test.

In this chapter I confine myself to a simple theory that pertains to three electrical properties of the micropipette glass electrodes: (a) the electrical resistance; (b) the liquid junction potential; and (c) the electrokinetic properties. Finally, I indicate briefly how these ultrafine-tipped pipette microelectrodes may be fabricated, and summarize some of our results [6].

7.2   ELECTRICAL RESISTANCE

The electrical resistance of the open-tipped micropipette electrode is easily calculated if we make a number of simplifying assumptions. With reference to Fig. 7.1, let the barrel length be l with an internal radius of $a(0)$. Let us assume the tip region to be uniformly conical with an included angle θ. Designating the axial coordinant z, the radius in the conical

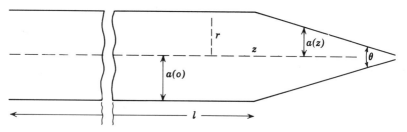

Fig. 7.1. See text for explanation.

region is $a(z)$. If we now assume a constant specific resistance, γ, of the fillant electrolyte, and if we neglect any conductivity through the glass walls, we may write the total resistance Γ as

$$\Gamma = \frac{1}{\pi} \int \frac{\gamma}{a^2(z)} \, dz, \tag{7.1}$$

where integration is to be carried out over the entire length of the pipette. This gives

$$\Gamma = \frac{\gamma l}{\pi a^2(0)} + \frac{\gamma \cot \theta/2}{\pi} \left[\frac{1}{a(t)} - \frac{1}{a(0)} \right], \tag{7.2}$$

where $a(t)$ is the radius at the tip. If we now assume that $a(0)$ is much larger than $a(t)$, the total resistance is approximated by

$$\Gamma \cong \frac{\gamma \cot \theta/2}{\pi a(t)} \tag{7.3}$$

and is seen to be directly proportional to the cotangent of $\theta/2$ and inversely proportional to the tip radius. In making this simple derivation, we have also neglected surface conductance of the glass. Furthermore, it is valid only for low-frequency measurements. One can easily ascertain with $\theta = 4°$ (an average angle for our electrodes) and $\gamma = 5 \, \Omega$ cm that a measured resistance of $10^7 \, \Omega$ implies a tip radius of 2.5×10^{-6} cm (250 Å).

It should be noted here that long-shank electrodes having a very gradual taper (so that the included angle is very small) give rise to a large electrical resistance. Moreover, such electrodes may possess undesirable mechanical properties by being both fragile and flexible.

7.3 LIQUID JUNCTION POTENTIAL

It is well known that surface of many glasses may exchange cations with bathing solution. This is certainly the case with the Pyrex that is generally used in fabricating pipette microelectrodes. Of course, at large radii this surface charge would have an insignificant effect on the electrical properties at the junction between the fillant solution and the test solution. With small radii, however, we might expect that the bulk solution would contain an excess positive charge. Let us consider the surface to have a constant charge density $\epsilon\sigma$, where σ contains a sign of the charge and ϵ is the electronic charge. For simplicity, let us consider only uni-univalent electrolytes. Let us also assume zero net current flow and zero convective flow of the electrolyte solution. We then perform our analysis in two parts: (a) a quasiequilibrium that relates to the potential difference between the tip region and the barrel region; and (b) an irreversible part relating to the liquid junction at the tip. In general the charge density ρ will be a function $\rho(r,z)$. However, for simplicity, let us assume that it is only $\rho(z)$. With these assumptions, electrical neutrality dictates that

$$\iint \rho(z)r\,dr\,dz = -a(z)\epsilon \int \sigma\,dz, \qquad (7.4)$$

or at any position z

$$c_+(z) - c_-(z) + \frac{2\sigma}{a(z)} = 0, \qquad (7.5)$$

where

$$\rho(z) = \epsilon[c_+(z) - c_-(z)]. \qquad (7.6)$$

Employing the Gibbs-Donnan equilibrium, we may write

$$c_+(z) = -\frac{\sigma}{a(z)} + \left[\left(\frac{\sigma}{a(z)}\right)^2 + c_+{}^2(0)\right]^{1/2} \cong c_+(0) - \frac{\sigma}{a(z)}, \qquad (7.7)$$

where the last approximation is valid for those cases in which $\sigma/a(z) \ll c_+(0)$. Utilizing the relevant equations of motion for each ion, namely

$$J_+ = RT\omega_+ \frac{dc_+(z)}{dz} + c_+(z)\omega_+ F \frac{d\psi(z)}{dz} \qquad (7.8)$$

$$J_- = RT\omega_- \frac{dc_-(z)}{dz} - c_-(z)\omega_- F \frac{d\psi(z)}{dz}$$

and the condition

$$J_+ - J_- = 0, \qquad (7.9)$$

we may substitute from (7.7) and integrate to obtain

$$-\frac{F}{RT}[\psi(t) - \psi(0)] = \frac{\omega_+ + \omega_-}{\omega_+ - \omega_-} \ln \frac{(\omega_+ + \omega_-)c_+(0) - [\sigma/a(t)](\omega_+ - \omega_-)}{(\omega_+ + \omega_-)c_+(0) - [\sigma/a(0)](\omega_+ - \omega_-)} \qquad (7.10)$$

J is the flux per unit area, R is the gas constant, T is the absolute temperature, ω denotes the mobilities, F is the Faraday, and ψ is the electrical potential. To a first approximation

$$\left|\frac{\sigma}{a(0)}(\omega_+ - \omega_-)\right| \ll (\omega_+ + \omega_-)c_+(0) \qquad (7.11)$$

so that the logarithmic term may be expanded to first order terms to give

$$\ln\left[1-\frac{\sigma}{c_+(0)}\frac{(\omega_+-\omega_-)}{a(t)(\omega_++\omega_-)}\right] \cong -\frac{\sigma}{a(t)c_+(0)}\frac{(\omega_+-\omega_-)}{(\omega_++\omega_-)}. \qquad (7.12)$$

Thus to this approximation the quasiequilibrium part of the potential is

$$\frac{F}{RT}[\psi(t)-\psi(0)] \cong \frac{\sigma}{a(t)c_+(0)}. \qquad (7.13)$$

It is seen to be a direct function of the surface charge density and inversely proportional to the tip radius.

We can treat the irreversible liquid junction by a variety of approaches. Because there are many other more limiting uncertainties, we choose the simplest approach, which assumes fractional mixing of the fillant electrolyte and the test solution. Without loss of generality, we will assume that the test solution and fillant solution are composed of the same electrolyte, but at differing concentrations. We then have

$$\begin{aligned} c_+(m) &= c_+(t)(1-\alpha)+c_+(\infty)\alpha \\ c_-(m) &= c_-(t)(1-\alpha)+c_-(\infty)\alpha, \end{aligned} \qquad (7.14)$$

where m denotes the region of mixing, t denotes the region within the microelectrode at the tip, and ∞ denotes the region remote from the tip in the test solution. Electrical neutrality dictates that

$$c_+(\infty)=c_-(\infty)=c(\infty). \qquad (7.15)$$

Again applying the equations of motion (7.8), one may integrate from the limits of $\alpha=0$ to $\alpha=1$ to obtain

$$\frac{F}{RT}[\psi(\infty)-\psi(t)]=-\frac{(\omega_+-\omega_-)[c(\infty)-c_+(0)]+(\omega_++\omega_-)[\sigma/a(t)]}{(\omega_++\omega_-)[c(\infty)-c_+(0)]+(\omega_+-\omega_-)[\sigma/a(t)]}$$
$$\times \ln\left[\frac{(\omega_++\omega_-)c(\infty)}{(\omega_++\omega_-)c_+(0)-(\omega_+-\omega_-)[\sigma/a(t)]}\right]. \qquad (7.16)$$

We have also employed (7.7).

In order to gain better understanding, let us examine the case for small differences in the electrolyte concentration between fillant solution and test solution and also where $\omega_+=\omega_-$. Equation (7.16) then reduces to

$$\frac{F}{RT}[\psi(\infty)-\psi(t)]=-\frac{\sigma}{a(t)c_+(0)}, \qquad (7.17)$$

which is seen to be of equal magnitude but opposite sign from that of (7.13). Under these circumstances, as well as with low surface-charge density, the sum of these two potentials (the quasiequilibrium and irreversible liquid junction) vanishes as is expected. Further study of (7.10) plus (7.16) however, allows one to conclude that with other things being equal the smaller the tip radius the greater is the magnitude of the junction potential. This magnitude is also related to the surface-charge density $\epsilon\sigma$ and, other things being equal, the expected potential difference is greater, the greater the magnitude of the surface charge. To avoid significant junction potential, one should work with reasonably large tips and electrolytes whose ionic mobilities are such as to counteract the ion-exchange character.

7.4 ELECTROKINETIC PROPERTIES

A markedly elementary derivation of the electrokinetic properties of a microelectrode is not very revealing, as one of the necessary assumptions is that the radius be large. I will employ a slightly more elaborate derivation, which still must be considered to be very much of an approximation. I follow in part the approach of Rice and Whitehead[7], and employ the Poisson-Boltzmann equation. Let us consider only the region at the tip and assume that ψ is a function only of the radial coordinate r and not of the axial coordinate z. This is valid only if the included angle of the conical section is small. We also assume that $\rho(r, z)$ is now only a function of r, $\rho(r)$. However, we expect that in any case both ψ and ρ are much stronger functions of r than of z, even for reasonable conical angles. With $\rho(r)$ given by (7.6), it may be written as

$$\rho(r) = -2c(0)\epsilon \sinh\frac{\epsilon\psi(r)}{kT},   (7.18)$$

where k is Boltzmann's constant. The usual analysis now confines itself to the cases where $\epsilon\psi \ll kT$ such that the hyperbolic sign may be expanded in a power series retaining only the first two terms. The Poisson-Boltzmann equation then becomes

$$\frac{1}{r}\frac{d}{dr}\left[r\frac{d\psi(r)}{dr}\right] = L_D\psi(r),   (7.19)$$

with

$$L_D = \frac{8\pi c(0)\epsilon^2}{DkT},   (7.20)$$

where L_D is the Debye length, a parameter related to the thickness of the diffuse double layer in the vicinity of a charge surface. The solution of (7.19) is then

$$\psi(r) = \psi(a)\frac{I_0(L_Dr)}{I_0(L_Da)}. \tag{7.21}$$

where I_0 is the zero-order modified Bessel function of the first kind. The charge density may be then written

$$\rho(r) = \frac{DL_D^2}{4\pi}\psi(a)\,\frac{I_0(L_Dr)}{I_0(L_Da)}. \tag{7.22}$$

One may obtain a more exact series solution of the Poisson-Boltzmann equation by retaining higher-order terms in the expansion of the hyperbolic site of (7.18). This is shown in the *Appendix* at the end of this chapter; but for the moment we note that it may differ significantly from the Bessel solution and under certain circumstances (i.e., when $\epsilon\psi$ is not much less than kT) the value of $\psi(r)$ may be radically different from that calculated from the Bessel solution. (This point will be discussed after evaluating the electrokinetic properties.)

The Naivier-Stokes equation specialized for solenoidal flow may be written

$$\eta\nabla^2\vec{u}(r, z, \phi) - \nabla p + \vec{F} = 0, \tag{7.23}$$

in which $\vec{u}(r, z, \phi)$ is the flow velocity, p is the pressure, and $\vec{F}$ is the body force per unit volume. We assume that for the components of the velocity $u_r = u_\phi = 0$ and $u_z = u_z(r)$. The zth component of the body force under consideration is

$$\vec{F}_z = E_z\rho(r). \tag{7.24}$$

With these assumptions, the solution of (7.23) is

$$u_z(r) = -\frac{1}{4\eta}\left(\frac{dp}{dz}\right)(a^2 - r^2) - \frac{\epsilon\psi(a)}{4\pi\eta}E_z\left[1 - \frac{I_0(L_Dr)}{I_0(L_Da)}\right], \tag{7.25}$$

in which we have also employed (7.22). From (7.25) we may note that if the pressure gradient dp/dz vanishes and the radius is large enough so that $L_Da \gg 1$, then

$$u_z(r) = -\frac{\epsilon\psi(a)}{4\pi\eta}E_z,$$

that is, the usual electrokinetic expression which indicates that the flow in response to an applied electric field is uniform throughout the radius of the capillary. If, however, $L_D a \ll 1$, one may easily show — on expanding the Bessel function and neglecting terms higher than the 4th power — that

$$u_z(r) = -\frac{\epsilon\psi(a)}{4\pi\eta}\frac{L_D}{4}(a^2 - r^2). \tag{7.27}$$

This shows that the velocity profile under these circumstances is parabolic. Thus we have Poiseuille flow, resulting from the fact that there is overlap of the diffuse double layer as the capillary radius becomes very small and leads to a more uniform charged density distribution as a function of r.

From (7.25) we can obtain the total volume flow through the capillary

$$V = 2\pi \int_0^a u_z(r) r \, dr, \tag{7.28}$$

which gives

$$V = -\frac{dp}{dz}\frac{a^2}{8\eta}A_t - \frac{\epsilon\psi(a)}{4\pi\eta}E_z A_t\left[1 - \frac{2 I_1(L_D a)}{L_D a I_0(L_D a)}\right], \tag{7.29}$$

where A_t is the area at the tip and I_1 is the first-order modified Bessel function of the first kind. We note that under conditions such that $V = 0$, the negative pressure gradient per unit electrical field strength is

$$\frac{-(dp/dz)}{E_z} = \frac{2\epsilon\psi(a)}{\pi a^2}\left[1 - \frac{2 I_1(L_D a)}{L_D a I_0(L_D a)}\right]. \tag{7.30}$$

Again, if $L_D a \gg 1$, we have the usual electrokinetic relation for this ratio, as the ratio of Bessel function vanishes under such conditions. We may also note that with $V = 0$ we can obtain the velocity profile by utilizing (7.30) together with (7.25). The result indicates that for large values of $L_D a$, both positive and negative flow exists. In other words, there may be internal circulatory flow.

Users of microelectrodes are perhaps more interested in the streaming potential and the electrokinetic volume flow under conditions of zero pressure gradient. To obtain these electrokinetic effects, we must compute the total current flow, which may be considered in two parts: (a) that which is convective as a result of fluid flow; and (b) that which is conductive because of an applied electric field. The convective part, i_1, is

$$i_1 = 2\pi \int_0^a u_z(r)\rho(r) r \, dr, \tag{7.31}$$

which on substituting from (7.25) and (7.22) gives

$$i_1 = \frac{dp}{dz}\frac{\epsilon\psi(a)}{4\pi\eta}A_t\left[1 - \frac{2\,I_1(L_Da)}{L_DaI_0(L_Da)}\right] - \left[\frac{\epsilon\psi(a)}{4\pi\eta}\right]^2\eta L_D{}^2E_zA_t\left[1 - \frac{2\,I_1(L_Da)}{L_DaI_0(L_Da)}\right.$$
$$\left. - \frac{I_1^2(L_Da)}{I_0^2(L_Da)}\right]. \qquad (7.32)$$

The conductive part, i_2, is simply

$$i_2 = \lambda E_zA_t, \qquad (7.33)$$

where λ is the specific conductivity of the fluid and is assumed uniform, $(\lambda = 1/\gamma)$. The total current is than

$$i = i_1 + i_2, \qquad (7.34)$$

and we may compute the streaming potential arising from flow with $i = 0$ as

$$\frac{E_z}{dp/dz} = -\frac{\epsilon\psi(a)}{4\pi\eta\lambda}g\,(L_Da, \alpha), \qquad (7.35)$$

in which

$$g(L_Da, \alpha) = \frac{[1 - 2\,I_1(L_Da)/L_DaI_0(L_Da)]}{1 - \alpha[1 - 2I_1(L_Da)/L_DaI_0(L_Da) - I_1^2(L_Da)/I_0^2(L_Da)]}, \qquad (7.36)$$

$$\alpha = \left[\frac{\epsilon\psi(a)}{4\pi\eta}\right]^2\frac{\eta L_D^2}{\lambda}$$

$$\frac{V}{i} = -\frac{\epsilon\psi(a)}{4\pi\eta\lambda}g\,(L_Da, \alpha). \qquad (7.37)$$

It is to be noted that these two ratios, (7.35) and (7.37), are equal. This is expected from the phenomenological theory of irreversible processes together with Onsager's reciprocal relations. The function $g(L_Da, \alpha)$ has been computed by Rice and Whitehead[7] and is an S-shaped function of the logarithm of L_Da, varying between zero for small L_Da and unity for large L_Da. Accordingly, the electrokinetic coefficients in the streaming potential and in electro-osmotic flow are smaller in fine capillaries than in large-bore capillaries. Thus errors in potential arising from possible pressure gradients in an experimental setup are smaller in the fine-tipped electrodes than in the conventional ones. Furthermore, because of the possible overlap of the diffuse double layer, the flow resulting from an applied electric field is less per unit area in the small-tipped microelectrodes than in the large ones. This may be interpreted as an abnormal increase in viscosity[8].

In concluding this section, we must reiterate that the foregoing theoretical considerations are highly oversimplified and represent models which are only partially representative of reality. The conclusions that have been drawn, therefore, must be guarded, and the results are subject to experimental testing.

Nothing has been stated in the foregoing concerning selectivity of the electrical properties according to the particular ions present. In fact there is nothing in the simple theory presented to suggest that any selectivity or specificity exists. However, I suggest that (a) if we were able to maintain enough convective flow at the tip of a microelectrode to overcome any back diffusion of ions from the test solution and (b) if the tip were behaving in an ion-exchange fashion, specificity could conceivably reside in the nature of the fillant electrolyte and be determined solely by this electrolyte. That is, the selectivity from one ion to the next could be in part dependent on the composition of the fillant. This suggestion is, however, largely intuitive and thus quite tentative. Undoubtedly as the tip becomes very small, the conductance through the glass becomes significantly important; this conductance would become the principal factor governing specificity of the electrical potential for the various cations.

7.5 FABRICATION OF FINE-TIPPED ELECTRODES AND SOME ILLUSTRATIVE RESULTS OF THEIR USE

We have found that it is rather easy to fabricate ultrafine-tipped electrodes the inside diameters of which can be as small as 100 Å as determined from resistance measurements. The technique employed is a modification of the conventional two-stage pull, the modification simply being the application of cooling jets of air onto the softened glass at the time of the forceful second pull. A simplified apparatus has been constructed to achieve this result; with appropriate adjustments of heater current, the amount of softening before the second pull, and the amount of cooling air, fine-tipped electrodes can be fabricated routinely.

We have used these devices to study the electrical potential in the sodium-transporting epithelial tissues of frog skin and toad bladder[2,9]. The electrical potential profile is found to be a monotonic function throughout the cellular region where transport is effected, as illustrated in Fig. 7.2. Furthermore, it appears that the results obtained by others — which show step function changes in the electrical potential as the microelectrode penetrates the tissue — are an artifact resulting from the elastic

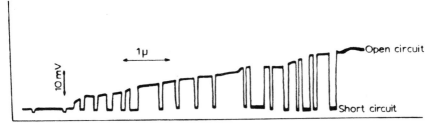

Fig. 7.2. Tracing of recorded electrical potential of microelectrode as advanced into toad bladder from mucosal surface. Rate of advance: 7μ/min. Preparation intermittently short-circuited. (*Chowdhury and Snell*, [2]. *Reprinted by permission of Elsevier*).

indentation of the cellular boundaries and possibly the cytoplasmic membranes, followed by sudden and intermittent penetration. An experiment substantiating this conclusion is illustrated in Fig. 7.3. In this experiment a conventional microelectrode having a tip diameter of the order of 5,000 Å was advanced to the point of deliberate and considerable tissue indentation. The advance mechanism was then stopped and a series of small finger taps were applied to the apparatus. The tension of the indentation was relieved as incremental penetration occurred. Figure 7.3 shows that the potential increased in steps closely associated with the applied taps (indicated by arrows).

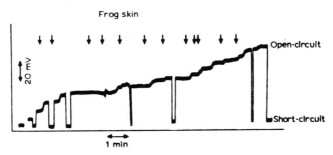

Fig. 7.3. See text for explanation. (*Chowdhury and Snell*, [9]. *Reprinted by permission of Elsevier*).

REFERENCES

[1] G. Ling, and R. W. Gerard, *J. Cellular Comp. Physiol.*, **34**, 383 (1949).
[2] T. K. Chowdhury, and F. M. Snell, *Biochim. Biophys. Acta*, **94**, 461 (1965).
[3] L. Engbaek and T. Hoshiko, *Acta Physiol. Scand.*, **39**, 348 (1957).
[4] G. Whittembury, *J. Gen. Physiol.*, **47**, 795 (1964).

[5] T. K. Chowdhury, and F. M. Snell, unpublished observations.
[6] F. M. Snell, and T. K. Chowdhury, in *Intracellular Transport*, Academic, New York, 1966.
[7] C. L. Rice, and R. Whitehead, *J. Phys. Chem.*, **69**, 4017 (1965).
[8] F. A. Morrison Jr., and J. F. Osterle, *J. Chem. Phys.*, **43**, 2111 (1965).
[9] T. K. Chowdhury, and F. M. Snell, *Biochim. Biophys. Acta*, **112**, 581 (1966).

APPENDIX

Solution of Poisson-Boltzmann Equation

For uni-univalent electrolytes the charge density is given by (7.18) and the Poisson-Boltzmann equation in cylindrical co-ordinates may be written as

$$\frac{1}{r}\frac{d}{dr}\left[r\frac{d\psi(r)}{dr}\right] = \frac{8\pi c_+(0)\epsilon}{D}\sinh\frac{\epsilon\psi(r)}{kT}. \tag{17.A-1}$$

The hyperbolic sine may be expanded to give

$$\frac{1}{r}\frac{d}{dr}\left[r\frac{d\psi(r)}{dr}\right] = L_D\,\psi(r)\left\{1+\frac{1}{3!}[\beta\psi(r)]^2+\frac{1}{5!}[\beta\psi(r)]^4+\ldots\right\}. \tag{17.A-2}$$

with $\beta = \epsilon/kT$. We now assume a solution of the form

$$\psi(r) = \sum_{i=0}^{\infty} A_i r^i \tag{17.A-3}$$

and compute the appropriate derivatives and substitute in (17.A-3) to obtain

$$\sum_{i=0}^{\infty} i^2 A_i r^{i-2} - L_D^2 \sum_{j=0}^{\infty} A_j r^j \left[1+\frac{1}{3!}\left(\beta\sum_{k=0}^{\infty} A_k r^k\right)^2+\frac{1}{5!}\left(\beta\sum_{K=0}^{\infty} A_K r^K\right)^4+\ldots\right].$$

The coefficients, A_i, may be determined in relation to A_0 by collecting terms in successive powers of r, giving

$$A_1 = 0$$

$$A_2 = \left(\frac{L_D}{2}\right)^2 \frac{1}{\beta}\sinh\beta A_0$$

$$A_3 = 0$$

$$A_4 = \left(\frac{L_D}{2}\right)^4 \frac{1}{4\beta}\sinh\beta A_0 \cosh\beta A_0$$

$$A_5 = 0$$

$$A_6 = \left(\frac{L_D}{2}\right)^6\left[\frac{1}{16\beta}\sinh\beta A_0 \cosh^2\beta A_0+\frac{1}{18\beta}\sinh^3\beta A_0\right]. \tag{17.A-5}$$

Thus the series solution is

$$\psi_s(r) = A_0\left[1 + \left(\frac{L_D r}{2}\right)^2 \frac{1}{A_0\beta} \sinh A_0\beta + \frac{1}{2!2!}\left(\frac{L_D r}{2}\right)^4\right.$$

$$\left.\times \frac{1}{A_0\beta} \sinh \beta A_0 \cosh \beta A_0 + \ldots\right]. \tag{17.A-6}$$

The comparable solution of (17.A-2) involving only the first term of the right hand side is

$$\psi_B(r) = A_0\left[1 + \left(\frac{L_D r}{2}\right)^2 + \frac{1}{2!2!}\left(\frac{L_D r}{2}\right)^4 + \ldots\right] = A_0 I_0(L_D r), \tag{17.A-7}$$

and as

$$\frac{1}{A_0\beta} \sinh A_0\beta = 1 + \frac{(A_0\beta)^2}{3!} + \frac{(A_0\beta)^4}{5!} + \ldots > 1 \tag{17.A-8}$$

and, likewise, the coefficients of higher terms are always larger for the complete series than for the Bessel series, we obtain

$$\frac{\psi_s(r) - \psi_B(r)}{A_0} > 0. \tag{17.A-9}$$

CHAPTER 8

Single and Coaxial Microelectrodes in the Study of the Retina

TSUNEO TOMITA

Department of Physiology, Keio University School of Medicine, Shinjuku-ku (Tokyo)

8.1 INTRODUCTION

There are, of course, many different types of microelectrodes that have been developed to meet special requirements. My own 20 years' experience in research with microelectrodes used on the retina has given me occasion to experiment with several of these different types.

When the depth recording in the isolated frog's retina was first undertaken in an attempt to analyze the electroretinogram (ERG) into components and to localize them in different retinal layers[1] the microelectrode did not have to be very small. For this purpose it was sufficient if the electrode could penetrate the retina without causing too much distortion of the tissue; as I recall, the microelectrode used in the above-noted experiment was as large as several microns – or even larger at the tip. Nevertheless, the conclusion drawn on such semimicroelectrodes survived exacting tests by several authors[2–6] who employed modern superfine micropipettes.

It was during the period of the investigations by Svaetichin and Brindley that the coaxial microelectrode was developed, and it subsequently provided a powerful tool for the study of the ERG[7]. As described in a later section, the coaxial microelectrode was designed for simultaneous recording of potentials at two adjacent points, and, if necessary, for additional recording of the potential difference between the two by a differential connection in the amplifier system. As a result of experiments using this technique, the most convincing evidence in support of our earlier conclusion concerning the localization of ERG components was provided[8].

With regard to the extracellular application of single microelectrodes to structures like the retina, which is a dense mass of small cells, it is difficult to obtain a good isolation of unit activity. However, the technique is adequate for the study of single ganglion cells which respond with impulse discharge; the all-or-none property of the impulse provides a means of differentiating one unit from another. Moreover, the extracellular microelectrode is easy to manipulate. The situation is different in the cells which constitute more distal retinal layers and which respond only with a slow and graded potential. In this case there is no decisive way to isolate single-unit activity from that of neighboring units insofar as extracellular recording is concerned. To my understanding, the difficulty in isolating unitary responses has been a barrier to further analysis of the ERG, which doubtless represents a sum of these unit activities. In order to surmount this barrier it was necessary to develop a technique of intracellular recording in small cells and a technique of marking the site of recording which would permit identification of single units. As a first step, the construction of very minute pipettes was undertaken.

8.2   MINUTE SINGLE PIPETTE FOR INTRACELLULAR RECORDING

8.2.1   Construction, Applicability, and Limitations
The minute pipette we now use for intracellular recording from within vertebrate retinas is constructed according to the Naka method, with which we became familiar during Naka's visit to this laboratory in 1959. Borosilicate glass tubing of approximately 0.7 mm outer and 0.5 mm inner diameter is heated at the center part by setting it in a U-shaped iridium-platinum ribbon of 50μ in thickness and 7 mm in width and applying current to the ribbon for an optimal length of time that is determined empirically. A few tenths of a second after the heating current is turned off, a sudden pull is exerted to obtain a pair of pipettes at one time. They are characterized by shallow tapering and small tip diameter (less than 0.1μ). The width of the heating ribbon is a determinant of the tapering angle toward the tip of the micropipette. An electron-microscopic illustration of such a pipette, which we owe to Dr. Naka, is provided in Fig. 8.1.

The pipettes are then filled with $2 M$ KCl* by boiling, and their tips are examined under a light microscope (600×). The tip appears blurred,

---

*The use of $2 M$ KCl rather than $3 M$ KCl resulted in a better yield of usable micropipettes.

I micron

Fig. 8.1. Electronmicroscopic picture of minute pipette of type currently being used for intracellular recording from cells in vertebrate retinas. (Courtesy of Naka.)

of course, as the size is far below the resolving power of any light microscope. In well-made pipettes however, the blurred image becomes fainter toward the tip and finally dissolves into the background. This kind of microscopic examination also serves for discarding pipettes that contain air bubbles. The pipettes thus selected are further checked with a megohmmeter. Those ready for use show a resistance around 200 MΩ with their tips immersed in 2 M KCl solution. Pipettes are prepared in the morning, kept in 2 M KCl solution, and used in the same afternoon. (Those kept overnight are less sharp and are no longer suitable for such purposes.)

The minute pipette was first applied to the frog's retina by Naka et al. [9] and shortly afterwards by Tomita et al.[10]. Among several kinds of responses determined by this means, the following three were most frequent: (a) impulse discharge at "on", "off", or "on-and-off" of light, accompanied by no distinct slow potential change; (b) impulse discharge similar to (a), but superimposed on distinct slow potential changes; and (c) slow sustained negative potentials. The first two were localized in the axons (optic nerve fibers) and soma-dendritic regions of ganglion cells, respectively, and the third was identified as the S-potential — which is known to originate in some structures in the bipolar cell layers which have not yet been positively identified and are tentatively termed "S-compartments." Despite thousands of trial penetrations beyond the S-potential level into the receptor layer, there was no indication of intracellular recording from single photoreceptors. The material was changed from the frog's retina to the carp's, as the cone inner segments in the carp appeared large enough to be impaled by a minute pipette. However, the experiment was without success.

While I was considering every possible means of solving this problem, construction of a building began just in front of ours, and I noticed hundreds of huge concrete piles being hammered into the ground, one after another. This gave me an idea which eventually developed into what we call the "jolting" technique.

8.2.2 An Auxiliary Technique: "Jolting"

I first tried to apply the principle of the pile driving to our minute pipette, but I soon realized the difficulty of driving such a slender pipette in purely longitudinal directions. (I have heard of similar attempts, present and past, from a number of microelectrode people, but few of these investigators reported a great deal of success; their reason is probably the same as mine.) Accordingly, the scheme was turned around 180° so that the retinal tissue was driven or *jolted* at a high acceleration against a vertically held minute pipette; this permitted intracellular recording from single carp's cones to be made for the first time.

The original model of the jolting device is schematically shown in the upper diagram in Fig. 8.2. It was remodeled from a classical electro-

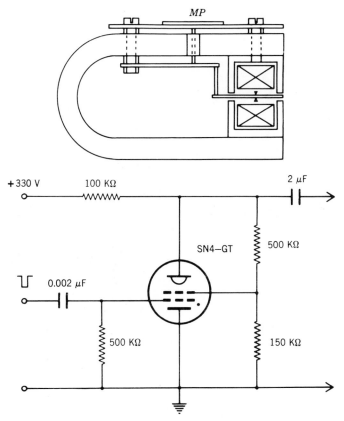

Fig. 8.2. Original device for "jolting" retina mounted on *MP*, which is made of chlorided silver plate serving as indifferent electrode (upper diagram). Lower drawing is circuit diagram of current pulse generator to activate jolter.

magnetic speaker, with its cone replaced by a metal strip. The "voice" coil through which strong current pulses are discharged was rewound with thicker wire. The lower diagram in Fig. 8.2 is the electronic circuit for discharging current pulses through the coil of some 100 amp in intensity and a few μ sec in duration. Since the fly-back pulse of the oscilloscope beam is used to trigger the discharge, the display on the oscilloscope screen of potentials from recording pipettes is not disturbed by pulse artifacts. Figure 8.3 shows some examples of mechanical responses to a current pulse and the method of their recording. As may be predicted from the mechanism, the response is a composite damped oscillation, the dominant frequency of which is determined by the mass and compliance of the metal strip. The schematic diagram in the lower left of Fig. 8.3 illustrates how the recording was made. One of a pair of tiny pieces from a razor blade is fixed with wax on the jolting table (*MP* in Fig. 8.2) with its cutting edge up, while the other piece is attached with the edge down

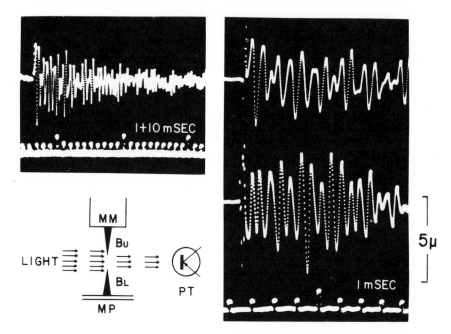

Fig. 8.3. Schematic illustration of arrangment for recording mechanical response of *MP* in Fig. 8.2 to a current pulse (lower left), and some sample records with it. Upper recordings are from one jolter with slow (left) and fast sweep (right); lower right recording is from another jolter of similar design. (Tomita[11]. Reprinted by permission of Cold Spring Harbor Laboratory of Quantitative Biology.)

to the arm of a micromanipulator to make a narrow slit between the two. The light beam, modulated by the mechanical movement of *MP*, is received by a phototransistor (*PT*) which converts the displacement to a potential change. Calibration is made simply by reading a shift of the dc level to a given displacement of the upper edge of the slit by aid of the micromanipulator to which this edge is attached.

The following calculation may serve for a rough estimate of the maximal acceleration which the retina undergoes by this device. To make the matter simple, the mechanical response of *MP* to a single current pulse is assumed to be a damped sinusoidal oscillation, that is,

$$s = e^{-\beta t} A \sin \omega t, \tag{8.1}$$

in which *s* is the displacement of *MP* from resting position, β is the damping factor, and *A* is the initial amplitude of damped oscillation.

As we are dealing with the maximal acceleration which is obviously found in the first sinusoidal wave, we can approximate

$$e^{-\beta t} = 1 \tag{8.2}$$

As the acceleration α is the second derivative of the displacement *s*,

$$\alpha = d^2 s / dt^2 = -A\omega^2 \sin \omega t \tag{8.3}$$

$$|\alpha_{max}| = A\omega^2 = A 4\pi^2 f^2. \tag{8.4}$$

From recordings in Fig. 8.3, let us assume $A = 2\mu = 2 \times 10^{-4}$ cm, and $f = 2 \times 10^3$/sec. Substituting these values in the above equation,

$$\alpha_{max} \simeq 32 \times 10^3 \text{ cm/sec}^2. \tag{8.5}$$

Or, since the gravity acceleration *g* is 980 cm/sec², we obtain a value of 32 *g* as an approximation.

Concerning some preliminary observations on single cone potentials in the carp (using the original model of jolter), reference will be made to Tomita[11].

The new model we are currently using is of a dynamic type, as schematically illustrated in Fig. 8.4. It is so designed that the retina on the jolting table is illuminated by light incident from underneath, while the recording micropipette is applied from the top. The new model has proved superior to the original one in that the characteristic frequency of the vibration in response to a single current pulse is about 10 kc, or five times higher than that of the original. Theoretically, a fivefold increase in frequency should increase the rate of acceleration by a factor of $5^2 = 25$, provided the amplitude remains the same. A test showed that the acceleration available from the new model was so strong that it had to be

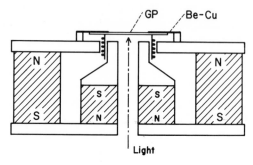

Fig. 8.4. New model of jolter, which permits light incident from underneath on vitreal surface while recording with microelectrode applied from the top. GP, glass plate on which the retina is mounted with receptor side up; Be-Cu, beryllium-copper diaphragm; N and S poles of permanent magnets. (Tomita et al.[12]. Reprinted by permission of Pergamon.)

attenuated to the optimal rate by means of a variable resistance connected in parallel with the input of the jolting device. Another advantage of the new model is a rapid damping; the vibration in response to a single current pulse falls off in a few msec, which is only one-tenth of the damping time of the original model.

The method and some results of intracellular recording obtained by Tomita et al.[12] from single cones in the carp will be discussed briefly. The retina detached from the pigment epithelium is mounted, with its receptor side up, on the jolter. A micropipette is advanced slowly into the retina in steps of a few microns by aid of a micromanipulator, while the jolter is activated at the beginning of each sweep of the oscilloscope beam. The retinal region where the recording is made is illuminated by a small light spot with every sweep of the oscilloscope beam. The use of a small light spot minimizes the confounding action of extraneous retinal potentials and also avoids bleaching of the remaining part of the retina. The artifact of the current pulse that jolts the retina is heard, through a connected speaker, as a "click" which usually sounds high-pitched but becomes low and tympanic whenever it is followed by a considerable negative or positive shift of the potential level. The change in the acoustical quality of the click is so distinct that it serves as an indication of the elicited response. When a response to light is noted after recording a resting potential at a depth corresponding to the receptor layer, the spectral response is recorded by scanning the spectrum in steps of 20 mμ.

Figure 8.5 shows three sample records of single-cone spectral responses which are maximally sensitive at different wavelengths. They represent those with the greatest signal-to-noise ratio obtained from hundreds of

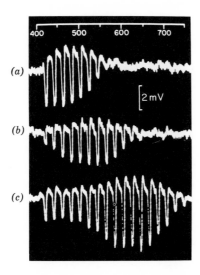

Fig. 8.5. Sample recordings from single cones, demonstrating three types of spectral responses found in carp retina. Scanning of spectrum is made in steps of 20 mμ with monochromatic light adjusted to equal quanta (2×10^5 photons per μ^2sec), and with duration of light of 0.3 second at each wavelength followed by intermission of 0.6 second. A downward deflection indicates negativity. Recording is made with CR-coupled amplifier having time constant of 0.5 second. Spectral scale is given in terms of millimicrons at top of figure. (Tomita et al.[12]. Reprinted by permission of Pergamon.)

recordings. The average recordings contained more noise as would be expected upon consideration of the relatively small amplitude of response (less than 5 mV) and the high resistance of the pipettes of minute-tip caliber, which usually measured more than 200 MΩ. As a result of the screening process, based solely on the greatest signal-to-noise ratio, 142 recordings were utilized in the statistical analysis. From each of the selected recordings the response amplitude at individual wavelengths was measured. The data were then normalized with the maximal response set at 100% and plotted on amplitude-wavelength coordinates. The adjacent points were connected by a straight line.

Figure 8.6 shows the result of photographically superimposing all of the 142 curves obtained in the above-described manner. The fact that the selected recordings were still contaminated with noise resulted in considerable variability in the data; consequently, the existence of subtypes was not immediately evident on only a cursory examination. However, the results of superimposing the response curves suggested three definite

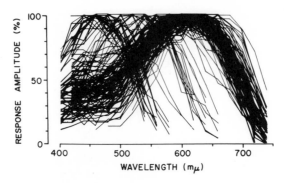

Fig. 8.6. Result of superimposing 142 spectral response curves selected from hundreds, based solely on greatest signal-to-noise ratio and normalized with maximal amplitude set at 100%. Further explanation of procedure is in the text. (Tomita et al.[12]. Reprinted by permission of Pergamon.)

groups; one peaking in the reg region, another at green, and still another at blue. The three groupings may be more distinct on the phase of the curve that falls toward the longer wavelengths. In this way, 105 curves out of 142 (74%) were included in the category of R-type, 14 curves (10%) in G-type, and 23 (16%) in B-type. From the curves of each category a histogram of the peaking wavelengths was constructed and the mean value along with the standard deviation were calculated. The result of this analysis is presented in Fig. 8.7, and the mean spectral response curves in Fig. 8.8. It is worthwhile to note that the maximally sensitive wavelengths of the responses are close to those of the difference spectra of single goldfish cones measured by Marks[13], who used a microspectrophotometer.

Several criteria have been used to identify the above-mentioned responses as single-cone potentials. First, they are observed only after recording a resting potential of 30–40 mV. Second, they are found at the depth corresponding to the layer of the cone inner segments. The only histological structures in the receptor layer that would permit intracellular recording are the cone inner segments. Thirdly, these potentials are differentiated from the S-potential not only by the depth at which they are recorded but also by having no substantial area effect.

Although the above criteria were considered sufficient, it was still desirable for the final identification of the single-cone response directly to observe the very site of the recording; this was accomplished recently by Kaneko and Hashimoto[14] by means of an electrode marking technique to be described below.

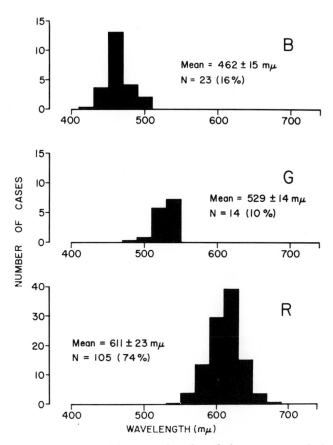

Fig. 8.7. Histograms of peaking wavelengths of three groups of single-cone spectra response curves. (Tomita et al.[12]. Reprinted by permission of Pergamon.)

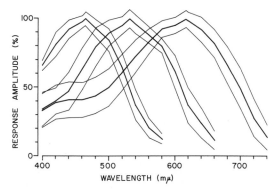

Fig. 8.8. Averaged response and standard deviation curves of three types of cones. (Tomita et al.[12]. Reprinted by permission of Pergamon.)

8.2.3   Identification of the Site of Recording by Marking

The microelectrode used by Kaneko and Hashimoto[14] is a minute pipette as described previously, but it is filled with 4% aqueous solution of Niagara Sky Blue 6B, a dye used by Potter et al.[15] for their study of the connections between cells of the developing squid. Pipettes are filled by boiling them in the dye solution for more than 10 min; they are then inspected under the microscope to select those filled with a uniform distribution of the dye. The advantage of this dye is that it is highly soluble in water but only slightly so in organic solvents, which permits the dye to be retained during the course of later procedures: dehydration in graded acetone-water mixtures, embedding in Epon, and so forth.

Whenever a potential is recorded by aid of the jolter and identified as a single-cone response by the previously outlined criteria, the dye is injected electrophoretically by the arrangement shown in Fig. 8.9. The lead-off wire of the recording electrode is connected through a 40 MΩ current-limiting resistor (R_L), to the negative terminal of a 135 V battery (B) by a foot-switch (FS). An important feature of this technique is that a small portion of the electrode tip, a micron or two in length, is broken by the applied high voltage—which results in a more rapid injection of the dye, but at a rate determined by the current-limiting resistor. The optimal duration of current for the injection is controlled visually under the dissection microscope. It usually takes less than a second to produce a tiny blue spot at the electrode tip. If the electrode is intracellular, the blue

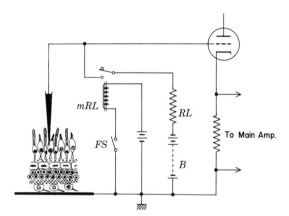

Fig. 8.9. Arrangement of marking single cells in retina with micropipette filled with 4% aqueous solution of Niagara Sky Blue 6B. B is a 135 V dry battery; RL, current limiting resistor of 40 MΩ: mRL, microrelay activated by foot switch (FS). Further explanation is in text. (Kaneko and Hashimoto[14]. Reprinted by permission of Pergamon.)

spot stays after the current is turned off. If the electrode is deliberately placed extracellularly as a control, a blue spot is observed during current but is soon diffused away and completely lost within seconds after the current is switched off.

A microphotograph showing a blue stain within the cone inner segment is illustrated in Fig. 8.10. The background stain is with 0.5% basic fuchsin alcoholic solution. In more than 100 preparations Kaneko and Hashi-moto[14] were unable to find a single section in which the stain appeared in structures other than the cone inner segment. The method is extremely reliable, and they are now applying the same technique to localize other unidentified responses within the retina.

8.3   COAXIAL MICROELECTRODE

8.3.1   General

In studies with microelectrodes it is often desirable to obtain simultaneous recording of potentials from two points lying close to each other, or from points separated by any desired small distance within the tissue. For this purpose the coaxial microelectrode has proved useful[7, 16]. A schematic illustration of the electrode is given in Fig. 8.11. The device

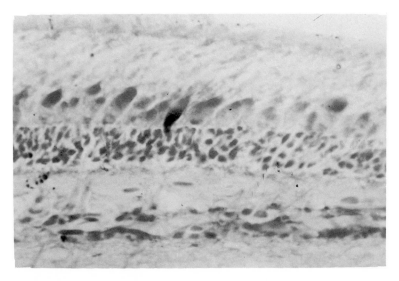

Fig. 8.10. Microphotograph showing cone marked blue at inner segment by arrangement in Fig. 8.9. (Kaneko and Hashimoto[14]. Reprinted by permission of Pergamon.)

consists of an inner superfine pipette filled with 3 M KCl solution, and a Ringer-filled outer pipette with a larger tip diameter ranging from a few microns to some 50 μ or even more, depending on the purpose of the experiment. A microphotograph of this type of electrode is shown in Fig. 8.12. The inner pipette is seen to protrude about 10 μ out of the tip of the outer pipette. Figure 8.13 shows our holder for the coaxial electrode, attached to a metal case (A) which contains the twin-triode used as the preamplifier tube. The entire assembly shown in this figure is attached to a micromanipulator that allows the electrode to be positioned in a given region of the tissue. The axial distance between the tips of the

Fig. 11. Schematic illustration of coaxial microelectrode.

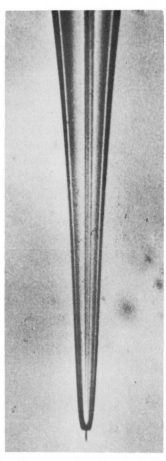

Fig. 8.12. Microphotograph of tip of coaxial microelectrode. (Tomita[16]. Reprinted by permission of Japan. J. Physiol.)

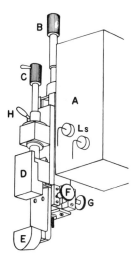

Fig. 8.13. Holder for coaxial microelectrode attached to metal case (A) which contains twin triode used as preamplifier tube. B and C, coarse and fine screws for advancement of inner pipette; D and E, insulator blocks on which inner and outer pipettes are attached by oiled clay; F and G, screws for alignment of outer pipette with inner pipette; H clamp for coarse screw B; Ls, lead-off wires to be connected to inner and outer pipettes by means of fine chlorided silver wires. (Tomita[16]. Reprinted by permission of Japan. J. Physiol.)

inner and outer pipettes is controlled by two screws, coarse (B) and fine (C), each rotation of the fine screw giving a displacement of 40 μ to the inner pipette. A similar device has been developed independently by Frank[17].

8.3.2   Principle and Circuit of Cross-Compensation

As discussed earlier[16, 18], one limitation of the coaxial micro-electrode is the interaction between the inner and outer pipettes. As the inner one penetrates through the whole length of the outer pipette, both are electrically connected by a capacitance across the wall of the inner pipette; therefore, they cannot be independent of each other. Because of this interpipette capacitance, any potential recorded from one pipette gives a transient potential change to the other. This may be illustrated by an equivalent circuit as shown in Fig. 8.14. The two circuits inserted on both sides of capacitance C are for the purpose of compensation—but let us forget about them for a moment. The inner pipette, which records an emf E_1, is connected through a high resistance of its own (R_1) to one of

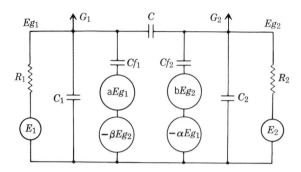

Fig. 8.14. Diagram showing principle of compensation for coaxial microelectrode. (For explanation, see text.) (Tomita[16]. Reprinted by permission of Japan. J. Physiol.)

the grids of the preamplifier tube G_1. The input capacitance of this side is represented by C_1. The outer pipette, which records another emf (E_2), is connected likewise to the other grid G_2, with C_2 and R_2 representing the capacitance and the resistance of this side, respectively. The interpipette capacitance C intervenes between the two pipettes. The interaction caused through this interpipette capacitance must be eliminated by some kind of cross-compensation.

The parts added to the equivalent circuit on both sides of the interpipette capacitance serve for the cross-compensation as well as for the compensation of the individual input capacitances C_1 and C_2. As interaction takes place by currents flowing from one side to the other across the interpipette capacitance C, it is necessary for elimination of the interaction to bypass just the same amount of current that flows through C. A pair of emf's ($-\beta Eg_2$ and $-\alpha Eg_1$ in Fig. 8.14) are responsible for this bypass action. Let us take $-\beta Eg_2$ as a model. Whenever Eg_2 yields current across C in the direction from G_2 to G_1, for instance, $-\beta Eg_2$ also yields current across Cf_1 in the direction from G_1 to ground. These two kinds of current will take the same absolute value when $Eg_2 \times C = -\beta Eg_2 \times Cf_1$. In other words, the effect of Eg_2 on Eg_1 is compensated under the condition that $\beta = C/Cf_1$. For the same reason, the effect of Eg_1 upon Eg_2 is also compensated when $\alpha = C/Cf_2$. Another pair of emf's connected in series with these, that is aEg_1 and bEg_2, are for the compensation of the individual input capacitances C_1 and C_2. This is accomplished in the way with which we are all familiar in the conventional single pipette technique.

Figure 8.15 shows the circuit of the model we are currently using, which is designed according to the above-described principle. One of the

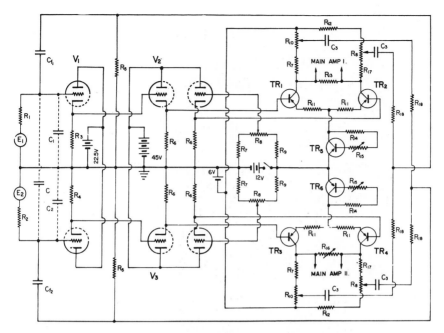

Fig. 8.15. Cross-compensation circuit diagram to eliminate capacitive coupling in both double-barreled and coaxial microelectrodes[18]. $V_1 - V_3$, 12AU7; $TR_1 - TR_6$. Sony 2SB50; Cf_1 and Cf_2, 15 $\mu\mu F$; C_3, 0.5 μF; R_3 and R_4, 200 kΩ; R_5, 500 kΩ; R_6, 20 kΩ; R_7, 5 kΩ; R_8, 2 kΩ; R_9, 30 kΩ; R_{10}, 1 kΩ; R_{11}, 100 Ω; R_{12}, 1 kΩ; R_{13}, 50 kΩ; R_{14}, 200 kΩ; R_{15}, 250 kΩ; R_{16}, 100 kΩ; R_{17}, 3 kΩ; R_{18}, 40 kΩ. Elements connected by dashed lines: equivalent circuit for a coaxial microelectrode. (See also Fig. 8.14.)

outputs is obtained between the collectors of TR_1 and TR_2, and the other output between those of TR_3 and TR_4.

The tests were made in steps as shown in Fig. 8.16. Square pulses of 20 mV amplitude and 1 msec duration were used. The resistance of the inner pipette in this test was 43 MΩ and the resistance of the outer pipette was 1.6 MΩ. Recordings in the left column are controls, obtained without compensation, and those in the right column are recordings with compensation. The arrangement for recording at each step is shown by the diagram on the right. When the inner pipette is directly fed by a pulse, small artifacts of a fast time course are induced in the outer pipette (Fig. 8.16aA; Eg_2). We can predict this time course from the resistance of the outer pipette, (i.e., 1.6 MΩ) and from the value of the interpipette capacitance C plus input capacitance C_2, which are estimated to be about 25 $\mu\mu F$ in total to make a time constant of about 40 μsec. The artifact is

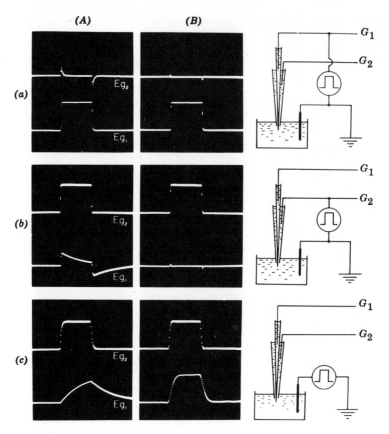

Fig. 8.16. Effect of compensation on coaxial microelectrode with circuit in Fig. 8.15. (Explanation in text.) (Tomita[18]. Reprinted by permission of *IEEE*)

minimized by adjusting the gain for $-\alpha Eg_1$ (Fig. 8.16aB). The effect of the outer pipette fed directly with a square pulse on the inner pipette (Fig. 8.16bA; Eg_1) is greater, of course, as the resistance of the inner pipette is much higher. The large effect also becomes negligibly small after adjusting the gain control for $-\beta Eg_2$ (Fig. 8.16bB). After the cross-compensation is complete in this way, the individual input capacitances are compensated by adjusting the values of aEg_1 and bEg_2. The left bottom recording (Fig. 8.16cA) is a control and the right bottom recording (Fig. 8.16cB) is the result of compensation.

A few words are appropriate here concerning the application of this design to a double-barreled microelectrode of the type of Coombs, Eccles and Fatt[19], in which a capacitance intervenes between the two barrels. In this case also, it is very easy to eliminate the interaction. Figure 8.17 shows three recordings at three different stages of compensation; without compensation, with an appropriate compensation, and with an overcompensation.

8.3.3 Examples of Applications

Simultaneous Intra- and Extracellular Recording. It has been known since the work of Hartline, Wagner, and MacNichol[20] that the intracellularly recorded action potential from within single ommatidia in the lateral eye of the horseshoe crab consists of two distinct components: a slow sustained positive-going potential during illumination (termed the ommatidial action potential or OAP), and a train of positive-going impulse spikes superimposed on the OAP. The action potential can also be recorded extracellularly, but with a marked difference; the polarity of OAP is reversed to be negative, while spikes superimposed remain positive. The difference in the configuration of the action potential obtained separately by intra- and extracellular recording should be demonstrated more clearly by simultaneous recording, using a coaxial microelectrode. Figure 8.18 illustrates an example of such[16]. The lateral eye of the horseshoe crab was cut by a razor blade perpendicular to the cornea to expose ommatidia at the cut surface. A coaxial microelectrode, the inner and outer pipettes of which were set flush at the tips, was introduced into one of the exposed ommatidia to a position at which a best extracellular recording was found from both pipettes. At this positioning of the outer pipette, the inner pipette alone was protruded about 20 μ out from the tip of the outer pipette to impale the underlying cell. Figure 8.18, which was obtained under such conditions, shows clearly that the OAP reverses its polarity on the opposite sides of the cell membrane, but that the impulse spikes are positive irrespective of

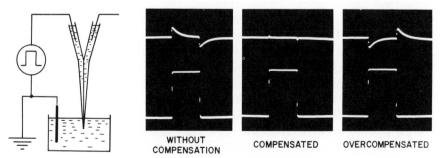

Fig. 8.17. Effect of compensation on double-barreled microelectrode with circuit in Fig. 8.15. (Tomita[18]. Reprinted by permission of *IEEE*)

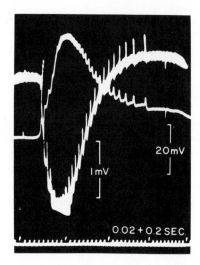

Fig. 8.18. Simultaneous intra- and extracellular recording of response to illumination of cell within ommatidium of lateral eye of horseshoe crab. Extracellular tracing starts upper and shows negative deflection twice crossing intracellular tracing, which shows positive deflection. Impulse spikes superimposed on two tracings are synchronous and both positive. (Tomita [16]. Reprinted by permission of *Japan. J. Physiol.*)

whether the recording is intra- or extracellular. The conclusion is that the battery for the OAP is situated across the cell membrane at the recording site, while the impulse spikes are generated in structures at a distance from the recording site. Figure 8.19 summarizes the conclusion in a schematical way. The localization of the site of impulse initiation in the axon hillock in this figure was based on another observation [21].

Isolation of the Receptor Potential by Fractional Recording. There was a long-term discrepancy among investigators' findings concerning the contribution of the receptor potential to the vertebrate ERG [22]. A new approach to this problem was taken by Murakami and Kaneko [23], who used the coaxial microelectrode. Figure 8.20 shows the arrangement, and Fig. 8.21 depicts a series of fractional recordings of the ERG at various depths in the frog's isolated retina (mounted receptor side up on the indifferent electrode). As seen from the arrangement, Channel I of the amplifier records the ERG fraction between the outer and inner pipettes of the coaxial microelectrode (upper tracing in each record of Fig. 8.21), while Channel II shows the ERG fraction between the inner pipette of the coaxial microelectrode and the indifferent electrode (lower tracing). Each fraction changes the size and configuration as the inner pipette

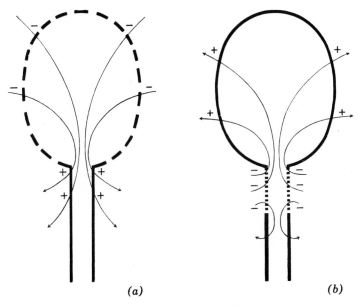

(a)   (b)

Fig. 8.19. Working model of cell or aggregate of cells in single ommatidia in lateral eye of horseshoe crab. Left diagram (a) shows outflow of current at axon hillock owing to OAP produced by depolarization of cell membrane; right diagram (b) shows initiation of propagated impulses at axon hillock by outwardly directed OAP current shown in (a). (Tomita[21]. Reprinted by permission of the American Physiological Society.)

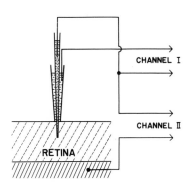

Fig. 8.20. Arrangement for recording ERG with coaxial microelectrode in retina. Channel I records potential between inner and outer pipettes; Channel II records potential between inner pipette and indifferent electrode. (Murakami and Kaneko[23]. Reprinted by permission of Pergamon.)

143

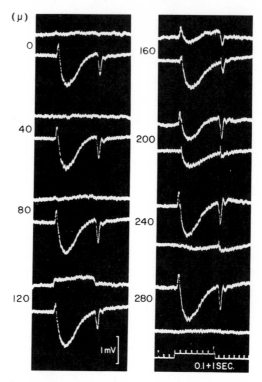

Fig. 8.21. Depth recording of frog's ERG from receptor side with arrangement shown in Fig. 8.20. Outer pipette was positioned on receptor surface; inner pipette penetrated into retina in steps of $40\,\mu$. Upper tracing of each recording was obtained through Channel I, and lower tracing through Channel II. Depth of inner pipette is shown in microns on left of each column.

traverses the retina (Fig. 8.21), but the sum of the two fractions is unchanged and equals the transretinal gross ERG.

Let us confine our attention to the bottom recording in the left column of Fig. 8.21. This recording was obtained with the inner pipette located at a depth corresponding approximately to the outer plexiform layer, so that the upper tracing records potential changes across the receptor layer while the lower tracing shows those across the remaining retinal layers (the bipolar and ganglion cell layers). It is reasonable to assume that the response in the upper tracing in this record represents the receptor potential. This is a graded sustained positive potential and is identified as a constituent of PIII of the ERG upon consideration of the reversal of polarity because of recording from the receptor side of the retina. It should be noted that the remaining ERG fraction in the lower tracing in the same record still shows a considerable size of a-wave, which manifests itself as an initial positive transient and which doubtless contributes to $PIII$. Apparently, there are at least two $PIII$ subcomponents, one arising in the receptor layer (distal $PIII$) and the other in more proximal layers (proximal $PIII$). Murakami and Kaneko[23] were successful in demonstrating that these $PIII$ subcomponents differ in latency and also in sensitivity to chemicals.

Depth Recording of the S-Potential. The S-potential is a graded sustained response to light recorded from within vertebrate retinas by means of microelectrodes. It is classified[24] into two major types; one is the luminosity type (L-type), the response polarity of which is negative irrespective of the wavelength of stimulating light (Fig. 8.22a), and the other is the chromaticity type (C-type), which changes the response polarity depending on the wavelength (Fig. 8.22b and c).

When the S-potential was first observed in the fish retina by Svaetichin[25], he believed that the recording was made intracellularly from single cones, and termed it the cone action potential. However, his conclusion was challenged by an experiment with coaxial microelectrodes[26]. Figure 8.23 illustrates a result of depth recording in the carp retina by means of a coaxial microelectrode with its inner pipette protruding 50 μ out from the tip of the outer pipette. Keeping this distance of 50 μ fixed, both pipettes were inserted into the retina from the receptor side in steps of 35 μ. In order to differentiate tracings through the inner pipette from those through the outer pipette, artificial pips were superimposed on the tracings through the inner pipette to mark the moments of "on" and "off" of light.

The top thick tracing in Fig. 8.23a is a control record from the retina surface. It actually consists of two nearly identical tracings, one from the outer pipette at the distal retinal surface and the other from the inner pipette at a depth of 50 μ in the retina. The middle tracing in Fig. 8.23a is from the outer pipette at a depth of 140 μ; the bottom tracing is from the inner pipette at 140 + 50 μ. The subsequent recordings $(b$–$d)$ were obtained after every 35 μ step-advance of the microelectrode. It is noted that in the first two recordings (a and b) the inner pipette picks up a larger response than does the outer pipette, but in c and d the response from the outer pipette is dominant. Apparently, as the coaxial microelectrode was advanced into the retina, the inner pipette first reached the site of the maximal response, followed by the outer pipette, which was 50 μ behind the inner one. The depth for this maximal response was definitely proximal to the receptor layer and was sufficient to exclude the cones as being the origin of this response. It was considered improbable that the

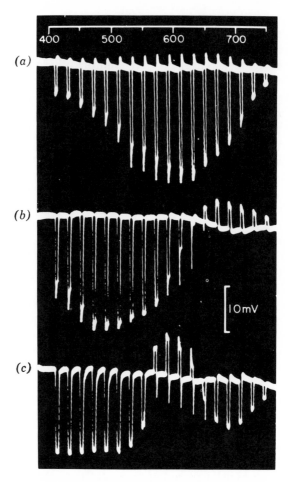

Fig. 8.22. Three types of spectral response curves of S potentials in carp retina: (a) luminosity type; (b) biphasic chromaticity type; (c) triphasic chromaticity type. (Tomita[22]. Reprinted by permission of J. Opt. Soc. Am.)

outer pipette of about 5 μ tip diameter could penetrate into single cones of the diameter of only 10 μ or so. The exclusion of the cones as the origin of the response was made conclusive by later experiments, including those with the marking technique at the recording site[27–31]. The terminology was also changed from the cone action potential to the S-potential. The nature and origin of the S-potential have been studied extensively since that time, but opinions of investigators are still at variance.

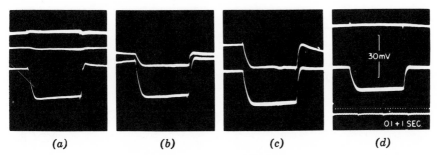

Fig. 8.23. Depth recording of S potential from receptor side of carp retina by means of coaxial microelectrode with inner pipette protruding a fixed length of $50\,\mu$ out of tip of outer pipette. Micrometer reading for outer pipette from distal retinal surface: $140\,\mu$ in (a), $175\,\mu$ in (b), $210\,\mu$ in (c), and $245\,\mu$ in (d). Tracings through inner pipette are superimposed by a pip at "on" and "off" of light for differentiation from tracings through outer pipette. (Further explanation in text.) (Tomita[26]. Reprinted by permission of *Japan. J. Physiol.*)

Impedance Measurement of the S-Compartment. As described in the preceding section, the nature and origin of the S-potential are still an open question. The S-compartment is a tentative term given to the recording site of the S-potential under such circumstances. One unusual property of the S-potential, besides several others reported[11], is that the resistance of the S-compartment remains the same during and after responding to illumination[32]. Tomita and Kaneko[33] were interested in repeating their experiment, but with the use of coaxial microelectrodes and the arrangement shown on the left in Fig. 8.24. The circuit on the right illustrates the principle for which we are indebted to Dr. F. A. Dodge of Rockefeller University. As $E_1 > E_2$ in the circuit and also R_L, a resistor of 100 MΩ inserted as a current limiter, dominates all the other resistances connected in series, current i in the circuit is expressed approximately by:

$$i = \frac{E_1}{R_L}. \qquad (8.6)$$

The value e in Fig. 8.24, or the potential recorded by the inner pipette of the coaxial microelectrode, then becomes:

$$e = E_2 - (R_x + R_c)\frac{E_1}{R_L}. \qquad (8.7)$$

If e is brought to zero by adjusting E_2, we obtain

$$R_x + R_c = \frac{E_2}{E_1}R_L \qquad (8.8)$$

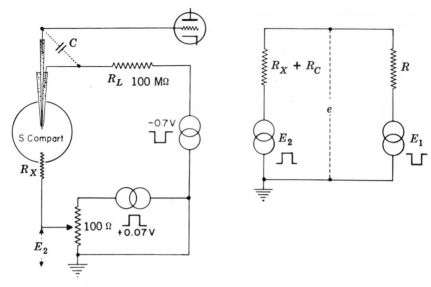

Fig. 8.24. Arrangement for resistance measurement of S compartment (left) and its equivalent circuit diagram showing the principle (right). S Compart., S compartment or recording site of the S potential; R_x, resistance of S compartment; R_L, resistor of 100 MΩ as current limiter; C, coupling capacitance between inner and outer pipette, which is eliminated by the cross-compensation circuit in Fig. 8.15; R, R_L plus trans-pipette resistance of outer pipette; R_c, coupling resistance between current and recording systems; E_1, source of square pulses applied to S compartment through outer pipette; E_2, source of square pulses identical in phase with those of E_1 but with polarity reversed; e, potential between inner pipette and the ground. (Tomita and Kaneko[33]. Reprinted by permission of Pergamon.)

If, on the other hand, E_2 is kept constant and R_c remains stable enough, we find that

$$\frac{de}{dR_x} = -\frac{E_1}{R_L} \qquad (8.9)$$

or

$$\Delta R_x = -\frac{R_L}{E_1}\Delta e. \qquad (8.10)$$

An advantage of the use of the coaxial microelectrode for the resistance measurement is that because the inner pipette protrudes a certain length out from the tip of the outer pipette, the coupling resistance between the current pipette (outer) and recording pipette (inner) is made very small compared with the single electrode (which serves for both current and recording) and also with the double-barreled microelectrode (in which one pipette is for current and the other is for recording). It has been shown that the resistive coupling of the coaxial microelectrode can be made one order below that of the double-barreled microelectrode of comparable size[33].

In order to insert both pipettes into single S-compartments and to keep the response stable throughout the resistance measurement, the outer pipette was made as small as $1\,\mu$ or less. However, with this tip size it was difficult to set the inner pipette to protrude an optimal length out from the tip of the outer one—even with the aid of a microscope. The final adjustment was possible only by observing electrical events with the arrangement in Fig. 8.25. In this arrangement current pulses are applied to the outer pipette through a current limiter of $100\,\text{M}\Omega$ while one observes V_i and V_0 displayed on a two-channel oscilloscope. While the tip of the inner pipette is inside the outer pipette the amplitude of V_i is, of course, comparable with that of V_0, but when the inner pipette protrudes out from the tip of the outer pipette, the amplitude of V_i shows a drastic decrease—corresponding to the decrease in the coupling resistance. With

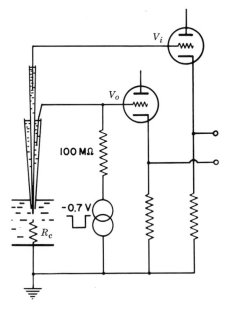

Fig. 8.25. Arrangement for final adjustment of intracellular coaxial microelectrode. R_c, coupling resistance between current and recording systems; V_i, potential induced across the inner pipette plus R_c; V_0, potential induced across the outer pipette plus R_c. Resistor of $100\,\text{M}\Omega$ is used as current limiter. (Tomita and Kaneko[33]. Reprinted by permission of Pergamon.)

further protrusion, however, a point is eventually reached at which a sudden increase in the amplitude of V_0 appears, signalling that the tip of the outer pipette has been choked up by the wall of the protruding inner pipette. The adjustment is complete when the inner pipette is set somewhat behind this final stage of protrusion.

The material was the carp's retina, detached from the pigment epithelium and mounted on a chlorided silver plate serving as the indifferent electrode. A coaxial microelectrode set in the manner described was advanced by means of a micromanipulator, until the electrode tip touched the retina surface. Pulses were sent from both E_1 and E_2 of Fig. 8.24, and the resulting pulses from the inner pipette were balanced (or e was brought to zero) by adjusting E_2. The electrode was then slowly advanced into the retina, while a 2 seconds illumination followed by 2 seconds dark interval was repeated. The moment of penetration of the electrode into an S-compartment was signalled by the appearance of a large S-potential in response to illumination, as well as by an unbalance of pulse on the record. From the amplitude of the unbalanced pulses, e, the effective resistance of the S-compartment was calculated, using (8.10). The results were widely scattered from nearly none up to 3 MΩ, but those between 500 and 1,500 kΩ were found to be common, confirming the observation of Watanabe et al.[32].

The measurement of resistance during one complete cycle of a 2 seconds light and 2 seconds dark interval was not always easy because of the large amplitude of the S-potentials. It was often necessary to limit the response amplitude to some 10 mV by reducing the intensity of the light stimulus. This was done in some experiments, including the one illustrated in Fig. 8.26, by inserting a red filter (Matsuda V-R68) into the passage of the light beam. The filter was useful not only for reducing the size of the response but also for quick screening of the chromaticity-type responses of the positive polarity for red light.

Figure 8.26 illustrates the recording of an L-type response while applying pulses balanced out in darkness. The potential from the inner pipette was led to both channels of a two-channel amplifier, one of which was dc coupled and of low gain in order to reproduce the original configuration (lower tracing). The other was CR coupled (time constant = 0.1 sec) and of high gain in order to emphasize any small change of resistance (upper tracing). Recording was made on moving film with a Grass kymograph camera. It can be noticed in the recording that the rising and falling phases of the response at "on" and "off" of light are associated with no discernible change of resistance, but that there is a gradual increase of resistance in the course of 2 seconds illumination – as evidenced by the growth of pulses toward the negative side. In the course

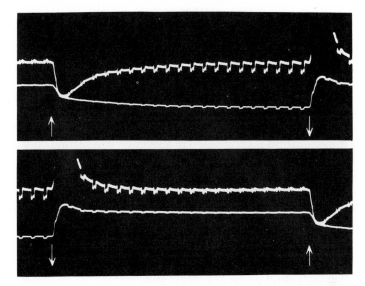

Fig. 8.26. Resistance change in S compartment (L-type, carp) during one complete cycle of 2 seconds diffuse illumination over retina and 2 seconds dark interval. Red light through filter (Matsuda V-R68) in front of tungsten filament lamp was used for illumination. Top recording continues to bottom one with some overlap. Besides dc recording (lower tracing), simultaneous recording with CR-coupled amplifier (time constant 0.1 sec) of high gain (about 5 times higher than the dc amplifier) was made in order to emphasize resistance change (upper tracing). Amplitude of pulses recorded with dc amplifier after withdrawal of electrode from compartment is shown in right bottom corner. Moments of light "on" and "off" are indicated by arrows up and down. (Tomita and Kaneko[33]. Reprinted by permission of Pergamon.)

of the following 2 seconds dark interval a gradual recovery or decrease of resistance toward the initial value of observation occurs. The experiment was repeated with the polarities of pulses reversed for E_1 and E_2 with the same results. In the case illustrated in Fig. 8.26 the observed maximal resistance change ΔR_x was 280 kΩ, but the value varied widely from one S-compartment to another. It ranged from nearly zero to several hundred kilohms, although there was no case in which Δe changed in the opposite direction.

The result in the case of C-type response was the same as in the L-type, provided the response was made negative by selecting an appropriate wavelength for the stimulating light. On the other hand, at wavelengths producing a positive response, Δe appeared in the opposite direction; there was a gradual decrease of resistance during illumination and a gradual recovery during the following interval. The direction of the resistance change was thus unequivocally predictable from the polarity of the response.

In conclusion, the S-compartment shows a small and gradual resistance change during response to light, the change being in the direction of increase if the response is negative and in the direction of decrease if the response is positive. This relation holds irrespective of the response types, L-type or C-type. The change is so slow that its detection is difficult if illumination is terminated in a fraction of a second, as was done in the experiment of Watanabe[32]. Such a slow resistance change cannot be the cause of the S-potential, but must be a kind of result of the S-potential. In this respect it is clear that the S-potential is quite different in nature from the kind of electrical activity usually ascribed to nerve cells.

REFERENCES

[1] T. Tomita, *Japan. J. Physiol.*, **1**, 110 (1950).
[2] D. Ottoson, and G. Svaetichin, *Cold Spring Harbor Symp. Quant. Biol.*, **17**, 165 (1952).
[3] G. Svaetichin, *Acta Physiol. Scand.*, **39**, Suppl. 134, 57 (1956).
[4] G. S. Brindley, *J. Physiol.*, **134**, 339 (1956).
[5] G. S. Brindley, *J. Physiol.*, **134**, 353 (1956).
[6] G. S. Brindley, *J. Physiol.*, **134**, 360 (1956).
[7] T. Tomita, and Y. Torihama, *Japan. J. Physiol.*, **6**, 118 (1956).
[8] T. Tomita, M. Murakami, and Y. Hashimoto, *J. Gen. Physiol.*, **43**, Pt. 2, 81 (1960).
[9] K. Naka, S. Inoma, Y. Kosugi, and C. Tong, *Japan. J. Physiol.*, **10**, 436 (1960).
[10] T. Tomita, M. Murakami, Y. Hashimoto and Y. Sasaki, in *The Visual System: Neurophysiology and Psychophysics*, (R. Jung and H. Kornhuber, eds.) Springer, Berlin, 1961.
[11] T. Tomita, *Cold Spring Harbor Symp. Quant. Biol.*, **30**, 559 (1965).
[12] T. Tomita, A. Kaneko, M. Murakami and E. L. Pautler, *Vision Res.*, **7**, 519 (1967).
[13] W. B. Marks, *J. Physiol.*, **178**, 14 (1965).
[14] A. Kaneko, and H. Hashimoto, *Vision Res.*, **7**, 847 (1967).
[15] D. D. Potter, E. J. Furshpan, and E. S. Lennox, *Proc. Nat. Acad. Sci.*, **55**, 328 (1966).
[16] T. Tomita, *Japan, J. Physiol.*, **6**, 327 (1956).
[17] K. Frank, *IRE Trans. Med. Electron.*, **ME-6**, 85 (1959).
[18] T. Tomita, *IRE Trans. Bio-Med. Electron.*, **BME-9**, 138 (1962).
[19] J. S. Coombs, J. C. Eccles and P. Fatt, *J. Physiol.*, **130**, 326 (1955).
[20] H. K. Hartline, H. G. Wagner and E. F. MacNichol, Jr., *Cold Spring Harbor Symp. Quant. Biol.* **17**, 125 (1952).

[21] T. Tomita, *J. Neurophysiol.*, **20**, 245 (1957).

[22] T. Tomita, *J. Opt. Soc. Am.*, **53**, 49 (1963).

[23] M. Murakami, and A. Kaneko, *Vision Res.*, **6**, 627 (1966).

[24] G. Svaetichin, *Acta Physiol. Scand.*, **39**, Suppl. 134, 17 (1956).

[25] G. Svaetichin, *Acta Physiol. Scand.*, **29**, Suppl. 106, 565 (1953).

[26] T. Tomita, *Japan. J. Physiol.*, **7**, 80 (1957).

[27] T. Tomita, T. Tosaka, K. Watanabe, and Y. Sato, *Japan. J. Physiol.*, **8**, 41 (1958).

[28] E. F. MacNichol Jr., and G. Svaetichin, *Am. J. Ophtalmol.*, **46**, Pt. 2, 26 (1958).

[29] G. Mitarai, *Proc. Japan. Acad.*, **34**, 299 (1958).

[30] T. Oikawa, T. Ogawa, and K. Motokawa, *J. Neurophysiol.*, **22**, 102 (1959).

[31] T. Tomita, M. Murakami, Y. Sato, and Y. Hashimoto, *Japan. J. Physiol.*, **9**, 63 (1959).

[32] K. Watanabe, T. Tosaka, and T. Yokota, *Japan. J. Physiol.*, **10**, 132 (1960).

[33] T. Tomita, and A. Kaneko, *Med. Electron. Biol. Engng.*, **3**, 367 (1965).

CHAPTER 9

The Difference of Electric Potentials and the Partition of Ions between the Medium and the Vacuole of the Alga Nitella

GREGOR A. KURELLA

Department of Biophysics, Moscow State University

9.1 INTRODUCTION

The fact that green algae of the family *Characeae* have no cortical cells makes them a highly preferred material for the study of selective ion absorption, mobility of protoplasma, bioelectrogenesis, and the dependence of photosynthesis on these processes.

Both internodal cells and those of lateral shoots are large coenocytes. Behind a sufficiently thick cellulose-pectin wall is a layer of immobile ectoplasma. Close to this wall, there is a cylindric chromathophore consisting of regularly arranged chloroplasts (Fig. 9.1). The underlying layer of cytoplasm—the endoplasm—is in normal, unexcited cells in constant ordered rotational motion. Counterflows of endoplasm, moving at a speed of 50–60 μ/sec are separated by two indifferent zones. In young cells, at a magnification of 300–400×, inclusions that are carried away by the flow of moving cytoplasm and the pulsating boundary between cytoplasm and vacuole are clearly visible. The bulk of the cell is occupied by a central vacuole.

The greater part of the research of our laboratory has been concentrated on *Nitella flexilis* and *N. mucronata A. Br./Miquel*, the latter species being cultured for many years on various artificial media and on agar substrate, that is, on media of exactly know composition. (*N. flexilis* is cultured in common aquaria.)

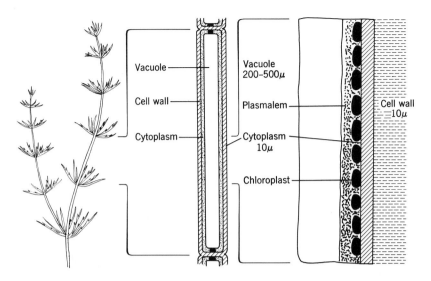

Fig. 9.1. Morphological structure of *Nitella* internode. (From L. Vorobjev, and G. A. Kurella, *Abhandlungen der Deutschen Akademie der Wissenschaften zu Berline* (*Klasse für Medizin*) **4**, 309 (1966).

9.2 METHODS

A word should be said about the general microelectrode techniques that we had to develop on our own in the early 1950's and that probably differ in some respects from the techniques of other laboratories.

9.2.1 Techniques for Drawing Microelectrodes

For our "intracellular capillary glass microelectrodes" (salt bridges of microscopic sizes), we use various glasses of the Pyrex type[1]. In some cases, however, we use chemically resistant aluminosilicate glasses of experimental compositions, which have a softening temperature somewhat higher than that of Pyrex. Capillaries with an outer diameter of 1.1 ± 0.1 mm (this standard was developed in the process of work and may, of course, be changed at will) are drawn from tubes with a diameter of 8–12 mm by the Perfiliev puller[2]. This puller permits capillaries to be made of any parameter and configuration, depending on the shape of the initial tubing. On one such installation it is possible to draw capillaries in which the relationship between inner and outer diameter remains the same as in the initial tube: they are, as we say, "drawn in similitude." On another installation a uniform shrinkage of the capillary's wall may be produced during drawing; in other words, it is possible to obtain thick-walled capillaries from thin-walled tubes. The second installation, being somewhat more complicated in adjustment, is less frequently used; it appears to be easier to select tubes with the required relationship of diameters. On both devices, more than 120 linear meters of standard capillaries of the given diameter (1.1 mm) are obtained from an initial tube with an outer diameter of about 10 mm. Waste of glass is negligible.

The capillary microelectrodes themselves are drawn on an "automatic forge" constructed by Bysov and Chernyshov[3]. Bysov noticed (and his observation was confirmed in practice) that the thicker the capillary's walls, the finer is the tip of the capillary microelectrode drawn on the forge. Two microelectrodes are obtained simultaneously, the upper with a short neck and the lower with a long neck. Microelectrodes made of standard capillaries are standard, too. In this case, also, waste of glass is minimal, and if necessary, one man can make up to 600 microelectrodes in one working day.

To verify the effectiveness of our methods, six microelectrodes were taken at random and their tips tested under an electronic microscope; in three of them the diameter of the tip was of the order of $0.1\ \mu$; in the three others it was no greater than $0.3\ \mu$. The microelectrodes of this lot were made of capillaries with a diameter relationship of $1:3$. Such electrodes penetrate very smoothly through cellular teguments, but are characterized by a correspondingly high electrical resistance. When filled with $2.5\ M$ solutions of KCl or NH_4Cl their resistance is of the order of $10^8\ \Omega$; thus they are suitable for investigations of slowly changing potential differences (e.g., resting potentials). The input resistance of electrometers will, of course, be high and result in a correspondingly high time constant.

Such microelectrodes are used in investigation of resting potentials of isolated single muscle fibers[4–8], bioelectrical phenomena on the membrane of cell nuclei[9], and in work with giant neurons of molluscs.

Microelectrodes of extremely high resistance are in most cases unsuitable for investigations of action potentials. However, on the same devices, it is possible without any complicated adjustment to prepare microelectrodes with a resistance of the order of $5–40\ M\Omega$ from capillaries with a diameter relationship of $1:2$ or $1:1.5$.

9.2.2   Techniques for Filling Microelectrodes

Different techniques for filling microelectrodes with electrolyte solutions were employed during different periods of our investigations[10]. At present we most often use the method of immersing the glass capillaries in boiling methanol or ethanol, subsequently replaced by electrolyte solutions. We prefer to work with electrodes kept but a short time in

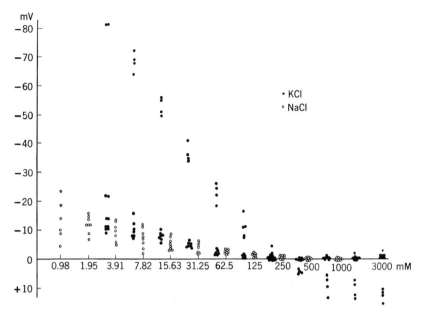

Fig. 9.2. Dependence of tip potential on concentration of outer test solution. The four microelectrodes are filled with 118-mM KCl solution. Test solutions contain either KCl (●) or NaCl (○). *Kurella* [11].

electrolyte solutions; the potential of the tip is always lower in such microelectrodes.

These so-called "tip potentials" are a source of trouble to all concerned with microelectrode techniques. Our interpretation of their origin[11] differs somewhat from that of most investigators[12]. Special investigations have shown that the tip potential is primarily the result of the cationic function of glass. Suffice it to say that the mean value of electrode potentials depends on the type of glass rather than on its configuration. A noticeable increase in one or another function of the electrode is observed and a corresponding rise of its own potential is obvious after a prolonged soaking in electrolyte solutions. Electrodes made of soft glasses have a more marked function, while in electrodes made of hard glasses, that is, of Jena-therm and Pyrex type, the potassium and sodium function together is expressed to a greater extent.

9.2.3 Tip Potential

In our opinion the tip potential is generated on the side wall of the capillary and shunted by the resistance along the channel of the micro-electrode. This view explains the correlation between electrode resistance and tip potential described by many authors. Microelectrodes made of quartz have practically negligible potentials. On the other hand, a capillary microelectrode made of glasses usually employed for hydrogen electrodes can be used as a microelectrode with hydrogen function even if its channel is not sealed—provided its channel section is small enough and the resistance along the channel is correspondingly high. An adequate correction will of course be necessary, and the same applies to micro-electrodes made of glasses with pronounced sodium or potassium functions. Sealed microelectrodes, reversible for one or another ion, but made from corresponding kinds of glass, are, naturally, preferable.

To all appearances, junction potentials (diffusion or streaming potentials by their nature) play a quite insignificant part in real capillary microelectrodes with an outer diameter of less than 0.5μ. This situation exists because ionic diffusion through the channel of such a capillary is determined by the section of the channel rather than by the difference in mobility of the given ions in water. With a diameter relationship of $1:3$ at the tip and an outer diameter of 0.1μ, the diameter of the channel will be of the order of 0.033μ.

The tip potentials of microelectrodes are usually checked by bringing the microelectrode into contact with an external medium (for instance, with Ringer's solution), and those potentials will depend essentially on the medium's composition and concentration. In thicker electrodes, there is no guarantee against contact potentials being generated upon the contact of the microelectrode tip with a polyelectrolyte gel—as, for instance, when the properties of the gel affect the ionic mobility of the electrolyte which fills the microelectrode during its diffusion in the gel phase [13]. I think, however, that in this case, too, the best way to prevent the appearance of such contact potentials is to reduce diffusion to a minimum, that is, to use extremely fine microelectrodes with a minimum section of the channel [14].

9.2.4 Micromanipulators

For investigations of constant potential differences we use small reference electrodes (silver-chloride or calomel) in which the micro-electrode is held by a small clamp on its end. Such combined half-cell electrode holders are conveniently mounted on the micromanipulator. The half-cells have a stable potential—especially the calomel ones, in some of which the potential was found to remain constant for several months without refilling.

Depending on the problem under investigation, different types of micromanipulators are employed. If the work is done at magnifications up to 120× (i.e., when steroscopic microscopes can be used), the microelectrode is often introduced vertically; at magnifications as high as 300–400×, the microelectrodes must be inserted horizontally (owing to the restricted range of the objective).

On micromanipulators of Reinert's type, microscrews are used for insertion in a vertical direction. The same manipulator can be used for horizontal insertion of microelectrodes in the vacuole of such large cells as *Nitella* by shifting its shoulders along a sliding plate. If the pushing effort is rough enough, the electrode is practically always thrust directly into the vacuole. As the resting potential remains unchanged after such an insertion (with the microelectrode implanted in the cell) even on the sixth day, it may be assumed that the electrode passes successfully through all the barriers inside the cell and is not subsequently covered by a cap.

Smooth insertion of microelectrodes (at angles different from the horizontal) into diverse objects including *Nitella* is achieved by means of hydraulic pushing devices which can be mounted either directly on the microscope table or on the shoulder of the micromanipulator. This simple system consists of a 10- or 20-ml all-glass syringe with a half-cell electrodeholder mounted on its glass piston and a combined 1-ml syringe with a backward spring. The piston of the delivering syringe is shifted by means of a micrometric screw, the two syringes being connected by a polyethylene tube. Depending on the relationship of the piston sectional areas, the delivery of the microelectrode per turn of the microscrew (one step $= 0.5$ mm) ranges from 5 to $50\,\mu$ without idle running or complemental work — provided there are no bubbles in the system and the fluid is degassed. Microelectrodes inserted in *Nitella* or other cells by means of such a pushing device enter the cell smoothly.

9.3 MEASUREMENTS OF CELLULAR POTENTIALS

9.3.1 With One Microelectrode

The first potential difference is observed at the initial contact of the microelectrode with the cell wall. Under our conditions it is of the order of 39–44 mV and is conditionally termed "potential of the cell wall." It seems to be caused by the fact that the cellulose-pectinic wall of the cell resembles an ion-exchange resin which possesses predominantly cationic properties in equilibrium with a strongly diluted solution, that is, with the medium. The microelectrode itself is filled with 2.5 M KCl or NH_4Cl. A chain is formed in which the measured potential difference is determined by the value of the potential jump between the boundary of the electrolyte solution filling the microelectrode and the wall phase and by the jump of the potential at the boundary of the wall phase and the external solution. This potential difference is distorted by diffusion potentials the values of which depend on the activity gradient and the constants of self-diffusion of ions in the phase of the wall[16]. It also depends on several other factors, including the possible changes of the microelectrode's own potential owing to the changed resistance of the microelectrode channel.

During a further smooth advance of the microelectrode through the membrane, the "wall potential" usually remains unchanged. When the tip of the microelectrode comes out of the wall, the value of the potential difference decreases by 10–20 mV, presumably because the tip of the microelectrode coming out of the wall displaces the protoplast. In this situation the measured value will be determined by the potential difference between the external medium and the medium filling the space between the wall and the protoplasm. It seems that an additional interphase exists or is generated on the surface of the protoplasm separating the microelectrode from the cytoplasm phase. Whether this interphase is the plasmalemma still remains an open question. However, we do know that this seemingly elastic-enough film is a product of cytoplasm and is capable of new formation. In addition, it was found that at a slow advance of the microelectrode the level of the potential may remain constant.

The next jump of the potential, displayed only on a more rapid advance of the microelectrode, results from the puncture of this interphase. Its value varies between 115–140 mV. Another result of such a puncture may be an observable local reaction of cytoplasm, namely, the arrest of cytoplasmic motion at the place of insertion of the microelectrode. This reaction seems to be related to an increase of viscosity; it could be seen how the layer of moving cytoplasma bypasses a "cloud" of arrested cytoplasm. We interpret the second jump of the potential as being the cytoplasm potential. Apparently, at a higher rate of advance, the microelectrode breaks through the interphase mentioned above. However, at each stop this phase begins immediately to cover its tip; this is accompanied by a drop of the potential to a level 15–20 mV below the wall potential. In these cases a dense "cap" that envelopes the microelectrode is always observed.

The formation of a "cap" around the inserted portion of the microelectrode was described by Walker[17]. However, the present author did not observe a simultaneous drop of the potential value to a definite level, and could merely record a certain decrease of potential difference. With a rapid and more profound insertion, contact is probably established between microelectrode and cell sap because in this case the process of overgrowing of the tip is more prolonged. There is, however, no guarantee that rapid insertion will ensure contact between microelectrode and cell sap, for it is not certain that the microelectrode always breaks through so elastic a formation as the tonoplast. Consequently, we feel that the difference between the mean values of potential differences (measured with microelectrodes inserted into the cytoplasm and into the vacuole) obtained in our experiments with smooth introduction—which does not exceed 5 mV, although Kishimoto mentions 8 mV[18]—perhaps does not reflect the true picture. A further perfecting of the methods of measuring possible jumps of the potential on the tonoplast of *Nitella* cells is required.

9.3.2   With Two or More Microelectrodes

Several devices of the type described above are used for the introduction of two or more microelectrodes into a cell of *Nitella*.

A double hydraulic micromanipulator has been devised for the insertion of two microelectrodes into a single spherical cell (e.g., into cytoplasm and nucleus of a cell of the salivary gland of *Drosophyla* larvae). The mechanisms of this micromanipulator are capable of shifting autonomously in three planes with an accuracy of 5 μ per one full rotation of the pushing screw. Sagital delivery, too, is provided—allowing a shift of the microelectrode by 10 mm. The angle between the axes of the microelectrodes may also be roughly modified. Besides being double, this micromanipulator has the advantage that—in contrast to the pneumatic micromanipulator of Fonbrune[19]—any possibility of "complemental work" of the executive instrument is precluded; the instrument is delivered in a strictly sagital direction without any "yawning" along a complicated curve.

Depending on the purpose of the investigation, different electronic devices are employed. For investigations of resting potentials and other constant signals, we use devices with a relatively great time constant but high resistant input. We began our work with a homemade electrometer devised according to Dubridge–Bardt[11,20]; later, L. N. Vorobiev developed a three-channel electrometer for three electrodes. Both these devices have an input resistance of $10^{12}\,\Omega$ and zero drift not exceeding 0.1 mV/hr; in work on *Nitella* we often use slightly modified Soviet factory pH-meters with a self-recording potentiometer. These meters' sensitivity has been brought up to 250 μV per scale division. Such pH-meters have a negligible drift and the input resistance is sufficient for work with usual capillary glass electrodes. For work with microelectrodes of extremely high resistance and requiring instruments with a minimum zero drift, we have, in a number of cases, used an English electrometer with a vibrating capacitor of the "Vibron" type, model 33B, with an input resistance of 10^{13} Ω. At present we are using for these purposes the more reliable electrometer Orion-Keti (TR-1501) made in Hungary; it has an input resistance above 10^{14} Ω.

Action potentials are studied by means of cathode followers developed by Bysov and Bongardt[21], or simpler followers with a somewhat higher time constant but which are sufficient for investigations of action potentials of *Nitella*; the steepness of their ascent phase does not exceed 200 mV/sec. Factory electrometric amplifiers with a relative low time constant (U–1–2) are also suitable for these purposes[22].

9.4 MEASUREMENT OF INTRACELLULAR IONIC ACTIVITY

Formerly we used only material of animal origin, mainly single fibers of skeletal muscle of anurous amphibians. However, when we started on our study of intracellular ionic activity by using different reversible electrodes, our attention turned to plant cells – in particular to cells of *Characeae algae*. Among the many reasons for this change was the dilemma with which we were immediately confronted. The concept of mean activity of a solution, more or less synonymously formulated for aqueous electrolyte solutions (especially for pure aqueous solutions of binary electrolytes), becomes ambiguous when one speaks of colloidal solutions. In particular this factor applies to polyelectrolytic solutions and heavily swelling poly-electrolytic gels, that is, to media which by their physicochemical properties resemble cytoplasm.

I should like to illustrate this phenomenon by the following example: the pH of an aqueous solution is a concrete conception, but when we have to deal with very dilute solutions of polyelectrolytes with H-ions as counterions, it is no longer appropriate to say that the readings of an immersed microelectrode (ideally reversible for hydrogen) correspond to the mean hydrogen-ion activity of the solution. The medium is not homogeneous, hydrogen ions being unevenly distributed. Taking this into account, we selected for our investigations of intracellular pH values and their changes caused by external influences or excitation the large cells of algae with a big central vacuole. We assumed that such cell sap must assuredly present a more homogeneous medium that would cytoplasm. Already many investigators have realized, perhaps without full comprehension, that somewhere within the cell, far from the outer membrane where the potential drop called resting potential is located, there can be local ion activities other than those close to the inner boundary of the membrane (e.g., within mitochondria, nuclei, chloroplasts, etc.). Such ion activities, therefore, exert no direct effect on the potential-difference value measured on the membrane. In determinations of ionic concentration, usually by flame photometry, the concept of mean concentration of any one species of ion to be compared with the measured potential difference involves all the contents of the cell without analysis of their distribution within the cell. When confronted in practice with such phenomena, some researchers are perplexed by the results obtained and seek in theory the causes of the deviation of experimental values. It seems to us that in our investigations we were able to reveal some of the causes of these phenomena [4-9, 23]. However, as these investigations were made on isolated single muscle fibers, nuclei of neurons, and salivary gland cells, I shall not describe our findings here.

9.4.1 Intracellular pH

For studies of intracellular pH, we primarily use glass microelectrodes filled with antimony, that is, antimony microelectrodes [24]. Such electrodes probably have several defects (it is generally thought that their readings are affected by the pO_2 of the medium), but they also have a definite advantage over glass microelectrodes reversible for hydrogen [25]. Their working surface is confined to the buttend, so that when they are inserted into the vacuole of *Nitella* they react on the pH of cell sap only. We use hydrogen glass electrodes also, but with them one is always confronted with the problem of reliable insulation of the lateral surface. Extensive methodical work carried out by G. A. Popov and V. K. Andrianov [24, 26, 27] has shown that the readings of antimony microelectrodes are dependable enough. Some insignificant errors are, of course, possible in determinations of absolute pH values, but the changes of pH can be measured quite reliably.

Depending on the lot of antimony, the glass, and the method of fabrication, the function of separate antimony microelectrodes varies between 52 and 57.5 mV per unit of pH within the range from 3.5 to 8. Above pH 8 the function begins to decrease, but if there is a graduation graph the microelectrode can be used even at still higher values of pH. Our microelectrode was always calibrated before and after use in a series of five buffer solutions.

We have discovered that the limit of current density for antimony microelectrodes, beyond which processes are generated that lead to an increase of their already high resistance and to unsteadiness of the whole system, lies at $40-50 \mu A/cm^2$ [24]. At the time they were working, Buytendijk and Woerdmann [28] could not take this factor into account because the required electrometers were still nonexistent and they were unable to use antimony electrodes thicker than 200μ. Anybody can prepare a glass microelectrode filled with antimony thinner than 1μ, but work with such an electrode — considering its resistance (considerably more than $10^{11}\,\Omega$), the extreme value of the generated emf (above 500 mV), and the permissible current density with a working surface of no more than $0.5\,\mu^2$ — requires electrometers with an extra-high input resistance. No such electrometers with an input resistance higher than 10^{14} were available in the late 1950's; we, too, had to use electrodes filled with antimony, about $3\,\mu$ in diameter. The method was permissible with cells such as *Nitella*, but not with smaller cells of animal origin. This was a second reason for working with giant plant cells.

Our investigations of the properties of glass microelectrodes filled with antimony proved that they can be used, though with certain reservations, for studies of pH in vacuoles of *Nitella*. The misgivings expressed by some investigators[29] about a possible reduction of the oxide layer of antimony microelectrodes within the cell (so that there will be no reaction) have not been confirmed. It will be shown later that the same applies to silver-chloride microelectrodes. Fortunately, other researchers, too, have reached this conclusion independently of our work[30].

In investigations carried out with G. A. Popov[24,26] and V. K. Andrianov[27] (both at that time students working on their dissertations), we were able to show that the pH of cell sap of *Nitella flexilis* varies between pH 4.2 and 5.3, depending on season, age, and conditions of growth. However, in one and the same cell (under constant external conditions), no changes in time (up to 20 hr) occur in the pH of cell sap. This was observed with microelectrodes inserted permanently, as well as with microelectrodes introduced into the cell only during measurements. Two microelectrodes, one a capillary and one antimony, were used for pH measurements to avoid a simultaneous recording of the resting potential. The pH of cell sap in internodes of different-aged plants that are cultivated in media of pH 7.8 (in aquaria, in water with undefined ionic composition) varied between 4.8 and 5.8; in cells of the same age but which were cultivated in White's medium[31] of tenfold dilution (free of sucrose and other organic ingredients and with a pH of 5.6), the mean reaction of cell sap varied between pH 4.2 and pH 4.6.

At tenfold and hundredfold changes of osmotic pressure in the medium, achieved by addition of sucrose (which does not affect the pH of the medium), the pH of the cell shifts within 10–15 min by 0.33 and 0.55, respectively, toward higher acidity. The changes of the resting potential differed in cells of different cultures. In one of the cultures the resting potential increased regularly with increasing osmotic pressure[27], but later these phenomena could not be reproduced on cells from other cultures of *N. flexilis* and *N. mucronata*. As yet we have not been able to provide an explanation. We can merely state that in the first culture the dependence of the resting potential on potassium concentration was 38–39 mV per tenfold change if the change of osmotic pressure was not compensated, and 51–52 mV if it was compensated. In the other cultures, in which the resting potential did not respond to changes of osmotic pressure, the dependence of the resting potential on potassium concentration in the medium was about 51 mV, irrespective of osmotic pressure.

The dependence of the resting potential on potassium and sodium concentration in the medium has been investigated by many researchers [32–35], so that I need not dwell on the subject here. I should only like to observe that a tenfold increase in potassium concentration results in a change of the resting potential value and the shifting of pH of cell sap toward higher alkalinity by 0.25 ± 0.03 (the resting potential decreases by 51 ± 0.9 mV). The increase of potassium-ion concentration in the medium of plants cultured on a modified White's medium was achieved by introducing potassium sulfate, while lack in the primary solution was compensated by sucrose. At a hundredfold increase of potassium-ion concentration, the pH of cell sap shifted towards a higher alkalinity by 0.45 ± 0.04 and the resting potential decreased by 84 ± 1.3 mV.

Transitory changes of pH in the external medium ranging from pH 5.4 to 9.2 are not accompanied by any noticeable changes of intracellular pH, although the value of the resting potential is altered. These changes are, however, not adequate. Within the mentioned range of pH in the medium, the curve of dependence of the resting potential on pH passes through a maximum within the range of pH 7.6 to 7.8. At this concentration of hydrogen ions, the resting potential increased in a number of cells up to 210 mV; at pH 5 or 9, the resting potential of these cells approximated 120 mV.

If *Nitella* cells are kept for many days in media with a pH substantially differing from the initial one, the reaction of cell sap will shift very slightly in the same direction as the hydrogen-ion concentration of the medium. However, the reaction of cell sap is always by 1.5–3 units of pH more acid as compared with the medium.

Using the described method, in which a three-channeled electrometer measures in turn the potential differences between two of the three electrodes, we were able to follow all the changes in time both of the resting potential and the intracellular pH and to determine the sum of these two values; the changes occurring in the course of development of the action potential could be recorded, too. It was found that the pH of vacuole sap is practically unaffected by the action potential. The whole recording clearly registers the action potential, both in measuring potential differences between medium and vacuole and in recording the sum of potential differences made up of the resting potential and the potential of the antimony microelectrodes reversible for hydrogen. This indicates that antimony microelectrodes might be capable of responding to relatively rapid changes of pH, if any. The time constant of antimony microelectrodes was also verified in model experiments.

The speed of response of such an electrode also depends, of course, on the rapidity of renewal of the fluid layer around the microelectrode; however, it must be kept in mind that at rest (i.e., before the development of an action potential) both cell sap and cytoplasm are in constant rapid motion, 50–60 μ/sec, and that the layer of fluid at the tips of the microelectrodes is constantly renewed (a total stop of protoplasm streaming coincides with a peak of the action potential[36]). We considered these experiments necessary and the findings sufficient to warrant the conclusion that no relationship exists between development of action potential and any hydrogen-ion transport between medium and vacuole.

Microelectrodes of this type, made of quartz and filled with platinum (with a smooth surface) or gold—a preparation which is rather complicated—are being successfully used in our department by E. I. Efimzew and F. F. Litvin for intracellular determinations of pO_2 and its changes during photosynthesis (by amperometry)[37, 38]. Such platinum microelectrodes are also used in investigations of intracellular red-ox potentials and their changes during photosynthesis. In these investigations, too, *Nitella* cells are used as experimental material.

As already mentioned with respect to all our experiments, we have never discovered any significant jump of the potential that could be related to the tonoplast. In our first experiments we could not distinguish a clear-cut boundary between cytoplasm and vacuole; however, later on, using greater magnifications and working with younger cells, we were able to ascertain that noticeable jumps occur only when the capillary microelectrode inserted into the cytoplasm is already covered by a cap of precipitating cytoplasm. A further advance of the microelectrode and its breaking through the formed capsula might produce an abrupt increase of potential difference up to the original cytoplasm potential, but this has nothing to do with the jump of the potential at the tonoplast.

In our further studies of this problem we applied the method of microinjection of different salts, substances and distilled water into the vacuole (Fig. 9.3). Experiments carried out by A. I. Konoshenko[39] showed that microinjections through capillaries of the microelectrode type produce no changes in the value of resting potentials at the boundary between cytoplasm and vacuole, provided that all precautions are taken against the destruction of the cell by the microinjection itself. Potential difference between vacuole and external medium was measured by another

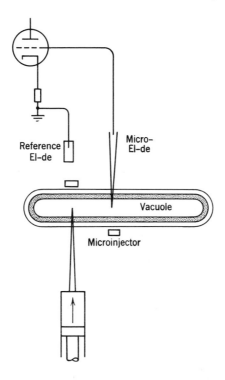

Fig. 9.3. Microinjection of KCl and NaCl in cell vacuole of *Nitella*. (From L. Vorobjev, and G. A. Kurella, *Abhandlungen der Deutschen Akademie Der Wissenschaften zu Berlin (Klasse für Medizin)* **4**, 309 (1966).)

capillary microelectrode. Flocculent precipitation in the vacuole was observed only when the introduced solutions contained calcium ions; if large amounts of calcium ions were added, the cell simply died off—and the resting potential value showed a corresponding gradual decrease.

During the last few years some investigators have reported a small jump observed on the tonoplast[35]. In a number of cases this could be explained by a changed value of the microelectrode potential, because the workers used microelectrodes having a diameter of the order of $10\,\mu$. The conditions of ion diffusion from so large a channel might be modified during the passage of the tip of the microelectrode from cytoplasm into vacuole. Recently some other papers have been published showing that in some marine algae jumps of the potential at the boundary between cytoplasm and vacuole may reach even higher values than those across the cell surface[40]. Therefore, we can no longer affirm with the same assurance that no jump occurs in fresh-water species of the genus

Nitella used in our experiments. There seems to be no full guarantee that the microelectrode smoothly advanced does pierce the tonoplast. If roughly advanced, the microelectrode's tip thrusts directly into the vacuole.

Even this single example convinces us once again that plant cells are far more complicated than cells of animal origin, which have no walls possessing ion-exchange properties (and, probably, selective properties, too), no tonoplast, and not so large a central vacuole.

9.4.2 Intracellular Chloride Activity

Prompted by our experience with antimony microelectrodes, we began working with glass electrodes filled with silver in order to develop a method for intracellular determination of chloride activity. Earlier attempts had been made at producing microelectrodes filled with silver or its alloys to obtain low-resistance microelectrodes[41] and to measure chloride activity[42]. But Mauro, who prepared a chlorinated-silver microelectrode of sufficient thickness, was disappointed by the result obtained; starting on the assumption of the homogeneity of intracellular contents, and a Donnan distribution of potassium and chloride ions between the axoplasm of giant squid axons and the medium, he expected to measure a zero potential difference between two symmetrical silver-chloride microelectrodes, one of which he inserted into the axon. But the actually measured potential difference amounted to 50% of the resting potential value. On the basis of these experiments, Mauro came to the conclusion that inserted silver-chloride microelectrodes are reduced within the cell and lose their chloride function. These considerations were accepted by other authors, or at least were cited in other papers[29], and this seems to have discouraged many investigators who proposed to measure the intracellular activity of chloride.

In our attempts at creating a glass microelectrode filled with silver we were confronted with many difficulties arising from the purely technical problem of soldering melted silver and its alloys into softened glass, a problem well known to all experimental physicists[43]. L. N. Vorobiev discovered an ingenious way out of our difficulty: the following method was devised for the determination of chloride activity in the vacuole sap of *Nitella* (Fig. 9.4).

One takes a dry capillary microelectrode and inserts it into the vacuole through a dried spot on the surface of the cell. After a few minutes, under the action of capillary forces and turgor, a small quantity of cell sap penetrates into the microelectrode. Through the broad end of the dry microelectrode, a fine silver wire is introduced, its tip tapered by electrolysis and reliably chlorinated. The function of such silver-chloride

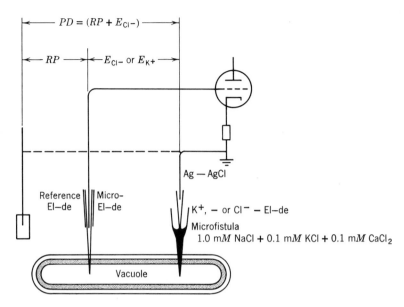

Fig. 9.4. Determination of chloride (or potassium) activity in vacuole of *Nitella*. (From L. Vorobjev, and G. A. Kurella, *Abhandlungen der Deutschen Akademie der Wissenschaften zu Berlin* (*Klasse für Medizin*) **4**, 309 (1966)).

electrodes is tested before and after each experiment. It was found to be equal to 55 mV per tenfold change of activity of the KCl solution within the range of 10^{-1} to $10^{-3} M$. The wire is pushed through the dry micro-electrode until it stops. When the wire comes into contact with the meniscus of cell sap in the neck of the capillary, chloride activity in the sample can be determined with sufficient stability. That the microelec-trode actually contains pure vacuole sap can be easily verified: If a usual microelectrode filled with KCl is introduced through the thick end of the dry microelectrode and brought into contact with the sample, the normal resting potential of the cell will be measured in relation to the reference electrode in the medium. The potential difference between such a concentric microelectrode and the second usual microelectrode, in-serted into the vacuole, is equal to zero. But if the contact is established between the sample and the silver-chloride microelectrode, then in relation to the reference electrode in the medium the measured potential difference will be equal to the sum of the value of the resting potential and the value of the potential on the chloride electrode. This procedure has been named the "microfistula" method. Naturally, it would be useless to attempt to employ it for investigations of the dynamics of changes of chloride activity in the vacuole, as there is practically no more exchange between the sample in the "fistula" and the bulk of cell sap.

9.4.3   Intracellular Potassium Activity

Studies of potassium activity in the cell sap of *Nitella* were carried out by means of sealed microelectrodes made of glasses melted by J. B. Koltunov[44, 45]. Using the terminology of G. Eisenman[46], these glasses may be designated as follows: $KAS_{21.6-4.5}$; $KAS_{18.6-6}$; $KAS_{16-4.5}$; $KAS_{19.2-5.1}$; and KAS_{20-5}. The best results were obtained with $KAS_{21.6-4.5}$. The selectivity coefficient K_{Na}^{K} of separate microelectrodes was up to 12–15. The effect of H^+, Ca^{2+}, Mg^{2+}, and Na^+ within the given limits of concentrations was negligible. The function of the microelectrode was checked before and after each experiment: it was 50–53 mV per tenfold change of KCl activity within 10^{-1} to $10^{-3} M$.

At first, in order to avoid complications associated with the problem of reliable insulation of the portion of the microelectrode not intended to be active, we also used the microfistula method in working with potassium microelectrodes. Later, having tried various methods of insulation, we began to introduce microelectrodes with a potassium function directly into the vacuole. The data obtained by these two methods were in good agreement. Paraffin wax, ceresin, silicon lac, and polystyrene were used for insulation. Polystyrene or silicon lac are also used today for insulation of intracellular glass microelectrodes with a hydrogen function. Furthermore, it was found that, owing to a gradual thickening of the microelectrode walls, this function is noticeably expressed only along the first $200-300\,\mu$, so that one can work quite safely on such relatively enormous cells as *Nitella* in which the microelectrode penetrates to a depth of $200\,\mu$.

The cells used for measurement of potassium-salt activity were placed either in artificial tapwater solutions of $1.0\,mM$ NaCl; $0.1\,mM$ KCl; $0.1\,mM$ $CaCl_2$, or in a standard nutrient solution with a twice greater concentration of K^+ and containing in addition Mg^{2+}, nitrates, phosphates, and trace elements. The intracellular potassium activity was not affected by an interchange of these solutions, while the resting potential in the standard solution was correspondingly lower. Potassium-salt activity measured by the "microfistula" method was $71.3 \pm 5.6\,mM$ (with a mean value of resting potential of 151.4 ± 2.7 mV). The direct introduction of the potassium electrode gave a mean value of 70.1 ± 3.9 mM (with a mean value of resting potential 155.3 ± 3.7 mV).

The mean activity of chlorides in the cell sap of plants cultured under similar conditions, measured by the method of microfistula with a silver-chloride electrode, was found to be equal to 125.6 ± 3.3 mM (with a mean value of resting potential 154.1 ± 2.3 mV). If the mean activity coefficient of alkaline chlorides in the vacuole is assumed to be equal to 0.76, the concentration of chloride in cell sap may be equal to 165 mM.

The conversion of potassium and chloride activities to concentrations shows that potassium in cell sap is most probably present in the form of chlorides and that our data agree well with those obtained by other authors who used chemical methods for determinations of ionic concentration in expressed cell contents [35,47,48].

Once one knows the summarized activity of chlorides and the activity of potassium salts, it is possible to give a rough estimate of the activity of sodium salts in cell sap. It should be of the order of 50 mM, because the sum of ionic concentrations of alkali-earth metals does not exceed 5–10 mM [49,50]. Measurements of the true values of sodium-salt activities by means of microelectrodes reversible for sodium are still in prospect.

As already mentioned, the described data refer to stationary activity values of the investigated salts. Some discrepancies with the data of other authors may be ascribed, *inter alia*, to the fact that no standardization of culture media has been carried out — although the composition greatly influences the composition of cell sap. We have discovered during the course of our control experiments that in plants cultured for several generations in media as far as possible free from one or another species of ions, the composition of cell sap becomes substantially altered.

Thus in media with a K^+ content known to be below 10^{-5} mM (all the potassium in the medium being substituted by sodium), cells are produced in which the activity of potassium salts in vacuole sap is three times lower and the chloride activity two times lower than in cells grown in the standard medium, while the mean value of the resting potential is almost normal (127 ± 4 mV). Cells grown in media containing no appreciable quantities of sodium ions determinable by flame photometry have a cell sap with twice as great activity of potassium salts and one-and-a-half times as great activity of chlorides as compared with vacuole sap of cells grown under standard conditions, the mean value of the potential being 125 ± 5.5 mV. The activity of potassium salts in vacuole sap was three times lower and sodium activity two-and-a-half times lower in cells grown in a medium free as far as possible of chloride, as compared with normal cells. The mean value of the resting potential in the third generation of these cells (i.e., in cells thrice transplanted on a new substrate and on the same medium free of chloride) was equal to 138 ± 5.2 and that of the fourth generation was equal to 173 ± 4 mV. No decisive conclusions can be drawn from these data about the mechanism determining the ionic composition of vacuole sap, because no unification of culture media has yet been achieved (such as we were able to achieve, for instance, in our work with isolated muscle fibers[5]) that would permit the creation of a standard background against which our results could be quantitatively analyzed and corresponding conclusions drawn. However, the data obtained by Su-jun Li and L. N. Vorobiev indicate that a conjunction exists between the influx of chloride and potassium into the vacuole. But in contrast to cells of animal origin, which contain very little chloride and in which chloride and potassium are distributed as a first approximation according to regularities described by Donnan, the vacuoles of plant cells contain chlorides in far greater quantities than the medium. (Incidentally, sodium, too, is present in quantities many times greater than in the medium.) In our experiments the standard external medium contained 1.3 mg ions chloride per liter, while the activity of chloride in the vacuole was one hundredfold higher.

It is important to note that in these experiments the mean resting potential value was near to normal, and no jump was observed that could be related to the tonoplast. We are more than convinced that in the cytoplasm, chloride activity is substantially lower than, and potassium activity about the same as, in vacuoles of cells grown under standard conditions. In all cases the resting potential of the cells, including plant cells, is determined by the mean activity of the potential determining ions (in particular potassium ions) immediately at the interphase of protoplasm and external medium. If this is so, then in cells grown in media free of one or another species of ions the mean activity of the potential determining ions should remain close to normal. Accordingly, it is clear that selective ion accumulation is determined by the properties of the polyelectrolytic structure of protoplasm rather than by the properties of the cell membrane. Investigations with electrodes reversible for hydrogen ions have shown that the method of direct introduction of reversible microelectrodes into the vacuole also permits study of the dynamics of activity changes caused by variations in the composition of the external medium — variations brought about by cell excitation and changes in illumination of photosynthetizing cells.

9.5 THE EFFECT OF ILLUMINATION ON THE RESTING POTENTIAL

A detailed investigation of the dynamics of activity changes in vacuoles of plant cells is still in prospect, but it is not without purpose that we mention illumination.

It has been firmly established by our work with V. K. Andrianov and

F. F. Litvin[51] that alternation of darkness and exposure of *N. flexilis* to white light and to red light (absorbed by chlorophyll only) of an intensity of the order of 2,000 lx results in an abrupt drop of the resting potential value to a new level. The process is rapidly reversible. A practically complete resemblance is observed between the shape of the curve of changes of the resting potential value in time under the action of light and the induction curves of photosynthesis. The rate of photosynthesis (oxygen evolution) was determined amperometrically by another group of investigators in our department, who used cells from the same culture[52].

The relationship of changes of resting potential value to intensity of incident light also bears a strong resemblance to the light curves of photosynthesis. We investigated the relationship of the resting potential value and of the magnitude of its changes to temperature in cells exposed to white light (intensity more than 4000 lx). It was found that the dependence of the resting potential value on temperature differs in cells kept in darkness and in those exposed to light; the value of Q_{10} within the temperature interval 6–16°C was equal to 1.03 in cells exposed to light, as compared with 1.1 when measured in darkness. An interesting fact is that the value Q_{10} for the effect of light — the fall of the resting potential value at illumination — is equal to 2.2–2.5, approaching the value of Q_{10} of photosynthesis in algae[53].

Thus we feel justified in concluding that a close relationship exists between the effect of illumination on the resting potential value and photosynthesis. If the cells are transferred into darkness, the initial value of the resting potential is restored; consequently, this is a reversible process, related to illumination only and not to temperature (for control, the temperature of the medium was always measured by light and in darkness and was found to remain stable within ±0.5°C). The concrete mechanism of this relation is as yet unknown. There is some evidence of marked changes of ion fluxes under the effect of light in the cells of another species of *Nitella*[54–56]. It may be reasonably assumed that the triggering of the mechanism of photosynthesis by illumination leads to sharp changes in exchange processes, accompanied by changes in the flow of ions and water across the cell surface and redistribution of ions and other substances between cell and medium and between the separate organoids within the cells, as well as to changes of gas composition of cell content [38]. This, evidently, results in a change of the activities of potential determining ions immediately at the inner boundary of the cell membrane, to which the jump of the resting potential is mainly confined (as evidenced by the change of the resting potential value at alternations of darkness and light and light and darkness). The verification of this assumption presents some methodical difficulties, as it involves investigations of the dynamics of activity not in the vacuole but in the cytoplasm and even in chloroplasts.

Activity changes in the vacuole are a mere consequence of the changes that take place in the cytoplasm. All data obtained till now on the changes of salt activities in cell sap, brought about by various influences exerted on the cell, show that these changes proceed at a very slow rate, whereas changes of the resting potential are far more rapid.

The subject of our investigations — green algae of the family *Characeae* — combines the functions of root-hair and green leaf of higher plants. By means of intracellular glass microelectrodes of different types, in combination with optical methods and the tracer method, this experimental material may furnish information of fundamental importance to the understanding of the principles of essential processes.

REFERENCES

[1] I. P. Kitaigorodskij, and S. I. Silvestrovitch, in *Manual on Glass Technology*, Vol 1, Gosisdat Literatury po Stroiteljstvu, Architekture i Stroimaterialam, Moscow 1963.

[2] B. W. Perfiljev, and D. R. Gabe, in *Capillary Methods on the Study of Microorganisms*, Acad. Sci. USSR, Moscow-Leningrad, 1961.

[3] A. L. Byzov, and W. I. Chernyshov, *Biofizika*, 6, 485 (1961).

[4] Zy-tiun Lian, and G. A. Kurella, *Biofizika*, 7, 700 (1962).

[5] Zy-tiun Lian, and G. A. Kurella, *Biofizika*, 8, 597 (1963).

[6] G. A. Kurella, and Zy-tiun Lian, *Biofizika*, 9, 78 (1964).

[7] G. A. Kurella, and Zy-tiun Lian, *Biofizika*, 10, 72 (1965).

[8] G. A. Kurella, and Zy-tiun Lian, in *Biophysics of the Cell*, Nauka, Moscow, 1965.

[9] W. F. Antonov, G. A. Kurella, I. F. Meschishen, and U. Ben-tzé, *Dokl. Akad. Nauk SSSR*, 161, 693 (1965).

[10] G. A. Kurella, *Biofizika*, 3, 243 (1958).

[11] G. A. Kurella, *Biofizika*, 3, 614 (1958).

[12] R. H. Adrian, *J. Physiol.*, 133, 631 (1956).

[13] J. Th. G. Overbeek, *Progr. Biophys. Biophys. Chem.*, 6, 58 (1956).

[14] G. A. Kurella, in *The Physico-Chemical Basis of the Origin of Resting Potential Difference, Trans. Moscow Soc. Naturalists*, 9, (Biol. Ser.) 74, (1964).

[15] N. T. Shelajev, in *New Soviet Optical Devices for Morphological Research*, Mashgiz, Moscow, 1955.

[16] D. Woermann, K.-F. Benhoeffer, and F. Helfferich, *Z. Phys. Chem.*, 8, 256 (1956).

[17] N. A. Walker, *Aust. J. Biol. Sci.*, 8, 476 (1955).

[18] U. Kishimoto, R. Nagai, and M. Tazawa, *Plant Cell Physiol.*, 5, 21 (1965).

[19] P. Fonbrune, in *Techniques de Micromanipulation*, Masson, Paris 1949.

[20] A. M. Bonch-Bruewitch, in *Use of Electronic Tubes in Experimental Physics*, Gosisdat Technico-Teoreticheskoj Literatury, Moscow, 1955.

[21] A. L. Byzov, and M. M. Bongardt, *Physiol. J. USSR*, **45**, 111 (1959).

[22] G. T. Shkurin, in *Manual on New Electronic Measuring Instruments*, Oboronisdat, Moscow, 1966.

[23] W. F. Antonov, G. A. Kurella, and L. G. Iaglova, *Biofizika*, **10**, 1087 (1965).

[24] G. A. Kurella, and G. A. Popov, *Biofizika*, **5**, 573 (1960).

[25] P. C. Caldwell, *J. Physiol.*, **124**, 1P (1954).

[26] L. N. Vorobiev, G. A. Kurella, and G. A. Popov, *Biofizika* **6**, 582 (1961).

[27] W. K. Andrianov, and G. A. Kurella, *Biofizika*, **8**, 457 (1963).

[28] F. J. J. Buytendijk, and M. W. Woerdmann, *Arch. Entwicklungsmech. Organ.* **112**, 387 (1927).

[29] R. C. Gesteland, B. Howland, J. Y. Lettvin, and W. H. Pitts, *Proc. IRE*, **47**, 1856 (1959).

[30] H. G. L. Coster, *Aust. J. Biol. Sci.*, **19**, 545 (1966).

[31] P. R. White, in *A Handbook of Plant Tissue Culture* (1943), (Russ. Translation), Foreign Lit., Moscow, 1949.

[32] G. E. Briggs, A. B. Hope, and R. N. Robertson, in *Electrolytes and Plant Cells*, Blackwell, Oxford, 1961.

[33] J. Dainty, *Ann. Rev. Physiol.*, **13**, 379 (1962).

[34] K. Oda, *Sci. Rep. Tohoku Univ., Fourth Ser. (Biol.)*, **27**, 159 (1961).

[35] R. M. Spanswick, and E. T. Williams, *J. Exp. Botany*, **15**, 193 (1964).

[36] R. M. Radvan, and L. N. Vorobiev, *Biofizika*, **10**, 889 (1965).

[37] E. I. Efimzew, and F. F. Litvin, in *Method of Preparation of Metallic Electrodes with Quartz Isolation* (9.x.1965 No. 103L471/31–16) Autor lizens (USSR) No. 190524 (1965).

[38] F. F. Litvin, and E. I. Efimzew, *Bull. Moscow Soc. Naturalists (Biol.)*, **4**, 140 (1966).

[39] A. I. Knooshenko, and L. N. Vorobiev, *Biofizika*, **10**, 703 (1965).

[40] J. Gutknecht, *Biol. Bull.*, **130**, 331 (1966).

[41] G. Svaetichin, *Acta Physiol. Scand.*, **24**, Suppl. 86, 5 (1951).

[42] A. Mauro, *Biol. Bull.*, **105**, 378 (1953).

[43] E. V. Angerer, in *Technische Kunstgriffe bei physikalischen Untersuchungen*, F. Vieweg & Sohn, Braunschweig, 1959.

[44] J. B. Koltunov, *Biofizika*, **8**, 619 (1963).

[45] J. B. Koltunov, in *Molecular Biophysics*, "Nauka," Moscow 1965.

[46] G. Eisenman, *Biophys. J.*, **2**, 259 (1962).

[47] M. Tazawa, *Plant Cell Physiol.*, **5**, 33 (1964).

[48] E. A. C. MacRobbie, *J. Gen. Physiol.*, **45**, 86 (1962).

[49] W. I. V. Osterhout, *Physiol. Rev.*, **16**, 216 (1936).

[50] M. Tazawa, and U. Kishimoto, *Plant Cell Physiol.*, **5**, 45 (1964).

[51] W. K. Andrianov, G. A. Kurella, and F. F. Litvin, *Biofizika*, **10**, 531 (1965).

[52] F. F. Litvin, I-tanj He, and E. I. Efimzew, *Physiol. Rastenij*, **12**, 364 (1965).

[53] E. Rabinowich, in *Photosynthesis*, Foreign Lit., Moscow, 1959.
[54] E. Barr, and T. C. Broyer, *Plant Physiol.*, **39**, 48 (1964).
[55] W. A. C. MacRobbie, *Biochim. biophys. Acta*, **94**, 64 (1965).
[56] F. A. Smith, *Biochim. Biophys. Acta*, **126**, 94 (1966).

CHAPTER 10

Dynamics of the Cell Membrane as an Electrochemical System

HARRY GRUNDFEST

Laboratory of Neurophysiology. Department of Neurology. College of Physicians and Surgeons. Columbia University

10.1 INTRODUCTION

A variety of techniques, one of the most important of which is recording with intracellular glass microcapillary electrodes, have in recent years greatly enriched our knowledge of the electrophysiology of excitable electrogenic cells. The ionic theory of bioelectrogenesis, initiated more than 60 years ago by Julius Bernstein[1, 2], has been placed on a realistic and quantitative plane largely through the work of Hodgkin and his colleagues[3–5], with the formulation of the sodium hypothesis[6], and the subsequent elucidation of spike electrogenesis of the squid giant axon[7]. The general validity of the ionic theory has been reinforced and extended as the range of cells and phenomena studied has enlarged[8–14]. Nevertheless, it has become evident at the same time that the rigorous application of electrochemical theory—in its classical form[15] as well as in various modifications[16–19]—has been far from adequate.

Perhaps we should not be unduly pessimistic about this state of affairs, for other areas in which workers are in quest of electrochemical theory are similarly affected. In order to account for the performance of a re-

The work in the author's laboratory is supported in part by grants from the Muscular Dystropy Associations of America; by U.S. Public Health Service Research Grants NB–03728, NB–03270, and Training Grant 5TI NB 5328 from the National Institute of Neurological Diseases and Blindness; and from the National Science Foundation (GB–2940).

latively simple, nonliving electrochemical system—the pH-sensitive glass electrode—three classes of theory are in contention[20]. The living cell membrane is a much more complex electrochemical system. A satisfactory theoretical treatment, therefore, will need to be appropriately more complex. The present account aims to indicate the order of these complexities in the membrane and some of the varieties of phenomena they engender.

10.2 THE MEMBRANE AS A HETEROGENEOUS SYSTEM

10.2.1 Structural Heterogeneity

The complexity of cells as electrochemical systems arises in part because of the heterogeneity of the cell membrane complex. The membrane is believed to be formed predominantly of a bimolecular leaflet of lipids and/or phospholipids, to the polar hydrophilic groups of which are attached proteinaceous macromolecules[21]. Many of the latter appear to be enzymes. The electrophysiological evidence indicates strongly that several varieties of specifically ion-permselective sites are dispersed over the leaflet. The sites are heterogeneous as to charge, effective pore size, chemical affinity, and reactivity to various stimuli. Despite the overwhelming evidence for these varieties of heterogeneity, no means are at present available to demonstrate and localize the morphological correlates, either by electronmicroscopy or by the techniques of macromolecular chemistry.

10.2.2 Functional Heterogeneity

The heterogeneity of the membrane as an electrochemical system is foreshadowed in the variety of functions that the different parts of the membrane must perform, each of which (Fig. 10.1) demands different properties. Thus a generalized sensory neuron has a specialized input component that receives information from the environment or from receptor cells and transduces this information into an electrical manifestation. A middle, conductile component is sensitive to the new, electrical signal, formulates a message containing this information, and propagates it to the output of the neuron. At the nerve terminals the electrical message is again translated into still another signal, intelligible to an appropriate site on the next cell in a transmissional chain. In general this output signal is a chemical agent (transmitter) which is highly specific for the immediate input of the next cell and may be completely unintelligible to other closely neighboring sites on that same cell.

The input component reacts to specific stimuli which may be chemical,

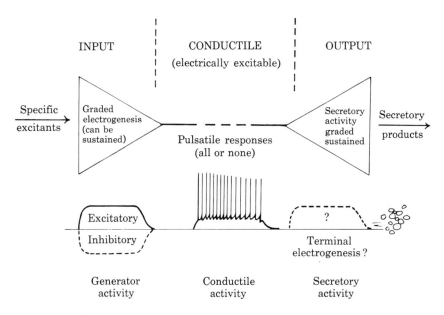

Fig. 10.1. Diagrammatic representation of functional components and electrical responses of receptor or correlational neuron. Electrically inexcitable input produces electrogenesis graded in proportion to its specific stimulus and usually sustained as long as the latter is applied. Possibility of hyperpolarizing electrogenesis, which may be produced by inhibitory synaptic membrane, is shown — but is not further considered. Depolarization at input, operating upon conductile electrically excitable component, can evoke spikes in latter. Spikes are encoded by frequency and number in proportion to generator depolarization. These pulsatile signals, propagated to the output, command secretory activity there that is roughly proportional to information encoded in message of pulses and sustained as long as message demands. Transmitter which is released at output can initiate synaptic transfer by operating upon input of another cell. Possibility of special output electrogenesis is indicated, but is not further considered. (Grundfest[22].)

mechanical, photic, or thermal. Except in the as yet insufficiently understood electroreceptors of electric fishes, the sensing and transducing elements are electrically inexcitable[22–25]. This property permits the input electrogenesis to be of either sign, linear with and graded in proportion to the stimulus, and lasting as long as the stimulus is applied. The depolarizing electrogenesis of the generator potential in sensory neurons — or of the excitatory postsynaptic potential (EPSP), which is initiated by transmissional activity — can be large enough to stimulate the voltage-sensitive (i.e., electrically excitable) middle component. The latter includes the classical spike-generating conductile variety of excitable membrane, but other varieties of electrically excitable responses are now also known[8, 11]. The spike is a triggered (i.e., nonlinear), all-or-none (digitalized) electrogenic activity. It is a specialized response necessary to overcome the lossiness of the membrane as part of an electrically conducting cable. As in other digitalized communication lines, the message is encoded by variation in the frequency and number of the impulses. Propagating to the output terminals of the nerve fiber, the message initiates activity in the output component of the membrane. This must be a secretory response which results in the release of the specific chemical transmitter agent that activates the input membrane portion of the next cell of the transmissional chain. The secretory capability of the neuron relates this cell type to receptor, gland, and neurosecretory cells[24, 26, 27]. However, there is as yet little information available regarding the nature of secretory activity in general.

10.2.3   Electrophysiological Consequences of Heterogeneity

Several salient features of the many consequences of functional heterogeneity can be represented by a few diagrammatic generalizations. The current-voltage (I–E) characteristics of the electrogenic components are shown in this way in Fig. 10.2. Electrically inexcitable elements behave as linear (ohmic) components in response to applied currents. However, when they are excited by their specific (adequate) stimuli, the slope of the I–E characteristic changes in the direction of decreased resistance. Thus the membrane element must have undergone an increase in permeability for one or several ionic species. The change may or may not be accompanied by an electrical sign that represents a change of the membrane potential from its resting value. The change may be in either direction, the sign and magnitude depending on the electrochemical conditions for the relevant ionic species.

The resistance of the electrically excitable components, however, is changed by applied current. The change may be in either direction — to decrease or increase permeability for one or for several ion species — and it may be induced by flow of inward (hyperpolarizing) currents as well as outward (depolarizing) ones. Accordingly, there are four classes of nonlinear effects induced in electrically excitable membrane. Increased permeability (activation processes) and decreased permeability (inactivation processes) may be caused by either depolarization or hyperpolarization. Each of the processes may involve different ion species. Furthermore, different changes may occur simultaneously or sequentially; in the latter case they may develop in different order in different cells, and the number of possible overt, phenomenological manifestations is large.

The various changes in membrane permeability may be represented

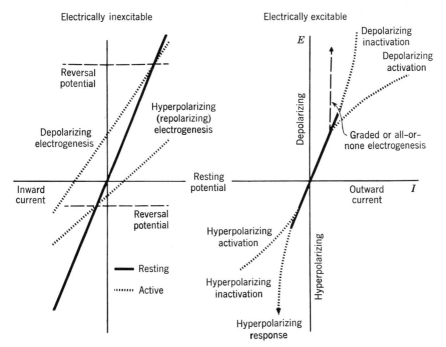

Fig. 10.2. I–E relations of electrogenic membranes. *Electrically inexcitable* membranes behave as ohmic resistances, E changing linearly with I. However, slope of relation changes during activation of membrane by appropriate stimulus. Broken lines represent active membrane of (depolarizing) EPSP's and of (hyper-polarizing) IPSP's respectively. Recorded amplitudes of the responses (given by differences between resting I–E line and that during activity) change with change in membrane potential. Thus a characteristic feature of electrically inexcitable electrogenesis is its change in sign when E exceeds reversal potential specified by intersection of resistance lines for active and passive membrane.

Electrically excitable membranes exhibit nonlinear behavior characterized by one or by several varieties of conductance changes. On stimulation with de-polarizing currents many but not all cells respond with conductance increase (depolarizing activation) for Na, Ca, or Mg—which introduces new inside-positive emf and causes graded or all-or-none depolarizing electrogenesis (cf. Figs. 10.11, 10.13 and 10.14). Increased conductance for K or Cl (inappro-priately termed "rectification") that is evoked by depolarizing stimuli (cf. Figs. 10.13, 10.14, 10.17 and 10.18) or for any ions by hyperpolarizing currents (cf. Figs. 10.26–10.28.) is usually not regenerative. However, it can become regenera-tive when electrochemical conditions permit (cf. Figs. 10.15–10.19). Decreased conductance or inactivation (likewise inappropriately termed "anomalous rectifi-cation") may be evoked by depolarizing or hyperpolarizing currents and it can also be regenerative (cf. Figs. 10.24–10.29). Various ionic processes which cause nonlinear relations exhibit different degrees of time variance (cf. Figs. 10.12, 10.15–10.19, 10.22 and 10.27–10.29). Effects of pharmacological inacti-vation are not shown in this diagram. (Grundfest [9]. Reprinted by permission of the Federation of American societies for Experimental Biology.)

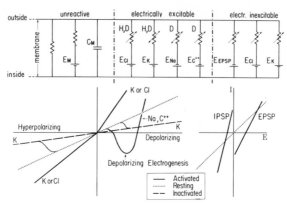

Fig. 10.3. Ionic batteries of electrogenesis. Above: Equivalent circuit diagram. Emf's contributed by electrogenic ion pumps are omitted. Below: Current-voltage (I–E) characteristics of electrically excitable components (left) and of electrically inexcitable (right), in voltage-clamp presentations. Origins are set at resting potential (E_M).

Membrane capacity (C_M) and invariant conductive component represent major, unreactive portion of membrane. Conductive component is subdivided into ion permselective element with E_M as average emf and nonselective element symbolized by resistance without emf. Reactive components are represented by variable resistances in series with different ionic batteries depending on different permselectivities. E_K and E_{Cl} in general are close to E_M, but E_{Na} and E_C^{2+} (Ca or Mg in various cells) are shown as inside positive. Permselective electrically excitable channels respond to depolarizing (D) and/or hyperpolarizing (H) stimuli with activation ($\nearrow$) or inactivation ($\swarrow$). Electrically inexcitable depolarizing electrogenesis of receptive and synaptic membrane is indicated by inside-positive battery (E_{EPSP}). Inhibitory synaptic electrogenesis involves increased conductance for either Cl(E_{Cl}) or K(E_K).

Unreactive electrically inexcitable electrogenic components have linear (ohmic) I–E characteristics, but activation of electrically inexcitable components by specific stimuli increases slope (indicating higher conductance in voltage clamp presentation). Depolarizing electrogenesis translates characteristic to right. Diagram shows inhibitory electrogenesis (IPSP) as hyperpolarizing and characteristic translated to the left. As membrane is polarized by applied currents, resting and active characteristics approach crossing beyond which the sign of recorded electrically inexcitable response is reversed relative to steady membrane potential. Reversal potential approximates equilibrium (Nernst) potential of ionic batteries that cause electrogenesis.

I–E characteristic of electrically excitable components exhibits nonlinearities which result from transition of resting membrane conductance to higher or lower values. Only the conductance increase caused by Na or C^{2+} activation shifts the characteristic significantly along the voltage axis. Three nonlinear regions with negative slope characteristics are shown. They mark transition from E_M to E_{Na} or E_C^{2+} by activation processes and from resting conductance to lower conductance by depolarizing and hyperpolarizing K inactivation respectively (Grundfest[14]. Reprinted by permission of the Federation of American societies for Experimental Biology) further by an equivalent circuit (Fig. 10.3). Each variety of ionic perm-selectivity forms a concentration battery and the relative permeabilities are represented as internal resistances. Unreactive but permselective components have invariant resistances, while the capacity to react to appropriate stimuli is denoted by variable resistances. The electrically inexcitable components respond to specific stimuli, which may increase their permeability (activator agents) or block the responsiveness (in-activator agents). The electrically excitable components respond to depolarization (D) and/or hyperpolarization (H) with either activation ($\nearrow$) or inactivation ($\swarrow$). All these effects may result in changes in the lumped equivalent (Thevenin) potential (E_M) which appears across an invariant membrane capacity (C_M).

The voltage clamp presentations of the I–E characteristics in Fig. 10.3 serve to re-emphasize the differences between electrically excitable (left) and electrically inexcitable (right) electrogenically reactive compo-nents. The latter behave as linear systems over the whole range of the characteristic. In some cases the range of measurements has been about ± 100 mV from E_M. Upon activation, the slope increases — indicating an increased conductance in the voltage clamp presentation. The charac-teristic line tends to shift so that its intercept on the voltage axis is a fraction of the emf of the relevant ionic system. For the generator membrane or for that of excitatory synapses, this emf is E_{EPSP}, positive to E_M. For the inhibitory synapses it is either E_{Cl} or E_K, which may lie at or on either side of E_M.

The greatest complexity of electrically excitable components (seen in Fig. 10.2) is also reflected in the voltage clamp presentations (Fig. 10.3). Activation processes for some ions (normally K and/or Cl) do not intro-duce a large translocation of the characteristic, but the increase in conductance is indicated by an increased slope. A dramatic effect is produced, however, when the activation process is for an ion (Na, C^{2+}) with a strongly inside-positive but normally occult emf. The characteristic is translated along the voltage axis to an intercept related to the new emf, and also has a high slope. The connection between the steady-state resting and active characteristic lines is through a region of negative slope, where the current flow is inward — despite an increasing shift to inside positivity. It is this condition that gives rise to the explosively regenerative all-or-none characteristic of spike electrogenesis[7]. In-activation processes that are induced either by depolarization or by hyperpolarization decrease the slope, but in general the characteristic line maintains the same origin as in the resting state. The result is that an increase in the membrane potential in either direction causes a decrease in the current. The transitions of the characteristic from the resting to

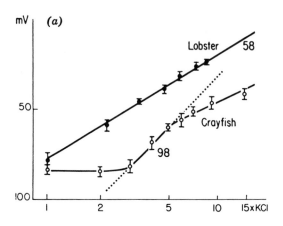

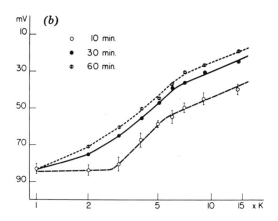

Fig. 10.4. Sensitivity of membrane of crayfish muscle fibers to changing K_0 under various conditions.

(a) Comparison between crayfish and lobster muscle fibers under similar experimental conditions. Abscissa is in terms of logarithm of multiples of initial concentration of K_0 (5.3 mM for crayfish saline; 15.4 mM for lobster). Ordinate shows membrane potentials measured 10 min after each increase in K_0. Membrane of lobster fibers behaves nearly as an ideal electrode for K; that of crayfish fibers responds little until K_0 is increased 3- or 4-fold. Change in E_M with K_0

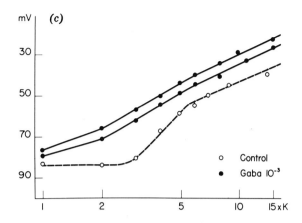

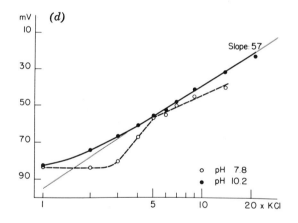

then became large (ca. 100 mV/decade K_0) and subsequently became about 40 mV/decade. (b) E_M-log K_0 characteristic of crayfish fibers became more linear and slope approached closer to 58 mV/decade when measurements of E_M were made 30 min and 1 h after each change in K_0. (c) Increased permeability to Cl following application of GABA resulted in a nearly ideal relation even when E_M was determined 10 min after a change in K_0. Data show two experiments with GABA. (d) Decreased permeability to Cl was effected by increasing pH of medium. This also was correlated with approach to ideal relation when E_M was determined 10 min after change in K_0. (Girardier, and Reuben[31].)

the inactivated state thus also represent negative slope regions. If a constant current sufficient to bring the membrane potential into the "forbidden" negative slope region is applied, there will be a regenerative shift to the inactivated state and a corresponding translation along the potential axis, owing to the larger IR drop. Activation processes that are initiated by hyperpolarization are also known, but their analysis is not yet as firm as are the data concerning depolarizing activations and the two varieties of inactivation processes.

Complex though the equivalent circuit of Fig. 10.3 may be as an electrochemical system, it still does not encompass other complex processes of which the membrane is capable. Ionic pumps, which themselves may contribute an emf[28–30], phase boundary potentials[17, 18], and the effects of differences in affinity and mobility of the solute ions in relation to the fixed charges of the membrane are not included.

10.2.4   Some Phenomenological Manifestations

The electrochemical complexity of the equivalent circuit of Fig. 10.3 manifests itself even in the behavior of the membrane as a K-electrode (Fig. 10.4). Lobster muscle fibers depolarize at a rate of about 58 mV/ decade increase in K (A), but under similar conditions the muscle fibers of the crayfish *Orconectes* exhibit a complex relation between E_M and log K[31]. This complexity can be eliminated if the measurements of E_M are made a long time after K_0 has been changed (B), or if the cell is treated with agents that change the permeability of the membrane for Cl (C, D). The complexity is caused by a marked time variance, which in turn may be ascribed to the interplay of the emf's of the Cl and K batteries of the system. Such interplays presumably arise from the kinetics of redistribution of the two ions.

Different cells differ widely in their permeability to various ions; for example, lobster axons are permeable to Cl (Fig. 10.5)—as are most cells that have been studied. Squid axons, on the other hand, and probably also fibers of the *Taenia coli* muscle of the guinea pig[14] are effectively as impermeable to Cl as they are to Na[32]. From the standpoint of evaluating theoretical conclusions, it is relevant that the E_M − log K_0 characteristic of squid axons has the same form as that of frog axons and muscle fibers (Fig. 10.6). The data on squid axons[33] could be fitted[6] to the constant field equation[16] by assuming high "permeability" coefficients (P_{Cl}) for Cl. Not only the osmotic data[32] but also several varieties of the electrophysiological data[28, 29, 34, 35] indicate that Cl permeability of the squid axon is as low as that for Na and that P_{Cl} must be correspondingly small.

Other electrophysiological tests, as well as the "good fit" of Fig.

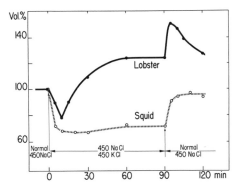

Fig. 10.5. Differences in permeability for Cl of lobster and squid axons. Ordinate: Relative volume changes following exposure of axons to medium made hyperosmotic by addition of 450 mM KCl. Squid axon (dotted line, open circles) promptly shrank and remained so until restored to isosmotic medium. Lobster axon (filled circles and solid line) shrank briefly, then swelled to new steady state. Further swelling occurred on returning to isosmotic medium. Differences denote lobster axon is permeable to Cl while squid axon is effectively as impermeable to Cl as to Na. (Freeman, and Grundfest[39], and unpublished data.)

10.6, are inadequate to evaluate permeability characteristics of the membrane. For example, lobster muscle fibers are effectively as impermeable to Cs as they are to Na, judging from osmotic data (Fig. 10.7), Nevertheless, they are depolarized by high concentrations of Cs almost as much as by K, Rb, and NH_4[36]. Another test, the use of "transport numbers" derived from electrophysiological data[37], is also inadequate when applied to crayfish muscle fibers[31], crayfish axons[38], and squid giant axons. The deficiencies of the electrophysiological criteria as tests for theoretical concepts stem from the basic inadequacy of the latter when derived from the postulates of a homogeneous electrochemical system.

As an incompletely understood reactive system the membrane often responds to challenges in various unexpected ways. For example (Fig. 10.8), when there is no permeant salt in the bathing medium, lobster axons that are subjected to a hyposmotic challenge lose a considerable part of their intracellular contents[39, 40]. This is caused by a large increase in permeability (Fig. 10.9) of the membrane for K[40]. Although the same increase in K permeability also occurs in the presence of a permeant anion (Cl), the loss of intracellular contents is small and may be negligible, presumably because of a compensating influx of salt. Similar findings have also been reported in frog muscle fibers[41].

Frog muscle fibers exhibit another complexity of behavior (Fig. 10.10):

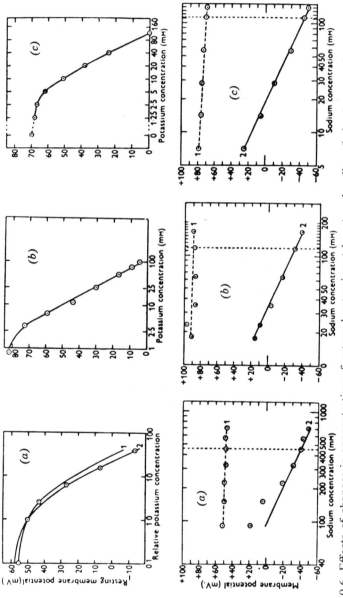

Fig. 10.6. Effects of changes in concentration of external potassium (above) and sodium (below) on membrane potentials of (a) squid axons, (b) frog muscle fibers and (c) frog axons. At low concentrations of K_0 the resting potential (E_M of Fig. 10.3) is strongly inside-negative. Depolarizations induced by increasing K can be fitted with constant field equation. For squid axon two values of P_{Cl} were assumed (0.3 and 0.45, respectively), but both are without experimental justification (see text). Changes in Na_0 did not markedly affect E_M (broken lines), but changed amplitudes of spikes in all 3 cells. Vertical dotted line indicates normal value of Na_0. The ordinates are expressed as $E_{outside} - E_{inside}$ so that negative values of spike amplitude signify overshoot of spike to inside positivity. Slopes of the 3 lines are those expected for a Na electrode condition. (Modified from Hodgkin [3] where references to original sources are given.)

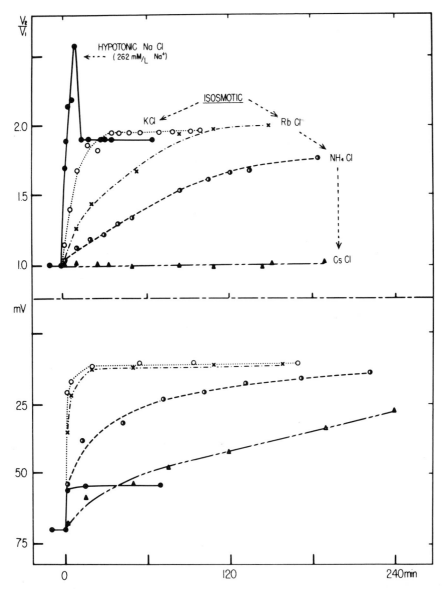

Fig. 10.7. Volumetric and electrophysiological estimates of permeability of lobster muscle fibers to monovalent cations. Top: Change in relative volume of muscle fibers at various times after exposure to different media. When medium was made hyposmotic by reducing NaCl level, fiber swelled rapidly, then reached steady volume almost double that of initial value. Substitution of NaCl with KCl, RbCl, or NH$_4$Cl to keep the medium isosmotic also caused the fiber to swell,

indicating introduced salt was permeating across membrane. Rates of swelling indicate differences in mobility of cations. Following substitution with CsCl, however, muscle fiber remained at initial volume, indicating membrane is effectively as impermeable to Cs as it is to Na.

Bottom: Membrane potentials of same fibers. Swelling caused by removal of Na was accompanied by rapid depolarization which is accounted for by the dilution of K_i through the entry of water into fiber. Enrichment of medium with K or Rb caused rapid depolarization to steady-state value of about $-10\,mV$. NH_4 also induced depolarization, but more slowly. However, Cs likewise depolarized the fiber. This electrophysiological indication that membrane is permeable to Cs is in contradiction to volumetric data. (Gainer, and Grundfest[36], and unpublished data.)

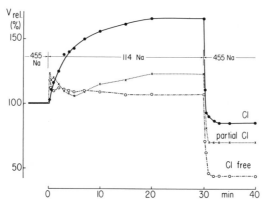

Fig. 10.8. Volume changes in lobster axons following hyposmotic challenges. Ordinate relative volume. Three experiments are shown, differing in condition only by partial or complete substitution of Cl with propionate in two of them. At time zero medium was made hyposmotic by reducing Na salt to 25% of control condition. This challenge lasted 30 min in each experiment. Swelling observed in presence of Cl during challenge was greatly diminished when Cl was reduced and still more so when absent. Shrinkages observed on removing challenges indicate loss to cell contents during hyposmotic challenge. Note that in absence of Cl there was a loss of about 60% of original volume. (Unpublished data, and Dunham, Freeman, and Grundfest[40]. Reprinted by permission of the *Biological Bulletin*.)

when the bathing medium contains no K, the muscle fibers respond to a hyperosmotic challenge with Na by an increased permeability for Na. Consequently, when the challenge is made by increase of NaCl in the medium, the salt is then permeant and enters the cell; that is, the reflection coefficient for NaCl changes from unity to nearly zero. The resultant change of volume is accompanied by a large depolarization. When the

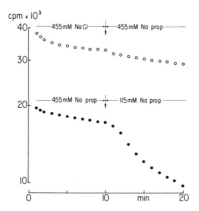

Fig. 10.9. K^{42} efflux induced in lobster axons by hyposmotic challenge. Above: Control experiment in which axon loaded with K^{42} in presence of Cl was transferred to a Cl-free but isosmotic medium. Substitution of propionate for Cl had little effect on rate of efflux of K^{42}. Below: axon equilibrated in Cl-free medium was challenged with hyposmotic Cl-free saline. Note marked increase in K^{42} efflux induced by hyposmotic challenge. (Unpublished data, and Dunham, Freeman, and Grundfest[40]. Reprinted by permission of the *Biological Bulletin*.)

hyperosmotic challenge is removed — but with the fibers still in a K-free medium — the cell swells but does not repolarize. Repolarization develops rapidly, however, when KCl is added. The quasi-steady-state membrane potential becomes somewhat more negative than the long-term resting potential. Presumably the hyperpolarization is due to electrogenic activity of the Na pump[12], and the behavior in the K-free hyperosmotic medium is due to inactivation of the Na pump.

10.3   THE MEMBRANE AS AN ELECTRODE SYSTEM

The foregoing examples have indicated some of the complexities of the membrane behavior that must be encompassed within a rigorous electrochemical theory of bioelectric phenomena. Nevertheless the existing theory can and does serve as an approximation under certain conditions, namely, when the bioelectric manifestations are due predominantly to a large change in conductance for one ion and this ion then becomes the predominant determinant of the Thevenin potential of the equivalent circuit (Fig. 10.3). Such a change occurs when the membrane is activated and becomes relatively highly permeable for one ion species (Fig. 10.11). The membrane potential then shifts in the direction of the ionic battery of high conductance. For example (Fig. 10.12), the inhibitory postsynap-

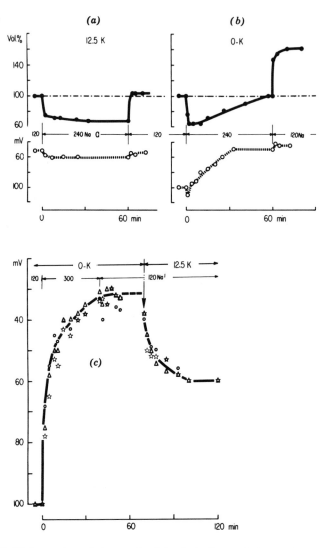

Fig. 10.10. Volumetric and electrophysiological effects of changes in permeability to Na induced in frog muscle fibers by presence or absence of K. (*a*) Fiber in saline containing 12.5 mM KCl was challenged with hyperosmotic solution (NaCl doubled). Shrinkage and correlated hyperpolarization from −54 mV to −62 mV were both reversible. (*b*) Another fiber similarly challenged, but in K-free saline. Transient shrinkage and marked swelling on removing challenge indicate entry of NaCl. Fiber hyperpolarized briefly and then depolarized rapidly to about −50 mV during challenge. Removal of latter did not cause repolarization. (*c*) Changes in membrane potential of 3 muscle fibers under same conditions as in

(b) except that fibers were challenged with 300 mM NaCl and depolarization was larger (to about -30 to -35 mV). Removal of challenge did not cause much repolarization, but addition of 12.5 mM KCl caused rapid shift of potential to about -60 mV, significantly more negative than resting potential of fibers equilibrated in 12.5 mM KCl (ca. -50 to -55 mV). (Modified from Grundfest[12], after unpublished data from Lopez, Reuben, and Grundfest[42].)

---

tic potentials (IPSP) of cat motoneurons, crayfish stretch receptors, and molluscan neurons are normally hyperpolarizing because the emf of the Cl battery (E_{Cl}) is somewhat negative to E_M. When the electrochemical conditions for E_{Cl} are changed so that the latter is positive to E_M, activation of the "inhibitory" membrane causes depolarization that is sufficient to excite the spike-generating component.

The ionic processes of electrically excitable and electrically inexcitable components exhibit a considerable degree of symmetry (Fig. 10.11) and this fact in itself must find correlates in the macromolecular structure of the membrane. However, the variety of overt manifestations of the different processes is much greater in the case of the electrically excitable components (Figs. 10.2 and 10.3), which the possibilities for both positive and negative feedback enlarge still more. In exploring the applicability of the ionic theory it will therefore suffice to consider only the responses of electrically excitable components of various cells.

10.3.1 Activation Responses of Electrically Excitable Membrane: Components of Spike Electrogenesis and "Anomalous" Spikes

In the spike electrogenesis of most cells both Na activation and K activation occur (Fig. 10.13) to cause the characteristic form of the electrogenesis[7]. However, one or the other of these processes can be eliminated or diminished to the point where it is negligible (Fig. 10.14). When the Na activation process is reduced, the depolarization that results from a shift from E_M to E_{Na} is reduced or abolished and the rest of the electrically excitable events that depend on this depolarization are also eliminated. On the other hand, when K activation is affected, the spike is prolonged because the pulse shaping effect of repolarization toward E_K is diminished.

Under normal conditions of animal cells there is an inside-positive electrochemical gradient for Na, Ca, and Mg (Figs. 10.3 and 10.10) so that depolarizing electrogenesis results from Na, Ca or Mg activation. It is, however, possible to cause cells to develop spikes that are caused by depolarizing electrogenesis by K or Cl activation. These responses can be readily demonstrated under favorable conditions.

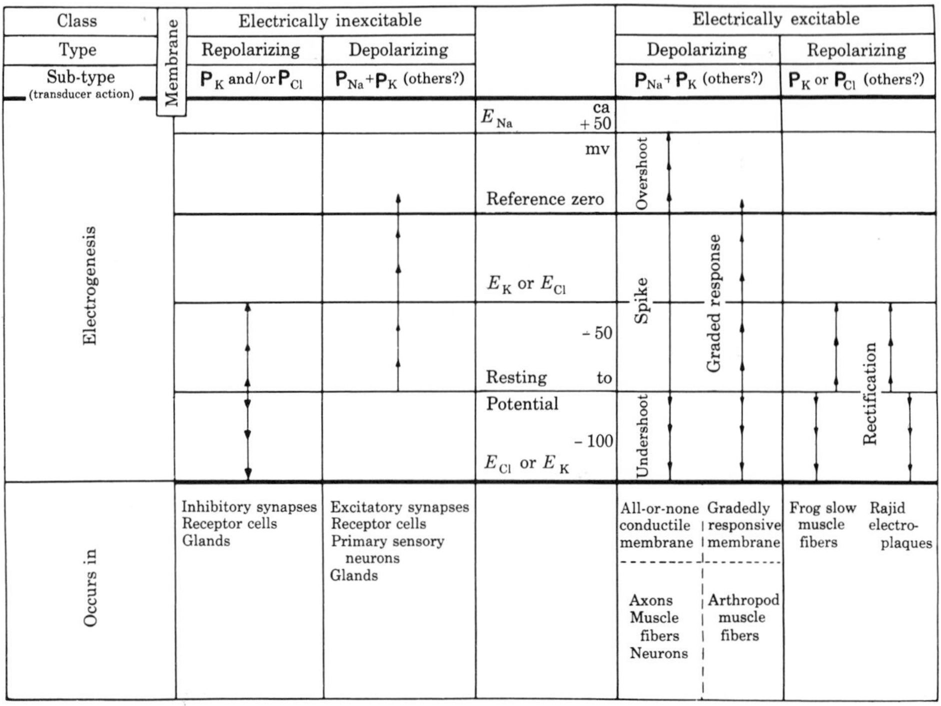

Class	Membrane	Electrically inexcitable			Electrically excitable	
Type		Repolarizing	Depolarizing		Depolarizing	Repolarizing
Sub-type (transducer action)		P_K and/or P_{Cl}	$P_{Na}+P_K$ (others?)		$P_{Na}+P_K$ (others?)	P_K or P_{Cl} (others?)
Electrogenesis				E_{Na} ca +50 mv Reference zero E_K or E_{Cl} −50 Resting to Potential −100 E_{Cl} or E_K (Overshoot, Spike, Graded response, Undershoot, Rectification)		
Occurs in		Inhibitory synapses Receptor cells Glands	Excitatory synapses Receptor cells Primary sensory neurons Glands		All-or-none conductile membrane / Gradedly responsive membrane --------------- Axons, Muscle fibers, Neurons | Arthropod muscle fibers	Frog slow muscle fibers Rajid electro-plaques

Fig. 10.11. Diagram of electrogenic manifestations of increased conductance (activation processes) for different ions in electrically inexcitable and electrically excitable membranes. Resting potential is represented as lying between E_K and E_{Cl}, the electrode potentials for K and Cl, respectively. The electrode potential for Na, E_{Na}, is indicated as about 50 mV positive to reference zero. The two classes of differently excitable membranes are each symmetrically subdivided into electrogenic types and subtypes characterized by nature of transducer actions which lead to increased permeability (P_K, P_{Cl}, P_{Na}) for different ion species. Types of cells or cell components in which different activities occur are also shown. Membranes could be further subdivided by various pharmacological reactions. Note that spike electrogenesis can occur without participation of K activation (Figs. 10.21 and 10.22), and that Na and K activation process of depolarizing electrically excitable membrane can be modified independently of one another (Fig. 10.14 and 10.21). (Grundfest[9]. Reprinted by permission of Springer–Verlag.)

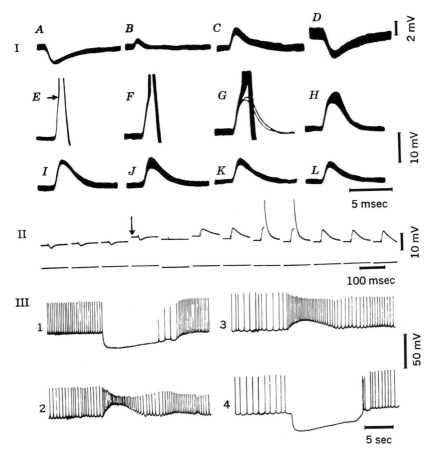

Fig. 10.12. Excitatory effects of normally "inhibitory" membrane resulting from change in electrochemical conditions for chloride and consequent inversion of electrogenesis from hyperpolarizing inhibitory action to depolarizing excitatory action. I. Cat motoneuron. *A*: IPSP was hyperpolarizing before iontophoretic injection of chloride into the cell. *B* and *C*: Depolarization after injection. *D*: Potential was again inverted to hyperpolarization during depolarization of the cell with applied current. *E*–*L*: Another sequence of records after large injection of chloride into neuron. Resting potential changed relatively little, so that large depolarizing IPSP's exceeded the critical firing level of the electrically excitable membrane components and elicited spikes (*E*–*G*). As chloride leaked out of cell, depolarizing electrogenesis of IPSP's became progressively smaller. (Modified from Ref. [43].) II. Crayfish stretch receptor. First three records show hyperpolarizing IPSP's. At arrow, normal saline medium was exchanged for chloride-free (glutamate) saline. At third subsequent record, IPSP became depolarizing and subsequently grew larger until it could evoke spike of receptor neuron. (Modified from Ref. [44]. III. Neuron in abdominal ganglion of *Helix*. (1) Repeti-

tive firing of cell was stopped by large hyperpolarization which was induced by application of acetylcholine. Latter agent activated inhibitory synaptic membrane which operates through Cl activation. (2 and 3) Chloride-containing saline was replaced with chloride-free medium. "Inhibitory" membrane then responded with depolarization to application of acetylcholine, and frequency of discharges was increased to such a degree that amplitudes of spikes diminished. (4) Electrogenesis again became hyperpolarizing and inhibitory after chloride-saline was reintroduced. (Modified from Kerkut, and Thomas [45].)

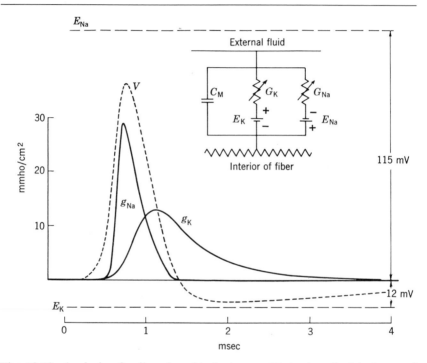

Fig. 10.13. Analysis of spike of squid giant axon (V, broken line) in terms of Hodgkin-Huxley theory. Abscissa is time in milliseconds, and ordinate on left is membrane conductance (G) or reciprocal of resistance, expressed in millimhos/cm^2 of membrane. Ordinate on right indicates potential of "sodium battery" (E_{na}) 115 mV positive to resting potential and of "potassium battery" (E_K) 12 mV negative to resting potential. Inset diagram shows equivalent circuit. Depolarizing stimulus initiates Na activation, an increase in conductance for the ion (G_{Na}). This change tends to carry membrane capacity (C_M) from resting potential (inside-negative) toward emf of sodium battery (E_{Na}) which is inside-positive. Process of Na inactivation is assumed to diminish G_{Na} rapidly from its maximum value. Slower rise in G_K activation outlasts G_{Na} and, after hastening repolarization of axon, is responsible for temporary hyperpolarization nearly to full value of potassium battery (E_K). (Modified from Hodgkin, and Huxley [7].)

196

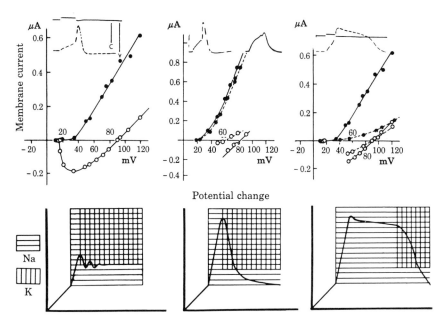

Fig. 10.14. Modifications of responses in electrically excitable membrane which result from independent variation in Na and K activation processes. Above: Effects of differential elimination of Na activation and K activation on *Onchidium* neurons. Voltage clamp experiments show peak initial current (open circles) and later outward current (filled circles) as functions of membrane potential. Solid lines connect control data; broken lines, after treatment with drugs. Insets show responses to brief stimuli. Left, control series, blanking of spike traces at 5 msec intervals; calibrations: $V = 100$ mV and $C = 5 \times 10^{-7}$ A. Center, neuron in preparation treated with 2% urethane. Early inward (Na) current was markedly diminished, while outward (K) current was unaffected. Response of cell was changed from spike to small graded potential. Blanking intervals on spike trace, 2 msec. Right, from experiment in which preparation had been exposed to tetraethylammonium ions (TEA). K current was diminished much more than was Na current, and response of cell became prolonged spike. Blanking signals at 5-msec intervals. Origins are at resting potential. (Ref. [8], modified from Ref. [46].) Below: Diagrams show how similar effects might be produced by changing relative times of onset of activation processes. Earlier onset of K activation would result in graded responses while delay in K activation would lead to prolonged spike. (Grundfest[8]. Reprinted by permission of the New York Academy of Sciences.)

K *Spikes.* Whereas most animal cells depolarize rapidly on exposure to high external K, the muscle fibers of the larval meat-worm (*Tenebrio molitor*) are insensitive to very high levels of K_0. Thus K_0 can be raised to levels at which E_K would be inside-positive, but E_M may remain inside-

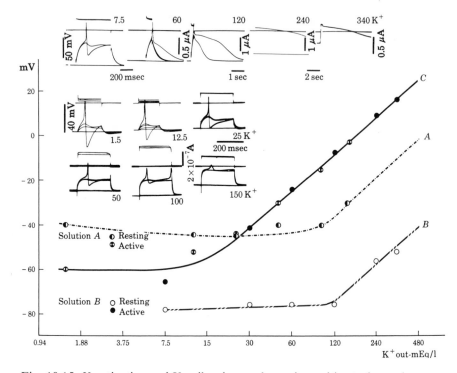

Fig. 10.15. K activation and K spikes in membrane insensitive to large changes in external K. Intracellular recordings from muscle fibers of *Tenebrio molitor* larvae. Graphs: ordinate shows membrane potentials; abscissa, K concentration of medium. One preparation (*A*) was soaked overnight in medium approximating frog Ringer's solution, but also containing 500 mM sucrose. Its muscle fibers had resting potentials of about − 40 mV, and this changed very little until external K exceeded 120 mEq/liter. Other preparation (*B*) was also soaked overnight, but in Na-free, high-Cl and high-Mg medium containing 7·5 mEq/liter K. Muscle fibers had larger resting potential (− 80 mV), which also remained unchanged until external K exceeded 120 mEq/liter.

Intracellularly applied depolarizing currents elicited responses shown in inset records. Six centered records are for different K levels used on preparation *A*. Top row is for five K levels in preparation K. Upper trace in each pair is zero potential and current monitor. Quartered and filled circles (*C*) indicate potentials of undershoots or plateaus of responses in respective records. For K levels above 15 mEq/liter, these changed with slope of 58 mV per decade increase in K, indicating membrane becomes a K electrode during activity. Prolonged over-shooting K spikes were evoked at highest levels of K in preparation *B*. They resemble in character Cl spikes shown in Fig. 10.19, but are caused by influx of K from external medium into muscle fibers. (Modified from Belton, and Grundfest[47].)

198

negative for several hours. However, a brief depolarizing stimulus which causes K activation shifts the membrane potential toward E_K, and a "K spike" then develops (Fig. 10.15). During the time that the K activation occurs the membrane behaves effectively as a K electrode; K spikes can also be induced even in cells that are depolarized by high K_0 (Fig. 10.16) if the membrane potential is restored to inside negativity by an applied current. A brief depolarization then also causes depolarizing K activation and during this time the emf shifts toward E_K. The K spikes of Figs. 10.15 and 10.16 are much more prolonged than those which are caused by Na activation. This indicates that the process of K inactivation is much slower than is Na inactivation. This is also demonstrated in the voltage clamp measurement of Fig. 10.16. The initial inward current that reflects the influx of K down its electrochemical gradient is maintained for a long time. In the squid axon and many other cells the increased Na conductance (g_{Na}; Fig. 10.13) is short-lived.

Cl *Spikes.* Although depolarizing Cl activation is the normal mode of spike electrogenesis in the fresh water algae *Chara* and *Nitella* [50, 51], a clear-cut instance of this electrically excitable process in animal cells has thus far been found only in the electroplaques of skates. The Cl activation can be demonstrated in the current-voltage characteristics (Figs. 10.17 and 10.18) as a time-variant increase in conductance of considerable magnitude [52, 53] that tends to shift the membrane potential about 5–7 mV positive to E_M. The threshold for the Cl activation, however, is at about 15 mV depolarization. Thus a depolarizing current, or the depolarizing electrogenesis of the synaptic membrane, does not cause spikes under normal ionic conditions. However, when the Cl of the external medium is replaced with an impermeant anion (as in Fig. 10.12), E_{Cl} becomes inside-positive. A brief depolarizing stimulus (Fig. 10.19) then evokes a larger prolonged all-or-none depolarization [54]. This Cl spike can be abolished by adding Cl to the medium. It is likely that the existence of depolarizing Cl activation has gone unnoticed in other animal cells, as the nonlinearity in the characteristics of the skate electroplaques (Fig. 10.17 and 10.18) that is due to Cl activation is made more evident by the absence of depolarizing Na and K activation.

In general, therefore, spike electrogenesis can be represented, as in Fig. 10.20a, by the occurrence of a negative slope region of the voltage-current characteristic created when depolarizing activation causes the membrane to shift from being an electrode for one emf (E_M of Fig. 10.3) to one for a more positive emf. The nature of the ion involved in the shift to the new emf is of secondary importance. Provided the new depolarizing emf is above threshold for the depolarizing activation process, there will

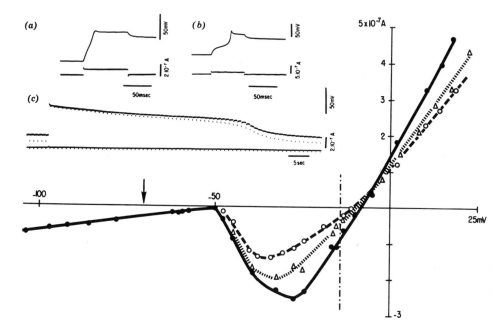

Fig. 10.16. Regenerative electrogenesis owing to K activation in puffer neurons bathed in isosmotic KCl solution and repolarized by applied inward currents, *a–c*: Prolonged K spikes evoked by depolarizing pulses. Holding potentials were −68 mV (*a*), −71 mV (*b*), and −60 mV (*c*). In latter recording, small, constant repetitive inward pulses were applied. Diminished amplitude of hyperpolarizations show that conductance increased during K spike. *Graph:* Voltage clamp data from another cell which was depolarized to −14 mV (broken vertical line) by isosmotic subsitution of KCl for NaCl. Holding potential was −70 mV (arrow). Conductance increase due to K activation by depolarizing stimuli represented by large change in slope of the I–E characteristic, was long lasting. Filled circles show peak current shortly after step changes in voltage. Triangles, measurements 80 msec after onset of depolarization. Open circles, 860 msec after onset. K equilibrium potential, point at which the 3 curves cross, is about −5 mV. The very gradual subsidence of K activation is denoted by decrease in high slope of characteristic. End of the K activation and of spike electrogenesis will restore the characteristic to resting slope. (Modified from Nakajima, and Kusano[48] and from Nakajima[49].)

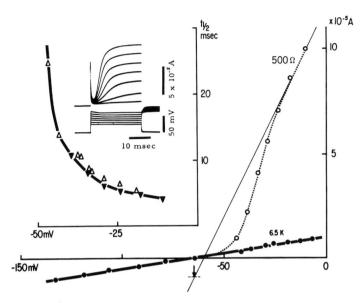

Fig. 10.17. Electrically excitable activity in electroplaque of skate (*Raja erinacea*) analyzed with voltage clamping. Records (upper left quadrant) show outward currents (top traces) evoked by depolarizing steps of various amplitudes (bottom traces). Increase in current to its steady state is relatively slow, but half-times ($t_{1/2}$) to reach the latter become shorter with increasing depolarization (inset graph). Main graph shows I–E characteristic in voltage clamp presentation. Heavy line and filled circles: ohmic relation of resting membrane. Dotted line and open circles: steady state attained during depolarizing activation. For large depolarizations, slope of the characteristic reaches limiting value (thin line) that represents decrease of resistance from 6·5 kΩ to 0·5 kΩ. Intercept of this line on voltage axis (ca. 7 mV positive to E_M) represents emf of the Cl battery of this cell. Arrow marks inward current that flowed after depolarizing pulse was terminated (undershoot in voltage clamp records), and represents dissipation of positive charge (of E_{Cl}) on membrane capacity. (Unpublished data and Hille, Bennett, and Grundfest[52]. Reprinted by permission of the *Biological Bulletin*.)

be an all-or-none response as the emf shifts from the original to the new value through the "forbidden zone" of the negative slope. The condition for the K spikes of Fig. 10.16 also depends on the existence of a negative slope region (Fig. 20c).

The new emf may dominate the Thevenin potential for various times, giving rise to gradations between a condition of "two stable states"[55] and a brief pulselike response that is normally seen in squid axons (Fig. 10.13) and many other cells in which depolarizing Na inactivation is present. In many cells the decay of the spike electrogenesis is accelerated

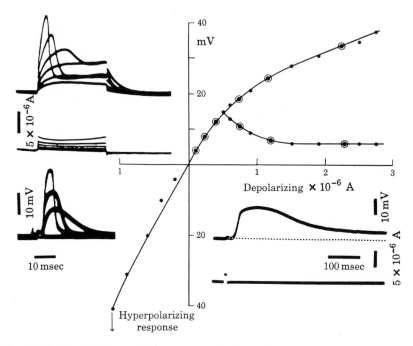

Fig. 10.18. Electrically excitable processes in skate electroplaques. Current clamp data. Origin of graph is resting potential. Large circles represent measurements from six superimposed records in upper left quadrant. Depolarizations (upper traces) elicited by 3 smallest currents (lower traces) fell on linear portion of I–E characteristic, but larger currents induced nonlinear time variant changes in potential so that characteristic developed two branches. In this presentation downward curvature of characteristic represents an increased conductance. In steady state, during largest conductance changes, potential fell to about 7 mV positive relative to E_M (cf. Fig. 10.17). Lower left quadrant: Superimposed PSP's in response to increasing neural stimulation. Steps of amplitude indicate that 3 nerve fibers were activated. Depolarization during smallest PSP was not sufficient to evoke electrically excitable Cl activation. Larger PSP's did evoke Cl activation and responses were shortened. Note their resemblance to potentials evoked by 2 largest depolarizing currents (above). Right lower quadrant: An iontophoretically applied jet of acetylcholine (signalled on lower trace) evoked long-lasting depolarization. Peak depolarization was limited to about 20 mV by onset of Cl activation. (Unpublished data and Cohen, Bennett, and Grundfest [53].)

by an increased conductance that leads to repolarizing electrogenesis and even to hyperpolarization (Fig. 10.13). The hyperpolarization itself may accelerate the decay by quenching the K activation.

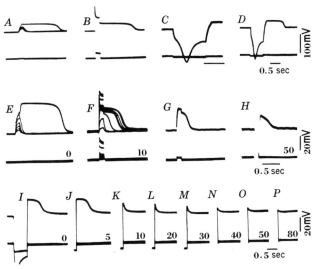

Fig. 10.19. Cl spikes and hyperpolarizing responses in skate electroplaques bathed in Cl-free saline. *A*, *B*, *E*: Long-lasting all-or-none responses triggered by depolarizing pulses. *C*, *D*, *I*: Inward currents evoked hyperpolarizing responses and, when the current was terminated, the cells produced "anode break" spikes. Spike triggered by depolarization (*E*) was diminished and finally abolished when Cl was reintroduced to concentration of 10 mM (*F*) and 50 mM (*G* and *H*). A very strong stimulus (outward current) was applied in the latter recording. Anode break spike (*I*) was also abolished by introducing Cl (*J–P*). (Unpublished data and Grundfest, Aljure, and Janiszewski[54]. Reprinted by permission of the Rockefeller University Press.)

Other "Anomalous" Spikes. The presence of depolarizing K activation is not an invariant condition even in normal spike electrogenesis[56]. In the spike of eel electroplaques, depolarizing Na activation is accompanied by depolarizing K inactivation (Fig. 10.21) and the increased conductance that is due to Na activation is superseded, as Na inactivation develops, by a decrease of conductance (Fig. 10.22) below that of the resting cell[57]. The Na and K processes are independent of one another and, as is the case with Na activation and K activation in other cells (Figs. 10.13 and 10.14), are differently affected by different agents[56, 58]. These pharmacological differences indicate that the ionic events in eel electroplaques also occur at separate sites of the membrane.

The foregoing description has dealt with analyses that have provided considerable information regarding ionic processes. The available data also serve to indicate the probable ionic processes of as yet incompletely analyzed systems. For example, the esophageal cells of *Ascaris* produce

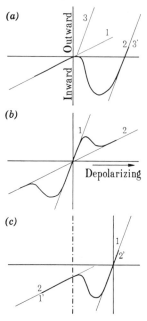

Fig. 10.20. Comparison of various negative slope characteristics involved in different types of electrogenesis. (*a*) The shift from the resting state (1) to higher conductance for a more positive ionic battery (2) results in negative slope characteristic that causes all-or-none response of normal spike electrogenesis. K or Cl activation (3) causes little change in membrane potential (cf. Figs. 10.17 and 10.18). If E_K or E_{Cl} is made positive to E_M (3′) "anomalous" spikes are evoked as seen in Figs. 10.19 and 10.23(*b*). Inactivation processes shift the characteristic from (1) to (2), but with little or no change in E_M. In either quadrant, transition is a negative-slope region giving rise to responses seen in Figs. 10.24 and 10.27. (*c*) Current-voltage characteristic of a cell depolarized by immersion in K-rich media. Original membrane potential is shown by broken line. Conductance (1) is higher then in the original state (2), but applying inward current causes transition to latter and hyperpolarizing response like that in Fig. 10.29. If the cell is kept hyperpolarized (1′), K activation is initiated by a depolarizing stimulus and the characteristic shifts temporarily to high conductance state (2′). Resulting K spikes are shown in Fig. 10.16. (Grundfest[14]. Reprinted by permission of the Federation of American Societies for Experimental Biology.)

diphasic spikes under some conditions and hyperpolarizing spikes under others[59]. These different responses can be accounted for qualitatively[13] by assuming a reasonable variety of interplay among their emf's—the resting potential (E_M), a depolarizing electrogenesis (E_D), and an electrogenesis that is hyperpolarizing and probably related to E_K (Fig. 10.23).

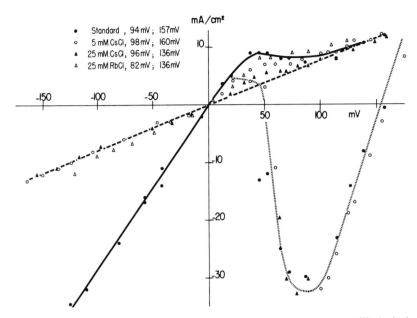

Fig. 10.21. Ionic currents in voltage clamped eel electroplaque. Filled circles show control conditions. Solid line in hyperpolarizing quadrant represents resting conductance state. With sufficient depolarization, characteristic develops time variant changes resolved into 2 branches. Initial inward current (dotted line) reflects influx of Na as the conductance for this ion increases by Na activation. Negative slope is characteristic of spike electrogenesis (Figs. 10.14 and 10.16). However, steady-state current after Na inactivation differs markedly from that observed in Fig.10.14 as it decreases with increasing depolarization. For large depolarizations the characteristic becomes linear, passing through the origin, but with a markedly reduced slope. Additions of Cs and Rb had relatively little effect on initial (Na) current, but altered steady-state characteristic, which then tended to have same low slope throughout as did the depolarized cell (broken line). Thus spike electrogenesis of eel electroplaque is without depolarizing K activation. K channels that are closed by depolarizing inactivation may also be closed by Cs or Rb (pharmacological K inactivation). Note that substitution of 5 mM CsCl for the KCl of the standard saline caused full pharmacological K inactivation in hyperpolarizing quadrant without significant effect on resting potential (94 → 98 mV) or spike amplitude (157 → 160 mV). (Nakamura, Nakajima, and Grundfest[56]. Reprinted by permission of the Rockefeller University Press.)

10.3.2   Electrically Excitable Inactivation Responses

Depolarizing Inactivation. As is seen in the voltage-current characteristic (Fig. 10.21), the transition from resting conductance of eel electroplaques to the state of depolarizing inactivation is through a region of

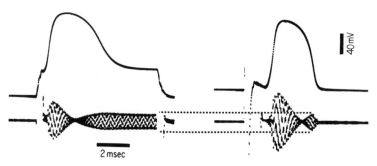

Fig. 10.22. Changes in impedance during spike electrogenesis of eel electroplaque. Simultaneous registration of spike (above) and ac. Wheatstone bridge output (below). Right: spike was evoked by neural stimulation and synaptic delay follows brief shock artifact. Postsynaptic potential and rising phase of spike are correlated with imbalance of bridge that represents decrease in impedance. Impedance returned to resting state shortly after peak of spike and there then followed phase of increased impedance during most of the falling phase. Left: spike was evoked by long-lasting depolarization. Bridge output shows that the peak of secondary phase of impedance increase was maintained as long as depolarizing current was applied. Thus these data confirm voltage clamp data of Fig. 10.21, namely, that sustained depolarization results in sustained increased resistance which results from depolarizing K inactivation. (Morlock, Benamy, and Grundfest[57]. Reprinted by permission of the Rockefeller University Press.)

negative slope. The similarity of the changes (shown diagramatically in Fig. 10.20a,b) may be expected to cause regenerative inactivation response (Fig. 10.24), and this has been observed in other cells as well[60]. However, there is one distinction between spikes and inactivation responses. The spike results from an increased conductance that in turn contributes a marked change of emf in the system (Fig. 10.20a). Once initiated, the spike is *autogenetic*[11] and can maintain itself as long as the depolarizing electrogenesis provides sufficient charge on the membrane capacity (Figs. 10.13–10.16, 10.19, 10.22 and 10.23). The inactivation response, however, does not involve a change in emf, but is caused by the increased voltage drop across a higher resistance (Figs. 10.21 and 10.24), and the latter is itself induced by the applied current. When the current is terminated, the inactivation response is also terminated (Fig. 10.24).

Hyperpolarizing Inactivation. Inactivation processes can also be initiated by hyperpolarizing stimuli (Figs. 10.2 and 10.3) and as shown diagrammatically in Fig. 20b; they also exhibit a region of negative slope, so that the also tend to be all-or-none effects. Hyperpolarizing inactivation responses have been observed in many cells and may exhibit considera-

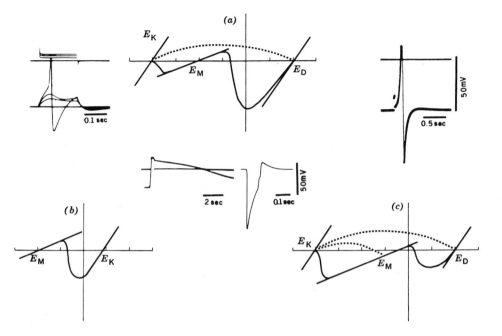

Fig. 10.23. Intracellularly recorded "anomalous" spikes in muscle fibers of *Tenebrio molitor* and in esophageal cells of *Ascaris lumbricoides*, and explanation of electrogenesis in diagrammatic voltage clamp presentations. (*a*) Under approximately normal ionic conditions $E_D > E_M > E_K$, where E_D is inside-positive ionic battery for depolarizing electrogenesis (Mg in *Tenebrio*). Depolarizing stimulus shifts characteristic from E_M toward E_D, giving rise to depolarizing electrogenesis of spikes. Depolarization induces K activation and a shift of the characteristic toward E_K (dotted line), giving rise to second phase of spike. Latter is particularly marked in recording from *Ascaris* cell (right). Hyperpolarization quenches K activation and potential returns to E_M. (*b*) *Tenebrio* muscle fibers can tolerate presence of as much as 340 mM K without appreciable change in E_M. However, $E_K > E_M$ and the response to brief stimulus (left center record) is a long-lasting K spike (as in Fig. 10.15). (*c*) *Ascaris* esophageal cells depolarize when solution contains increased concentration of Cl while E_K is relatively unchanged. Thus $E_M' \gg E_K$ and most or all of the electrogenesis is hyperpolarizing, occurring as the characteristic shifts from E_D to E_K or from E_M' to E_K. Hyperpolarizing spike is brief (as in *A*) because K activation is quenched by hyperpolarization. (Grundfest[13]. Reprinted by permission of the Rockefeller University Press.)

207

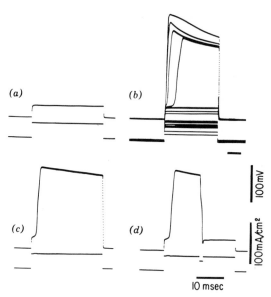

Fig. 10.24. Regenerative depolarizing inactivation responses in eel electroplaque immersed in isosmotic KCl. (*a*) current (lower trace) just subthreshold; (b) eight superimposed records with progressively increasing currents. Four largest stimuli evoked disproportionately larger depolarizations that developed more rapidly and to higher amplitudes with larger currents. Time base was slowed for these records. (*c*) and (*d*) current subthreshold (as in (*a*), but a brief small additional depolarization triggered large change in membrane potential. A brief inward current (*d*) abolished response which otherwise lasted as long as formerly subthreshold current was maintained (*c*). (Modified from Nakamura, Nakajima, and Grundfest[56].)

able variety in form. Eel electroplaques exhibit a simple type (Fig. 10.25). Once the disproportionately large change in membrane potential has been induced by a certain (threshold) level of applied current, there is little or no further change while the current is applied, and the membrane potential returns to the resting value as soon as the current is turned off. These results may be interpreted as being caused by a decreased conductance for one ion; and as the K conductance is the highest in the resting cell, it seems likely that both depolarizing and hyperpolarizing responses are due to K inactivation. However, the kinetics of the two types of inactivation may differ considerably. The transition into depolarizing K inactivation is rapid, whereas the completion of hyperpolarizing K inactivation is slower (Fig. 10.25).

The conditions under which hyperpolarizing inactivation responses are observed in different cells and the form of the responses may differ

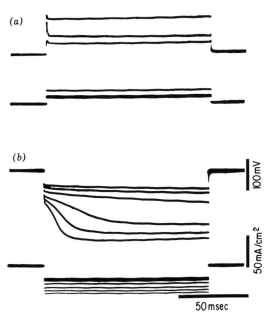

Fig. 10.25. Inactivation responses of eel electroplaque caused by (a) depolarizing inactivation and (b) hyperpolarizing inactivation. Unlike cell of Fig. 10.24, this electroplaque was not depolarized initially. (a) Three superimposed sweeps. Lower trace monitors outward (depolarizing) current which was increased in successive records. Larger currents induced disproportionately larger depolarizations indicative of increase in resistance owing to depolarizing K inactivation. Amplitudes of the inactivation responses are smaller than in Fig. 10.24 because resting resistance was higher (the cell had not been subjected to high K_0). (b) Six successive superimposed sweeps showing effects of increasing inward (hyperpolarizing) currents. With weak currents hyperpolarization (upper trace) tended to increase slowly, but the change became more rapid with stronger currents. Larger currents increased maximum hyperpolarization approximately proportionately. Once this large potential was induced by hyperpolarizing K inactivation it remained essentially steady. Note slower rate at which hyperpolarizing inactivation develops in comparison with depolarizing inactivation. (Unpublished data and Nakamura, and Grundfest[58].)

considerably. The large hyperpolarization induced in lobster muscle fibers by a threshold stimulus is not maintained, although the current remains constant[61]. Instead, there is a "spontaneous" return toward the resting level (Fig. 10.26), indicating that a second conductance change—an increase—must have developed during the applied current. The response thus has the form of a brief hyperpolarizing pulse followed by a more or less steady plateau. The pulse increases only slightly in

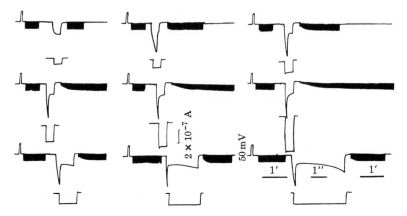

Fig. 10.26. Interplay of inactivation and activation processes during hyper-polarizing responses of lobster muscle fiber and subsequent persistent increased conductance. Each trace (ink-writer records) was registered in sequence of chart speeds. Initial fast: calibrating pulses of 50 mV. Slow: series of brief hyper-polarizing testing pulses of constant strength. Fast: during which hyperpolarizing current was applied. Slow: during which testing hyperpolarizing pulses were again applied. Below each recording is also shown record monitoring the applied current.

After hyperpolarizing response was evoked (middle record, upper row) membrane resistance remained lowered. This is indicated by decrease in ampli-tudes of hyperpolarizing testing pulses. As stimulating current was increased, early pulse-like part of hyperpolarizing responses became shorter, and membrane resistance subsequently decreased further and remained low for longer times. Conductance increase lasted more than 1 min with highest currents used (middle row). Note that only a small degree of after-depolarization is associated with relatively large conductance changes. During prolonged hyperpolarizing currents (lower row), membrane resistance began to rise, but after current was withdrawn membrane resistance fell below its resting value. (Reuben, Werman, and Grund-fest[61]. Reprinted by permission of the Rockefeller University Press.)

amplitude as the current increases and its duration becomes shorter. These effects indicate that the secondary conductance increase develops more rapidly with larger applied currents, persisting for some time after the current is terminated. During a long-maintained current the hyper-polarization may increase again and another brief pulselike phase may be induced. Thus the hyperpolarizing currents apparently cause several time-variant changes in conductance, one of which, the initial decrease, is due to K inactivation. The ionic nature of the secondary increase in conductance is not definitely known as yet, but it appears to be the result of hyperpolarizing Cl activation. In crayfish muscle fibers, hyperpolarizing Cl activation normally occurs at a very low threshold and this obscures a decrease in K conductance caused by hyperpolarizing K inactivation. However, when the fiber is bathed in a saline solution in which the Cl has been replaced with an impermeant anion, hyperpolarizing responses appear that are in general similar to those of the lobster fiber (Fig. 10.27).

A number of cells, notably axons of various forms, also exhibit hyperpolarizing responses, but only when the fibers have first been depolarized in the presence of high K (Figs. 10.28 and 10.29). These hyperpolarizing responses also have two general forms. Once the hyperpolarization has increased in toad and frog axons (Figs. 10.28 and 10.29) it remains at the higher value as long as the current is applied. The steady-state hyperpolarization is approximately proportional to the applied current (Fig. 10.29). As in eel electroplaques (Fig. 10.25), the hyperpolarizing response involves mainly one ionic process: K inactivation. In the squid axon, however (Fig. 10.28), the hyperpolarizing response has a pulselike component similar to that of the lobster (Fig. 10.26) or crayfish (Fig. 10.27) muscle fibers. The characteristics of the complex conductance changes responsible for the time-variant hyperpolarizing responses in the axon have not yet been analyzed.

One can readily account for the axons' having to be in a depolarized state when the hyperpolarizing current is applied[8]. At the normal resting potential the membrane conductance of the axons is low, either because all the reactive K channels are normally closed or because only

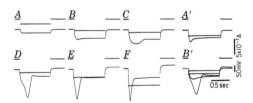

Fig. 10.27. Dependence of hyperpolarizing responses of crayfish muscle fiber on ionic conditions. $A–F$: Preparation was immersed for 12 h in medium in which all Cl was substituted by propionate. Small increase in inward current (monitored on upper traces) from C to D evoked disproportionately large but transient hyperpolarization. Subsequent applications of larger currents shortened large hyperpolarization, but amplitude was not increased proportionately. Although currents remained constant, membrane potential returned to less negative value, at which it remained as long as current was applied. B': similar hyperpolarizing responses were evoked after the preparation had been transferred to a medium in which NO_3 was the sole anion. A': hyperpolarizing response was abolished after preparation had been transferred to standard Cl saline. (Ozeki, Girardier, Brandt, Reuben, and Grundfest[62].)

TOAD

SQUID

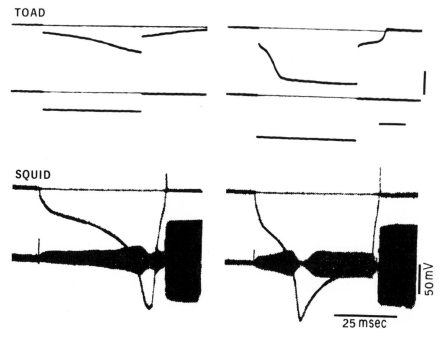

25 msec

50 mV

Fig. 10.28. Hyperpolarizing inactivation in axons of toad and squid. Fibers were depolarized by increasing K_0 and hyperpolarizing currents (about 50 msec in duration) were applied. Current was monitored on lower traces in top records. Wheatstone bridge output was monitored in lower set of records. When applied current was weak (left), hyperpolarizing response developed slowly. Response of toad axon resembles that of eel electroplaque (Fig. 10.25), while that of squid axon resembles response of lobster or crayfish muscle fibers (Figs. 10.26 and 10.27). Note, however, that inactivation processes of axons were made evident after depolarization of cells, whereas electroplaques and muscle fibers did not need to be depolarized in order to exhibit hyperpolarizing inactivation. (Modified from Tasaki[55].)

a few are open. Thus hyperpolarizing inactivation cannot develop or it is relatively insignificant. When the fibers are depolarized in K, the steady-state membrane conductance is high (Fig. 10.29). Application of a hyperpolarizing current then returns the conductance to the lower value and the consequent large voltage drop induced by the current constitutes the hyperpolarizing response. The fact that the response is all-or-none arises from the nature of the voltage-current characteristic (Fig. 10.20c). The high conductance branch of the depolarized cell is connected to the low conductance branch by a region of negative slope. The hyperpolariz-ing response of the initially depolarized cell is the converse of the K-

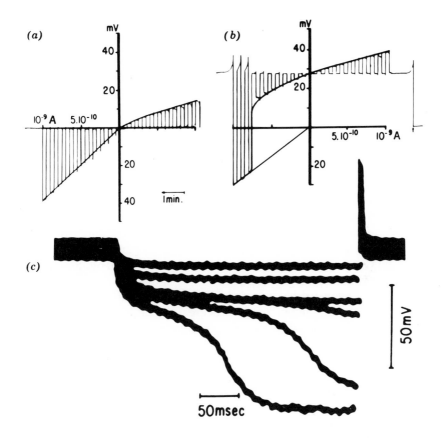

Fig. 10.29. Hyperpolarizing inactivation responses in frog axons. (a) and (b) I–E characteristics plotted out by periodic applications of current pulses whose amplitudes are shown on abscissa (depolarizing to the right). Ordinates: membrane potential in relative units, resting potential at origin. During measurements in (a), axon was in ordinary frog saline solution. E changed linearly with I for inward currents, but for outward currents developed curvature which betokens K activation. For measurements in (b), axon was depolarized by about 28 units following immersion in KCl-enriched saline. I–E characteristic shows a high conductance even for small hyperpolarizing current, but characteristics became increasingly nonlinear for inward currents 5×10^{-10} A and returned to low conductance state with currents 7.5×10^{-10} A. Extrapolation of this portion passes through origin. (c) Change in membrane potential during sufficiently large applied current is hyperpolarizing response. Note that form of hyperpolarizing response of frog axon resembles that of eel electroplaques (Fig. 10.25) or toad axons (Fig. 10.28). (Grundfest[11]. Reprinted by permission of Academic Press. Modified from Stämpfli[63, 64].)

spike that occurs if the original membrane potential is restored by an applied current and a brief depolarizing stimulus is given (Fig. 10.16). The difference is that the resulting change in the characteristic is in the direction of depolarization (Fig. 10.16, right or Fig. 10.20c). The change thus contributes a positive emf and a K-spike occurs as an autogenetic response. In the reverse condition, the new membrane potential can be maintained only as long as the hyperpolarizing current is applied.

10.4   CONCLUSION

The range of information that is contributed by electrophysiological studies provides a certain degree of insight into the functional capabilities of living membranes and into the diversitites of their performance as electrochemical systems. Although the structural data are not yet available, a theoretical accounting for bioelectrogenesis must take into consideration the heterogeneity of the system as revealed by electrophysiological and pharmacological data.

This heterogeneity has numerous facets. For example, some components of the membrane increase or decrease their permeability to certain ions on being subjected to a change in the electrical field; others are not sensitive to an electrical stimulus, but require more or less specific activation by chemical, mechanical, thermal, or photic energy sources. Both the electrically excitable and electrically inexcitable components may be specifically affected by certain agents and the degrees of specificity may vary widely. The degree of selectivity in permeability characteristics may also vary rather widely. All these varieties of responsiveness must have morphological correlates at the macromolecular and structural level.

Despite the obvious complexities, however, some degree of order does emerge and lends assurance to the premise that new approaches and new techniques will eventually provide a more rigorously quantitative electrochemical theory of bioelectrogenesis.

REFERENCES

[1] J. Bernstein, *Pfluegers Arch. ges. Physiol*, **92**, 521 (1902).
[2] J. Bernstein, *Elektrobiologie, F. Vieweg u. Sohn*, Braunschweig, 1912.
[3] A. L. Hodgkin, *Biol. Rev.*, **26**, 339 (1951).
[4] A. L. Hodgkin, *Proc. Roy. Soc. (London) Ser. B.*, **148**, 1 (1957).

[5] A. L. Hodgkin, *The Conduction of the Nervous Impulse*, Liverpool University Press, 1964.

[6] A. L. Hodgkin, and B. Katz, *J. Physiol.*, **108**, 37 (1949).

[7] A. L. Hodgkin, and A. F. Huxley, *J. Physiol.*, **117**, 500 (1952).

[8] H. Grundfest, *Ann. N. Y. Acad. Sci.*, **94**, 405 (1961).

[9] H. Grundfest, in *Properties of Membranes and Diseases of the Nervous System*, (M. D. Yahr, ed.), Springer, New York, 1962.

[10] H. Grundfest, in *The General Physiology of Cell Specialization*, (D. Mazia and A. Tyler, eds.), McGraw-Hill, New York, 1963.

[11] H. Grundfest, in *Advances in Comparative Physiology and Biochemistry*, vol. 2, (O. E. Lowenstein, ed.), Academic, New York, 1966.

[12] H. Grundfest, *Ann. N. Y. Acad. Sci.*, **137**, 901 (1966).

[13] H. Grundfest, *J. Gen. Physiol.*, **50**, 1955 (1967).

[14] H. Grundfest, *Federation Proc.*, **26**, 1613 (1967).

[15] D. A. MacInnes, *The Principles of Electrochemistry*, Dover, New York, 1961.

[16] D. E. Goldman, *J. Gen. Physiol.*, **27**, 37 (1943).

[17] T. Teorell, *Progr. Biophys. Biophys. Chem.* **3**, 305 (1953).

[18] T. Teorell, *Discussions Faraday Soc.* **21**, 9 (1956).

[19] T. Teorell, *Biophys. J.*, *Suppl.*, **2**, 27 (1962).

[20] R. G. Bates, in *Reference Electrodes*, (D. J. G. Ives and G. J. Janz, eds.), Academic, New York, 1961.

[21] J. F. Danielli, in *Recent Developments in Cell Physiology*, (J. A. Kitching, ed.), Butterworth, London, 1954.

[22] H. Grundfest, *Physiol. Rev.*, **37**, 337 (1957).

[23] H. Grundfest, in *Biophysics of Physiological and Pharmacological Actions*, (A. M. Shanes, ed.), American Association for the Advancement of Science, Washington, D.C. (1961).

[24] H. Grundfest, in *Essays on Physiological Evolution*, (J. W. S. Pringle, ed.), Pergamon Press, London, 1964.

[25] H. Grundfest, *Cold Spring Harbor Symp. Quant. Biol.*, **30**, 1 (1965).

[26] H. Grundfest, *Arch. Ital. Biol.*, **96**, 135 (1958).

[27] H. Grundfest, in *Nervous Inhibition* (E. Florey, ed.), Pergamon, London, 1961.

[28] H. Grundfest, C-Y Kao, and M. Altamirano, *J. Gen. Physiol.*, **38**, 245 (1954).

[29] H. Grundfest, in *Electrochemistry in Biology and Medicine*, (T. Shedlovsky, ed.), Wiley, New York, 1955.

[30] S. Nakajima, and K. Takahashi, *J. Physiol.*, **187**, 105 (1966).

[31] L. Girardier, J. P. Reuben, and H. Grundfest, unpublished data.

[32] A. R. Freeman, J. P. Reuben, P. W. Brandt, and H. Grundfest, *J. Gen. Physiol.*, **50**, 423 (1966).

[33] K. S. Cole, and H. J. Curtis, *J. Gen. Physiol.*, **24**, 551 (1941).

[34] C-Y Kao, and H. Grundfest, *Experientia*, **13**, 140 (1957).

[35] P. F. Baker, A. L. Hodgkin, and T. L. Shaw, *J. Physiol.*, **164**, 330 (1962).

[36] H. Gainer, and H. Grundfest, *J. Gen. Physiol.*, **51**, 399 (1968).

[37] A. L. Hodgkin, and P. Horowicz, *J. Physiol.*, **153**, 370 (1960).

[38] A. Strickholm, and B. G. Wallin, *J. Gen. Physiol.*, **50**, 1929 (1967).

[39] A. R. Freeman, and H. Grundfest, *Federation Proc.*, **25**, 570 (1966).

[40] P. B. Dunham, A. R. Freeman, and H. Grundfest, *Biol. Bull.* **133**, 462 (1967).

[41] J. P. Reuben, E. Lopez, P. W. Brandt, and H. Grundfest, *Science,* **142**, 246 (1963).

[42] E. Lopez, J. P. Reuben, and H. Grundfest, unpublished data.

[43] J. C. Eccles, *The Physiology of Nerve Cells*, Johns Hopkins Press, Baltimore 1957.

[44] S. Hagiwara, K. Kusano, and N. Saito, *J. Neurophysiol.*, **23**, 505 (1960).

[45] G. A. Kerkut, and R. C. Thomas, *Comp. Biochem. Physiol.*, **8**, 39 (1963).

[46] S. Hagiwara, and N. Saito, *J. Physiol.*, **148**, 161 (1959).

[47] P. Belton, and H. Grundfest, *Am. J. Physiol.*, **203**, 588 (1962).

[48] S. Nakajima, and K. Kusano, *J. Gen. Physiol.*, **49**, 613 (1966).

[49] S. Nakajima, *J. Gen. Physiol.*, **49**, 629 (1966).

[50] C. T. Gaffey, and L. J. Mullins, *J. Physiol.*, **144**, 505 (1958).

[51] L. J. Mullins, *Nature*, **196**, 986 (1962).

[52] B. Hille, M. V. L. Bennett, and H. Grundfest, *Biol. Bull.*, **192**, 407 (1965).

[53] B. Cohen, M. V. L. Bennett, and H. Grundfest, *Federation Proc.*, **20**, 339 (1961).

[54] H. Grundfest, E. Aljure, and L. Janiszewski, *J. Gen. Physiol.*, **45**, 598A (1962).

[55] I. Tasaki, *J. Physiol.*, **148**, 306 (1959).

[56] Y. Nakamura, S. Nakajima, and H. Grundfest, *J. Gen.Physiol.*, **49**, 321 (1965).

[57] N. Morlock, D. A. Benamy, and H. Grundfest, *J. Gen. Physiol.*, **52**, 22 (1968).

[58] Y. Nakamura, and H. Grundfest, *23rd Internat. Physiol. Cong. Abst.* **167** (1965).

[59] J. del Castillo, and T. Morales, *J. Gen. Physiol.*, **50**, 603 (1967).

[60] M. V. L. Bennett, and H. Grundfest, *J. Gen. Physiol.*, **50**, 141 (1966).

[61] J. P. Reuben, R. Werman, and H. Grundfest, *J. Gen. Physiol.*, **45**, 243 (1961).

[62] M. Ozeki, L. Girardier, P. W. Brandt, J. P. Reuben, and H. Grundfest, unpublished data.

[63] R. Stämpfli, *Helv. Physiol. Acta*, **17**, 127 (1958).

[64] R. Stämpfli, *Ann. N. Y. Acad. Sci.*, **81**, 265 (1959).

CHAPTER 11

Measurement and Control of Membrane Potential in Myelinated Nerve Fibers

B. FRANKENHAEUSER and A. B. VALLBO

The Nobel Institute for Neurophysiology, Karolinska Institutet, Stockholm, and the Department of Physiology, University of Umea

11.1 INTRODUCTION

The purpose of this report is to provide a short description of a technique developed for investigations of the activity in single nodes of myelinated nerve fibers. The membrane potential was measured without any appreciable attenuation, external electrodes only being applied to the nerve. Negative feedback was used to obtain a recording system with very high input impedance. The membrane potential was further stabilized and changed in rectangular steps that were achieved by an additional feedback system. Some of the basic findings obtained with these methods will be summarized here; in addition, a complete description of the analysis may be found in papers in *The Journal of Physiology* and *Acta Physiologica Scandinavica* (see references).

11.2 RECORDING CELL AND POTENTIAL CONTROL SYSTEM

A single, myelinated fiber from the sciatic nerve of a frog or a toad was dissected for a length of 5–10 mm. The nerve trunk was left intact proximally and distally to the region from which the single fiber was dissected. The preparation was then transferred in a recording cell made of Perspex, a cell having four pools separated by three partitions. The two center pools were 200–300 μ and the partitions were 150 μ thick, the end pools being considerably larger (about 2 cm in length). During the mounting of the fiber, the recording cell was filled with Ringer's solution well above the partitions. Petroleum jelly was applied on the partitions so that the fiber ran through three tunnels of petroleum jelly from one end pool to the other. The Ringer's solution was then carefully sucked away down to the level of the petroleum jelly seals so that four separate fluid pools were obtained while the fiber was entirely below the fluid surface.

Figure 11.1 shows a schematic diagram of the preparation in the recording cell. The pools were labeled A, B, C, and E as indicated in the figure. The potential across seal BC was recorded with a differential amplifier (labeled A_1 in the figure) with cathode follower input. The amplifier had a voltage gain of about 2,000×. The output was single-ended and placed through a cathode follower. Pool B was grounded and the output of the amplifier was connected to pool A. A signal at the input V_{CB} appeared amplified at the output V_{BA}, creating a negative feedback loop where the fiber forms a part of it. The changes in membrane potential were recorded as the potential developed across seal AB. These potential changes were very nearly equal to the potential changes across the membrane at the node in pool A, as will be shown. The system included a generator—the nodal membrane in pool A—and a loop of impedances. The impedance from a point inside the fiber under the node—point D—through the axoplasm and through the membrane to pool C will be referred to as Z_{DC}. The impedance from pool B to C, Z_{CB}, and the impedance from pool A to B, Z_{BA}, are accounted for mainly by the fluid layer between the nerve fiber and the petroleum-jelly seals. The amplifier was connected to this loop with the input between B and C and the output between A and B. The potential developed across these impedances will be referred to as V_{BA} and V_{CB}, for example, and the potential developed by the generator, that is, the nodal membrane, will be referred to as V_{in} or V_{DA}. The amplifier gain is a.

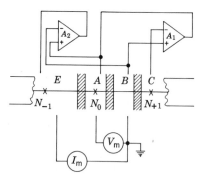

Fig. 11.1. Diagram of experimental arrangement, N_{-1}, N_0, N_{+1} are nodes of Ranvier; A_1 and A_2, feedback amplifiers; V_m, potential recording amplifier; I_m, current recording amplifier. A, B, C, and E solution pools insulated from each other by petroleum jelly seals (shaded regions).

It is obvious that for any change of the potentials in this system the following is valid:

$$V_{in} = V_{DA} = V_{DC} + V_{CB} + V_{BA},$$ (11.1)

and

$$V_{BA} = aV_{CB}.$$ (11.2)

Thus

$$V_{CB} = V_{BA}/a.$$ (11.3)

The current through the loop is:

$$\frac{V_{DC}}{Z_{DC}} = \frac{V_{CB}}{Z_{CB}}.$$ (11.4)

Thus

$$V_{DC} = \frac{V_{CB}Z_{DC}}{Z_{CB}}.$$ (11.5)

Substituting from (11.3) and (11.5) in (11.1) gives:

$$V_{BA} = V_{DA}\frac{aZ_{CB}}{Z_{DC} + Z_{CB}(a+1)}.$$ (11.6)

The impedance Z_{DC} is in the order of size of 60 MΩ. Z_{CB} was about 10 MΩ and the amplifier gain was 2000. Using these figures in (11.6), it follows that the potential changes between pool A and B are 99.65% of the changes in membrane potential. The input impedance of the system is quite high owing to the negative series feedback – in the order of size of 20,000 MΩ. Consequently the current drain from the fiber to pool C and to B is extremely small; pool C and point D are kept at a constant potential relative to ground. The only point which moves significantly with reference to ground during activity is the solution in pool A, that is, the outside of the node under investigation.

In order to excite the node, current pulses were injected from a low impedance stimulator with the output connected to pool E and B. Pool E was thus connected to ground through the low impedance of this stimulator. A pulse from the stimulator injected a current pulse into the node. During activity in the node there is almost no current spread within the fiber in either direction from the node. The current flowing in the internode from pool E to the point D inside the fiber is therefore proportional to the potential difference between pool E and B, and this current is independent of any changes in membrane potential at the node. The potential between E and B was measured with a separate amplifier, and it was possible to calculate the current forced through the node from this potential and from the resistance of the axis cylinder and the nodal membrane. With this method it was thus possible to record the potential changes across a single node, to apply current through the node, and to measure this current[1]. These properties are required for a voltage clamp technique, that is, for changing the membrane potential in steps and recording the ionic currents produced by the nodal membrane at any potential.

The membrane potential was controlled by a second feedback amplifier (labeled A_2 in Fig. 11.1). The input to this amplifier was the potential between pool A and B, that is, the potential across the nodal membrane — the output being connected to pool E. Any output potential drives a current through the node in the same way as does the stimulator. Thus both the input and the output of this amplifier were effectively across the membrane of the node; and as the feedback was negative, the amplifier worked as a voltage stabilizer, keeping the membrane potential at a value which could be determined by a reference potential. This reference potential in turn could be made to undergo rectangular steps. The current injected into the node by the second feedback amplifier (in order to keep the membrane potential at any value) was measured. This current is equal to current produced by the nodal membrane[2, 3].

11.3   ANALYSIS OF THE NODAL CURRENT UNDER VOLTAGE CLAMP CONDITIONS

An extensive analysis of the ionic currents in the myelinated nerve fiber has been performed during the last 10 years. The membrane currents associated with a potential step are treated as the sum of a capacitive current and an ionic current. The ionic current in turn is treated as the sum of a nonspecific leak current, an initial current, and a delayed current.

The capacitive current was not studied in detail because the voltage clamp technique is not suitable for reliable measurements of this current. The leak current was measured at anodal steps and at the equilibrium potential for the initial current. These measurements indicate a linear relation between the membrane potential and the leak current and, further, that the leak current does not change with time. Consequently it seems justifiable to describe the mechanism behind the leak current as a conductance which is independent of potential and time.

A step change in membrane potential in depolarizing direction was associated with an initial transient current, and this was followed by a delayed lasting outward current. A similar pattern has been found in the squid axon, and there it has been shown that the initial current is carried mainly by sodium and that the delayed current is carried mainly by potassium ions[4–8]. It seems likely that this could be the case also in the myelinated nerve fiber. Moreover, it was assumed that the ions were carried through the membrane as if they were moving passively down the electrochemical gradient. These assumptions were rigorously tested for the initial current. In Ringer's solution the equilibrium potential for the initial current was about 125 mV positive from the resting potential. If the initial current is a sodium current, then the equilibrium potential should depend upon the outside concentration of sodium ions. The experimentally measured change of this equilibrium potential should agree with the change in the sodium equilibrium potential as predicted by Nernst's equation. This point was first tested and it was found that the agreement was satisfactory. The initial current in the nodal membrane was also in other respects similar to the sodium current in the squid axon (e.g., in the smooth variation with the membrane potential of the amplitude and in the time course). There was one clear difference, however: for large pulses the peak current in the squid fiber is linearly related to the membrane potential, whereas the current plot was curved in this region in the toad fibers.

One reason for this effect could be that the sodium conductance is inactivated with such a short time constant at large pulses that it is not fully turned on before it is inactivated to a large extent. However, it could be shown that this was not the case. The sodium conductance can also be derived from measurements of the instantaneous current when the cathodal pulse is interrupted at the peak of the initial current. In this case the calculations of the sodium conductance are based on measurements of inward current only; when it is determined from the currents at cathodal steps, the calculation is based on measurements of inward current at pulses smaller than the equilibrium potential and outward currents at pulses larger than the equilibrium potential.

When the sodium conductance was calculated from the instantaneous currents, it was found that the conductance did not decrease to such a large extent at large pulses. As an example, at a cathodal step of 180 mV it was about 80% of its maximum value while it was only 40–50% of its maximum value when it was calculated from the peak initial current at cathodal steps. This finding indicated that the bending of the plot of the peak initial current to membrane potential depended on the direction of the current, but not upon the state of the membrane. It is known that a simple semipermeable membrane shows a kind of rectification such that at corresponding driving potentials when the ions, passing through the membrane, move from a solution of high ionic concentration to a solution of low concentration the currents are larger than when the ions move in the reverse direction. This rectification is explained by different ionic concentrations in the membrane. The "constant field equation" takes these concentration differences into account.

The next step in the treatment was therefore to apply the constant field equation to the experimental data. The permeability of the nodal membrane for the initial current as calculated with the constant field equation reached a ceiling value and decreased only slightly at large cathodal pulses; for example, at 180 mV it was as much as 80–90% of the maximum. Another finding which agrees with this treatment was that the value of the permeability calculated was independent of the potential at which the current was measured. This was not the case when the sodium-carrying mechanism was treated as a conductance.

The initial currents decreased in amplitude when solutions with low sodium concentrations were applied. This decrease was more pronounced in the region of inward currents than in the region of outward currents. The potential of current reversal (the equilibrium potential) was also affected by the change in sodium concentration. All these findings were satisfactorily and quantitatively accounted for by the constant field equation. The experimental findings thus strongly support the conclusion that the initial membrane currents are mainly carried by sodium and that sodium moves as in free diffusion in an electric field[2, 3, 9]. The question of how high the sodium specificity is was investigated further by Frankenhaeuser and Moore[10]; and it was shown that the ratio between the sodium permeability and the potassium permeability was about 20:1 for the mechanism carrying the initial current.

The sodium mechanism was described by empirical equations closely related to those used by Hodgkin and Huxley[4–7] for the squid fiber voltage clamp data. The permeability changes were described by rate constants that depend on membrane potential but are independent of time[11–13]. It was found that one of the rate constants describing the inactivation varied considerably from one fiber to another. In a report by Frankenhaeuser and Vallbo[14] it was shown that the variation in this rate constant was the main factor accounting for the variation from one fiber to another in their response to current pulses of long duration and linearly rising currents, that is, in the accommodation.

The delayed current was analyzed in a way similar to the initial current [15–18]. The findings justified the conclusion that the delayed current was mainly carried by potassium ions and that the mechanism could be satisfactorily described as a permeability. The potassium-permeability changes were described by empirical equations similar to those used for the sodium permeability. The detailed analysis of the delayed current was complicated by the fact that there was also a delayed small increase of the permeability for other ions, mainly sodium.

The voltage clamp analysis describes in quantitative form how the membrane currents depend on membrane potential, time, and ionic concentrations. The membrane potential was changed in steps during the clamp experiments and the currents did not affect the potential. During ordinary excitation of nerve the membrane currents change the membrane potential. Therefore the differential equations for the various currents are independent of each other in clamp experiments but depend on each other during excitation. The equations should in both cases describe all the major membrane currents. A computer program was written to solve the action potential and the excitability of the nerve model [19]. A large number of computations for normal and changed conditions was made, and the solutions were compared with experimental findings [14, 20]. These computations seem to indicate that the outcome of the experimental analysis was a satisfactory description of the mechanisms accounting for the excitability and the membrane action potential in the myelinated nerve fiber. Furthermore, it seems clear that the exact effect of a change of one constant cannot be very easily judged, because the expected effect is often cancelled to a large extent by secondary effects on other mechanisms.

REFERENCES

[1] B. Frankenhaeuser, *J. Physiol.*, **135**, 550 (1957).
[2] F. A. Dodge, and B. Frankenhaeuser, *J. Physiol.*, **143**, 76 (1958).
[3] F. A. Dodge, and B. Frankenhaeuser, *J. Physiol.*, **148**, 188 (1959).
[4] A. L. Hodgkin, and A. F. Huxley, *J. Physiol.*, **116**, 449 (1952).
[5] A. L. Hodgkin, and A. F. Huxley, *J. Physiol.*, **116**, 473 (1952).
[6] A. L. Hodgkin, and A. F. Huxley, *J. Physiol.*, **116**, 497 (1952).
[7] A. L. Hodgkin, and A. F. Huxley, *J. Physiol.*, **116**, 500 (1952).
[8] A. L. Hodgkin, A. F. Huxley, and B. Katz, *J. Physiol.*, **116**, 424 (1952).
[9] B. Frankenhaeuser, *J. Physiol.*, **152**, 159 (1960).
[10] B. Frankenhaeuser, and L. E. Moore, *J. Physiol.*, **169**, 438 (1963).
[11] B. Frankenhaeuser, *J. Physiol.*, **148**, 671 (1959).
[12] B. Frankenhaeuser, *J. Physiol.*, **151**, 491 (1960).
[13] B. Frankenhaeuser, *J. Physiol.*, **169**, 445 (1963).
[14] B. Frankenhaeuser and A. B. Vallbo, *Acta Physiol. Scand.*, **63**, 1 (1965).
[15] B. Frankenhaeuser, *J. Physiol.*, **160**, 40 (1962).
[16] B. Frankenhaeuser, *J. Physiol.*, **160**, 46 (1962).
[17] B. Frankenhaeuser, *J. Physiol.*, **160**, 54 (1962).
[18] B. Frankenhaeuser, *J. Physiol.*, **169**, 424 (1963).
[19] B. Frankenhaeuser and A. F. Huxley, *J. Physiol.*, **171**, 302 (1964).
[20] B. Frankenhaeuser, *J. Physiol.*, **180**, 780 (1965).

CHAPTER 12

Resistance Measurements by Means of Microelectrodes in Cardiac Tissues

EDOUARD CORABOEUF

Laboratoire de Physiologie comparée et de Physiologie cellulaire associé au CNRS, Faculté des Sciences, Orsay

I am particularly indebted to Prof. S. Weidmann for much useful information and advice and to Misses F. Denoit and E. Mercier for constant help in preparing the manuscript.

This study was supported in part by contract 6600438 and 394, D.G.R.S.T., Paris, France.

12.1 INTRODUCTION

The use of microelectrodes with an external diameter of less than 0.5 μ [1, 2] has been extended to cardiac tissue since 1949[3–12]. The technique of intracellular microelectrodes, moreover, is now so commonly used and has been so well described[13, 14] that is is superfluous for us to recapitulate at length. However, a few points are or can be more particularly related to cardiac applications, and for this reason we shall briefly mention them. We shall then consider the problem of resistance measurements of cardiac cell membrane by means of intracellular microelectrodes.

12.2 MICROELECTRODE TECHNIQUE APPLIED TO CARDIAC TISSUES

The main requirements met in applying the technique of microelectrodes to cardiac tissues arise from three facts: (a) cardiac cells are generally of rather small size, (b) if they are most often (as in the myocardium) free from connective tissue, they can be in other regions (e.g., nodal and conducting tissue) protected by thick connective sheath; and (c) they normally show rythmic mechanical activity, and the resulting contraction of the whole organ may be quite strong and of quite large amplitude.

12.2.1 The Size of Cardiac Cells

The size of cardiac cells depends on the species, the age and the nature of the tissue to which the cells belong. It is generally larger in mammals than in lower vertebrates: the cell diameter is between 10 and $20\,\mu$ in mammalian ventricular myocardium and around $5-7\,\mu$ in toad and frog ventricle. The diameter is smaller during the early stages of development. Moreover, the cylindrical shape of normal myocardial cells, with a length several times the diameter, does not exist in the very young embryo — in which cells are rather spherical. In a normal adult mammalian heart the smallest cells are found in the regions of the nodes ($3-4.5\,\mu$) and the largest ones in the lower part of the His bundle ($30-40\,\mu$).

The generally small size of cardiac cells and the presence in some cases of connective tissue require microelectrodes with external tip diameter of $0.3\,\mu$ or less (resistance between 15 and 30 MΩ). Even then, the penetration of a fiber through the connective tissue in sheep or calf Purkinje strands remains sometimes difficult to achieve.

12.2.2 Prevention of Mechanical Artifact and the Floating Electrode Technique

The problem of mechanical artifacts is of unequal importance, according to the tissue studied. With preparations of isolated Purkinje fibers the contraction is so weak that is is possible to penetrate the cells of such preparations with a fixed microelectrode and sometimes to maintain the tip of the electrode inside a cell for periods up to 10 or 12 h. With atrial tissue it is also possible to use fixed microelectrodes, the preparations being generally thin enough to be easily fixed in different points with pins or hooks. The situation is not fundamentally different with rather thick, nonperfused isolated ventricular preparations, contractions of which are weakened by some degrees of anoxia that develops in the depth of the tissue; such preparations are sometimes used in spite of the risk of diffusion of ions or toxic substances from the inner to the superficial layers of cells.

However, in most cases in which ventricular tissue has to be studied with intracellular microelectrodes, the problem of mechanical artifacts must be solved. In 1950-1952 different solutions (Fig. 12.1) were suggested to prevent movements of the tissue: immersion into gelatin[4] fixation by means of a ring without suction[15] or with suction[16], or fixation by means of a forklike holder[17]. A more physiologically suitable solution consists of making the electrode free enough to follow the natural movement of the tissue. Vaughan Williams[18] has suggested a system with a semiflexible microelectrode. The microelectrode is dipped blunt-end downward into a latex solution and immediately

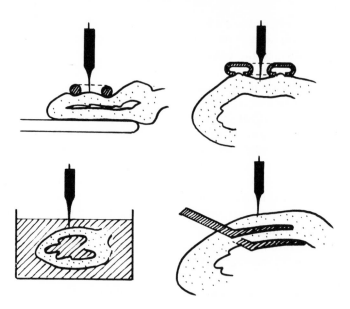

Fig. 12.1. Schematic representation of different solutions used to prevent mechanical artifacts when ventricle electrical activity must be studied with fixed microelectrodes. Lower left: immersion in gelatin[4]; upper left: fixation by means of ring[15]; upper right: fixation by suction[16]; lower right: use of fork-like holder[17].

withdrawn. When the latex is dried, it forms (after cutting of the glass tube) a sort of very flexible connection filled with KCl, allowing the use of the convenient Ag/AgCl or calomel junction. On the other hand, the system is sufficiently rigid to permit penetration of the electrode in a precise point of the preparation—but it is not flexible enough to follow the movement of a strongly beating ventricle.

The most flexible system, now quite classical, was described by Woodbury and Brady in 1956[19]. In this technique, only the fine tip of a conventional microelectrode is used (Fig. 12.2a and b) and a fine tungsten wire introduced into the blunt end of this tip establishes the electrical connection between the KCl solution filling the microelectrode and the cathode follower. The very small weight of the tip and the great flexibility of the wire allow the system to follow the movement of a large mammalian heart in situ, without the microelectrode leaving the cell which it has penetrated ("floating" microelectrodes). The penetration is generally successful as soon as the tip touches the tissue. To diminish drift in the baseline, it seems preferable to substitute for tungsten wire a fine chlorided silver wire (preheated in order to avoid metal elasticity).

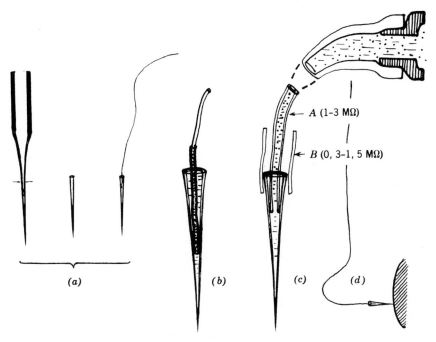

Fig. 12.2. (a) Preparation of "floating" microelectrodes, according to Woodbury and Brady[19]. (b) Detail of tip; tungsten wire may be replaced by chlorinated silver wire. (c) Modification by Benoit[21]; wire is replaced by fine inner (A) or outer (B) catheter filled with KCl. (d) Modification by Coraboeuf and Breton [20]; with 50–100μ wire, floating microelectrodes may be used to study vertically hung hearts. Wire may be replaced by microcatheter previously bent by heating in vicinity of the flame. (Reprinted by permission of the American Association for the Advancement of Science[19] and Masson[20,21].)

Using the Woodbury and Brady technique, with silver wire 20 μ in diameter and 12 cm long and with microelectrodes of standard taper (between 1/10 and 1/5) and resistance (15 to 25 MΩ), we obtained the following results. For tips 16–20 mm long (weight around 0.1 mg) the time during which the electrode remained in the same fiber of a perfused rat (at 24°C) or guinea-pig (at 32°C) ventricle lay between 1 and 15 min (30 penetrations; mean, 4 min). With a tip length of 9–12 mm, the time reached values ranging between 3 and 28 min (18 penetrations; mean, 9 min). With a tip length of 6–7 mm, we obtained on a few occasions good penetrations lasting more than 90 min. These results agree well with original observations of Woodbury and Brady on the frog. It is possible in a simple way to modify this technique to study the electrical activity of vertically hung hearts: the tip of a microelectrode is connected to a larger silver wire (diameter 50–100 μ) horizontally bent at its lower end (Fig. 12.2d)[20]. This technique gives good results with perfused rat (24°C) and guinea-pig hearts.

An interesting improvement of the "floating" electrode technique was developed in Orsay by Benoit[21]. It consists of using, instead of a metallic wire, a fine polyethylene tube pulled by hand in a flame and filled with KCl (resistance about 1 MΩ) (Fig. 12.2c). Using a catheter 20 cm long with an external diameter at the tip of 120 μ (internal diameter about 50 μ), we obtained good penetrations in a guinea-pig ventricle, the electrode remaining in the same cell for 12, 16, and 26 min in the three best attempts.

Another interesting improvement in recording intracellular potentials from the myocardium in situ on large animals (dogs) was described recently[22]. It consists of a miniature microelectrode holder with a vertical movement and with a slightly convex base to which the heart may be sutured. With this system not only surface but also deep cells (0.5 mm) may be impaled at will.

12.2.3 The Double-Barreled Microelectrode

It is well known that microelectrodes may be assembled in a more or less elaborate manner. Extracellular two-channel electrodes were used in 1940 to study potential gradients in the brain[23], and intracellular double-barreled microelectrodes used for polarizing and stimulating single cells were described[24–26] and applied to polarization of Purkinje fibers[27]. Construction of this type of electrode may be realized in different ways: one can bind together with wire, a rubber band, or a cement, the two tubes and stick them together in a flame[13, 28] or more simply hold them freely together and more or less bend them during heating so that they stick. It is also possible to twist them in the flame before pulling (Fig. 12.3). Drawing may be done easily by hand or by a puller.

An electron micrograph of double barreled electrode is shown in Fig. 12.4; in the case of single electrodes the smallest external diameter of a barrel seems to be about 0.07 μ[29]. The common resistance determined from the voltage recorded by one barrel when a current is passed through the other (the tip being dipped in Ringer's solution) is generally between 10 and 175 kΩ in our experiments (mean value of 20 determinations: 52 kΩ). The capacity between barrels is around 20 $\mu\mu$F —the same value as already given by Coombs[25], who indicates for common (or coupling) resistance a value of about 200 kΩ.

It is also possible to pull microelectrodes with 3, 5, 7, and 12 channels [30,31]. Such electrodes are used in nervous tissue or neuromuscular physiology studies, and in microelectrophoresis experiments[28]. Double

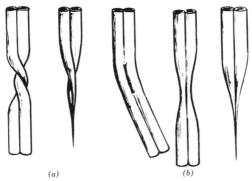

(a)   (b)

Fig. 12.3. Double-barreled microelectrodes are easily pulled after torsion (*a*) or after bending in the flame so that they stick (*b*).

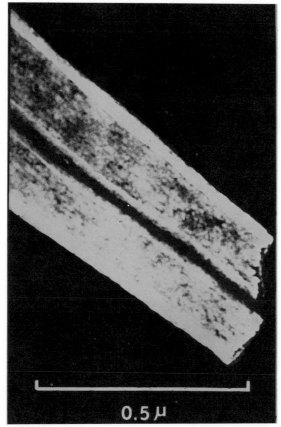

Fig. 12.4. Double-barreled microelectrode (electronmicrograph, courtesy of Professor J. Boistel).

or triple microelectrodes were recently used to measure intracellular pH of skeletal muscle[32] or as sodium-sensitive electrodes[33]. Coaxial microelectrodes were used by Tomita[34] and applied to measurement of transverse effect in myocardium by Tanaka and Sasaki[35].

Double microelectrodes may be used as fixed or as floating electrodes. In the latter case it is useful to remove KCl in each barrel at distance of 0.5–2 mm by suction, using a very fine catheter, then to wash carefully the blunt end of the electrode in bidistillated water to avoid shunt (Fig. 12.5). A small drop of oil or other equivalent substance can improve the insulation between the two barrels after the silver wires (or the poly- ethylene tubes) are inserted into each barrel. The mobility of the double- barreled floating microelectrodes is generally worse than that of single barreled electrodes for two reasons: (a) it is difficult to get tips so short and so light because of the necessity of emptying the upper part of the barrels at a reasonable length; and (b) the presence of two silver wires that must be carefully set wide apart makes the system more rigid. For these reasons mechanical artifacts are more important when strongly beating prepara- tions are concerned, and electrodes do not stay inside a fiber as long as a standard floating electrode can.

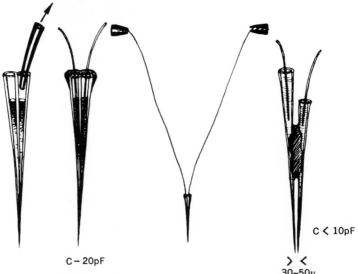

C – 20pF

C < 10pF

> <
30–50μ

Fig. 12.5. Left and middle: Double-barreled floating microelectrodes. Barrels are emptied by suction on 0·5–2 mm; when wires are inserted, blunt end of micro- capillaries is covered with oil or silicon. Wires are maintained wide apart to prevent occasional shunts (they may be replaced by microcatheters). Right: Cemented floating electrodes. (Gargouil, Tricoche, and Bernard[37]. Reprinted by permission of Gauthier-Villars.)

It is possible to group individual electrodes in different ways: For membrane resistance measurements it is useful to have electrode tips which are separated by a short and precisely determined distance; this may be achieved by fixing microelectrodes relative to each other with wax or cement[13]. Fixed electrodes[36] or floating electrodes (Fig. 12.5)[37] have been so used to polarize cardiac cells.

12.3 RESISTANCE MEASUREMENTS IN CARDIAC TISSUES

Since Weidman's experiments[38] showing that electrotonus in Purkinje fibers is accurately described by the cable equations, it is well established that this type of cardiac tissue shows membrane properties not fundamentally different from those of nerve or skeletal muscle. The problem is less clear in the case of auricular and ventricular myocardium, and we shall have to take into consideration the fact that this tissue must be considered as a functional syncytium in which the cablelike properties of the constituting fibers are masked by repeated branchings.

Membrane resistance measurements made by using intracellular microelectrodes may be achieved with one or two microelectrodes[39]. When one microelectrode is used, the resistance corresponding to the electrode and to the physiological medium is measured before (electrode + Ringer) and after (electrode + Ringer + cell) impalement of the fiber. The difference between the two values obtained corresponds to the input resistance of the cell and to a cytoplasmic component. When two intracellular microelectrodes are used, a current is applied by one of them through the membrane, and the resulting change in membrane potential is measured by the other. The current is generally applied during a rather long period (say 50–100 msec), so that the process of charging and discharging the membrane does not interfere appreciably with the measured potential change.

The change in membrane potential may be recorded at different distances from the polarizing electrode. Considering the cable properties of the fibers and the fact that the high resistance of the membrane acts as an obstacle to the current flow, it may be understood that an increase in membrane resistance produces a decrease of current in the vicinity of the electrode and at the same time makes that current spread further to cross the membrane. For this reason, a recording electrode localized far from the polarizing one will be more sensitive to resistance changes than if it were in its vicinity (see below). From this point of view, it is convenient to use distant electrodes. However, as the amplitude of electrotonic potentials decreasing rapidly with distance, it may become necessary to use close electrodes in tissues in which this decrease is the steepest, especially in ventricular tissue. Because of contraction-evoked difficulties in simultaneously maintaining two separate electrodes inside cells in strongly beating preparations, it may be of interest to use cemented or double-barreled floating electrodes or even single electrodes. However, results differ substantially when single or double electrodes are used, as will be indicated below.

If a current I_0 is applied through a microelectrode at the origin ($x = 0$) of a semi-infinite cable and if V_0 is the electrotonic potential produced at $x = 0$ by I_0, the polarization resistance or input resistance or effective resistance is given by

$$\frac{V_0}{I_0} = r_i \lambda = \sqrt{r_i r_m}, \tag{12.1}$$

$$\lambda = \sqrt{\frac{r_m}{r_i + r_e}} \simeq \sqrt{\frac{r_m}{r_i}}, \tag{12.2}$$

where r_i is the resistance of the core per unit length (Ω/cm); r_e is the resistance per unit length of the external fluid which can be neglected when $r_i \gg r_e$; r_m is the membrane resistance times unit length (Ω cm); and λ is the space constant. In this particular case the polarization resistance is known as the characteristic resistance, but more generally the polarization resistance depends upon the geometry of the fibers (particularly upon the type of termination and upon the number of branches). For example, in the case of an infinite linear cable impaled with a polarizing microelectrode, the polarization or effective resistance is one-half of the characteristic resistance (input resistance to a semi-infinite cable), as an infinite cable may be considered as two semi-infinite cables starting from the microelectrode. From r_i and r_m one can obtain the specific resistance of the myoplasm R_i (Ω cm) and the membrane dc resistance (resistance times unit area of membrane) R_m (Ω cm^2) from the following relations:

$$R_i = \pi a^2 r_i, \tag{12.3}$$

$$R_m = 2\pi a r_m, \tag{12.4}$$

where a is the radius of the fiber.

The measurement of membrane resistance with two microelectrodes is generally performed according to the technique used by Weidman [38,40], the current being delivered through a resistor of 50 MΩ or more to be independent of variations in electrode and membrane resistance. If a single electrode is used to pass current and to record voltage, the microelectrode is mounted in one arm of a bridge circuit[41]. The use of an external source of current is not indispensable; current may be supplied by the cell itself[42]. The microelectrode voltage clamp technique allowing measurements of membrane currents is applicable to "short" Purkinje fibers[43, 44], but its use remains difficult during the first milliseconds of the action potential. If depolarizing currents must be passed through a microelectrode, it seems preferable to use potassium citrate as a fillant instead of KCl[45].

When a cablelike fiber is polarized through a microelectrode (point polarization[46], the change in potential is not uniform along the fiber but decreases exponentially with distance — except in short cells or artificially shortened fibers (e.g., "short" Purkinje fibers[43]). As membrane conductances are voltage-dependent, they will be modified differently in the neighborhood of the electrode and far from it. It is thus possible to obtain an inward (sodium) current close to the electrode and an outward (potassium) current in a more distant area. Such opposite currents have a tendency to cancel effects. So, it will be necessary when triggering an excitation or a repolarization with a microelectrode to polarize the membrane well beyond the level required when the polarization is uniform.

12.3.1 Cablelike Structures (Purkinje Fibers)

Resting Membrane. In isolated Purkinje fibers (false tendon fibers), often deprived of branching over 1 cm or more, the drawback arising from the syncytial structure of cardiac tissue is reduced to a minimum. It is therefore possible to consider such preparations as cablelike fibers at least in species in which a characteristic Purkinje tissue is well developed (ungulates and, to a smaller extent, dogs, cats, or monkeys).

Weidmann[38] showed that electrotonic potentials can be recorded up to a distance of several millimeters. He found a space constant of 1.9 mm, a polarization resistance of about 0.5 MΩ, and an R_m value close of 2,000 Ω cm^2 (see Table 12.1 for these and other values). It may be noted that R_i (105 Ω cm) is about twice the resistivity of the Tyrode solution at 37°C (51 Ω cm). Vassale[47] finds a similar value for R_m at the same external potassium concentration (2.7 mM). R_m falls to about half its value when external potassium is doubled.

In Purkinje fibers it was shown[48–50] that the potential displacement produced by hyperpolarizing current pulses is slightly increased on replacing chloride with methylsulfate and is reduced with nitrate or iodide. The observed decrease in membrane conductance when chloride is replaced by an impermeant anion such as methylsulfate shows that chloride contributes only in a small part to the membrane conductance;

Table 12.1 Some Values of Electrical Constants of Cardiac Tissues Obtained with Intracellular Microelectrodes

Tissue	Species	τ_m (msec)	C_m (μF/cm²)	λ (mm)	R_{in} (kΩ)	R_m (Ω cm²)		Ref.
Purkinje fibers	Kid	19.5	12.4	1.9	478*	1,940	$[K]_0$ = 2.7 mM	[38]
	Calf, sheep		12.8†			1,680	$[K]_0$ = 2.7 mM	[47]
			2.4‡			850	$[K]_0$ = 5.4 mM	[47]
								[155]
	Sheep				69			[51]
					211			[51]
						2,200	"short" fibers	[52]
	Dog	12.9						[56]
Terminal Purkinje fibers	Dog	3.7		0.83	160			[56]
"P" cells	Rabbit			0.3–0.6	1068§			[87,122]
Atrial tissue	Frog	6.5		0.33		280		[71]
	Rat			0.13		220	after Ach	[71] [74]
Ventricular tissue	Frog				5,500	315‖		[36]
	Cat				6,400	489‖		[36]
	Frog				11,000			[107]

Species	τ_m	C_m	λ	R_{in}	R_m	Notes	Ref.
Toad	7.5				50		[80]
Mouse	4.2				38		[80]
			0.07		300	$x < 100\mu$	[35]
Rabbit			0.4–0.7		300	$x = 1$–1.5mm	[35]
			0.1–0.3 max		47§	"V" cells	[87,122]
					500–1,000		[147]
Rat					38.5§		[156]
					153¶		[156]
					1,450**		[156]
					14§§		[156]
Guinea pig					59§		[156]
					187¶		[156]
Chick embryo	9.6	20.0			7,000–12,000		[70]
Cultured	35.0			480††			[117]
heart cells				506‡‡			[119]

τ_m is the membrane time constant; C_m, membrane capacity; λ, space constant; R_{in}, input resistance; R_m, membrane dc resistance.
*Semicable (polarizing electrode close to a cut end).
†Measured from response to a current step.
‡Measured from the foot of the propagated action potential.
§Frank and Fuortes method.
§§Frank and Fuortes method, using the change in plateau amplitude in guinea pig.
¶V/I, double-barreled microelectrodes.
‖Estimated for a cell 150μ long, 12μ (frog), and 16μ (cat) in diameter.
**Single microelectrodes.
††Estimated for a cell 100μ long, 15μ wide, 5μ thick.
‡‡Estimated for a cell 200μ long, 15μ wide, 5μ thick.

235

in resting fibers, chloride ions could carry at the most 0.3 of the total membrane current. On papillary muscle, chloride removal results in an average decrease of effective conductance of about 11%.

In shortened Purkinje fibers the input resistance for hyperpolarizing currents increases by a factor of 3.7, $\frac{1}{2}$ h after ligation[51]. Clamping the voltage through a microelectrode in short Purkinje fibers, Deck and Trautwein[52] measured the membrane currents and calculated the resting conductances. They found a mean value of 0.46 mmho/cm² for resting g_K (corresponding to a membrane resistance R_m of 2.200 Ω cm²) and a mean value of 0.023 mmho/cm² for resting g_{Na}.

In the terminal Purkinje fibers the difficulty of obtaining theoretical electrotonic potential decays increases appreciably. The transition from the Purkinje fibers to myocardium is gradual[53,54], the terminal Purkinje fibers showing action potentials more or less intermediate between those recorded in false tendons and in ventricular muscle[55–57]. This transformation is accompanied by a progressive increase in the branching, and fibers become difficult or impossible to distinguish under direct microscopic observation. However, by moving the recording electrode in areas surrounding the current electrode, it is possible to find preferential pathways apparently composed of a number (8–10) of densely interconnected single fibers with a total diameter of 100–120μ, along which electrotonic potentials spread to a reasonable distance[56].

Except in the immediate vicinity of the tip of the polarizing electrode, the decline of electrotonic potential with distance seems to conform to the exponential decay. The mean value of λ found[56] is about half Weidmann's values. The characteristic resistance was 160 kΩ, a number apparently not very different from the theoretical value if the bundles are formed of about 10 small fibers.

Active or Depolarized Membrane. Weidmann[40] estimated the impedance changes during the course of action potential of Purkinje tissue of the kid by applying to the membrane through a microelectrode rectangular pulses of anodal current (duration, 40 msec; frequency, 12/sec) and by recording action potentials and electrotonic potentials by another microelectrode.

Considering the amplitudes V_1 and V_2 of two electrotonic potentials at two different times of the cardiac cycle (for instance, during the diastole and during the systole) and the corresponding membrane resistances R_1 and R_2, the equation may be written (according to Hodgkin and Huxley[58]):

$$\log_e V_1/V_2 = \log_e \sqrt{R_1/R_2} + \sqrt{(R_1/R_2)}\, l/\lambda, \qquad (12.5)$$

where l is the distance between microelectrodes.

It appears that during the spike the membrane resistance falls to about 1% of its early diastolic value and then increases progressively until the end of the plateau. The resistance comes back to the diastolic value 60–80 msec after the beginning of the response, and a value three times higher is reached just before the repolarization (Fig. 12.6a)[7]. Possible sources of errors affecting these measurements are listed by Weidmann

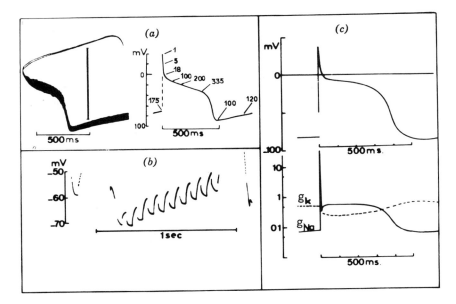

Fig. 12.6. (a) Left: Impedance changes in course of Purkinje fiber action potential appreciated from height of superimposed nonsynchronized potential changes caused by square pulses of polarizing anodal current. Vertical bar: 100 mV. (Redrawn from Weidmann[40]. With the permission of Professor Weidmann and the *J. Physiol.*) Right: Relative resistance values indicated for different phases of the action potential. (Weidmann[154]. Reprinted by permission of the *Ann. N.Y. Acad. Sci.*) (b) Diastolic impedance changes in rabbit sinus preparation. (Redrawn from Dudel and Trautwein[67]. With the permission of Springer-Verlag.) (c) Computed conductance changes during action potential of Purkinje fiber. (Noble [79]. Reprinted by permission of *Nature*.)

[40]: the fibers are not infinite in length and may have a diameter (50–100μ) which may become appreciable compared with λ when the value of this parameter decreases as is the case at the crest of the spike. It could be noticed also that if the pulses are long enough during the diastole (when the charging of the membrane capacity by the current pulse is almost complete), they may become too short during the plateau when the resistance increases appreciably.

The first measurements of the rectifier properties of Purkinje fibers, made by Weidmann[59], suggested that the permeability of the cardiac membrane to K ions does not increase significantly under the influence of long-lasting depolarizations. In fact, Carmeliet[49] (see also [60]), plotting the relative membrane resistance against membrane potential in sodium-free solution (that is to say, in absence of electrical response), observed that the resistance increased when the membrane was de-polarized at plateau values. (On the contrary, an increase in extracellular potassium concentration caused a drop in membrane resistance that may be interpreted as a rise of g_K [Fig. 12.7].) Similar experiments have been made by other workers[61,62]. For potential displacements of about 35 mV, the membrane resistance to depolarizing current is 6.7 times larger than that to hyperpolarizing current. These experiments and others[52] show that anomalous rectification is present in Purkinje fibers (Fig. 12.6c) but that during large depolarizations delayed rectifica-tion also appears. The threshold of the delayed rectification is sodium-dependent and increases when sodium is removed[63].

A positive chloride current seems to be responsible for the early rapid repolarization in Purkinje fibers[64], while a negative nonsodium current exists during the plateau[65].

On fibers arrested on the plateau level by lowering of the temperature [66], there does not seem to exist much resistance variation between dia-stole and plateau. Under the action of strophantin, membrane resistance increases except in extreme intoxication[67]. Resistance measurements were made by Reuter[68] on Purkinje fibers under the action of calcium and adrenaline. Adrenaline could increase the calcium influx when the membrane is depolarized.

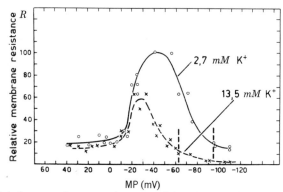

Fig. 12.7. Relative membrane resistance of Purkinje fiber measured for differ-ent membrane potentials in sodium-free solution with normal (2.7 mM) and high (13.5 mM) potassium concentration. (Redrawn from Carmeliet[49].)

12.3.2 Cardiac Syncytium

Atrial and Nodal Tissue. In the case of atrial or nodal tissue the variations of membrane resistance were measured during the slow diastolic depolarization characteristic of pacemaker cells in the sinus preparation of the rabbit's heart (Fig. 12.6*b*)[67, 69]. This depolarization is concomitant with (a) a progressive increase in resistance, as has already been demonstrated in Purkinje fibers by Weidmann[40], and (b) with a progressive decrease in potassium permeability. Similar results were obtained in cultured chick embryo cells beating spontaneously[70]. The variations of diastolic membrane resistance were also estimated or measured in normal conditions and during the action of acetylcholine or vagal inhibition in frog[71] and dog[72] and under the action of adrenaline in dog[73]. It is now well established that the vagal inhibition and the action of acetylcholine results in a decrease in diastolic membrane resistance because of an increase in potassium permeability.

Values for R_m and λ for frog auricle before and after application of acetylcholine[71] and for polarization resistance of rat atrial appendage [74] are indicated in Table 12.1. In this last tissue, λ is about $130\,\mu$ in the direction of the fibers and half this value perpendicular to fiber direction. It is greatly diminished by acetylcholine. As shown in Fig. 12.8, the steady-state voltage deflections produced in atrial tissue by current

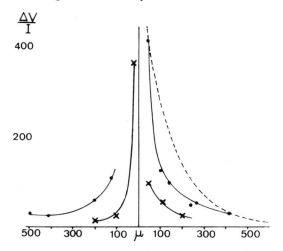

Fig. 12.8. Spatial decrement of voltage from polarizing microelectrode in rat atrial appendage. Recording electrode was moved parallel (filled circles) or perpendicular (crosses) to fiber direction. Abcissa: distance between polarizing and recording electrode; ordinate: potential change per unit current. For comparison, exponential is drawn, interrupted line, (Woodbury, and Crill[74]. Reprinted by permission of Professor Woodbury and Pergamon.)

pulses of 0.1 sec duration fall off quite rapidly with distance. The curve is not an exponential (an exponential curve passing by the first and last experimental points of right curve is indicated approximately by the interrupted line), but a Bessel function.

Recent work by Fozzard and Sleator[75] concerned the measurement of relative membrane resistance during the guinea-pig atrial action potential with two microelectrodes, one recording electrode of normal resistance (10–20 MΩ) and one current electrode of low resistance (1–5 MΩ) "sealed" on the surface 100–200 μ from the recording electrode. Assumptions made by the authors led them to discard the theoretical difficulties resulting from the geometry of the tissue and from the fact that the measurements concern slope conductances rather than chord conductances (see below). They assumed that the distance between polarizing and recording electrode is comprised between λ and 3λ, so that $R_m = kV^2$; that during the plateau of the action potential the membrane voltage is the same in any point of the preparation, making the longitudinal currents negligible; and that the plateau approximates a steady state voltage, so that $I_i = I_{Na} + I_K + I_{Cl} = 0$ and $1/R_m = g_m = g_K + g_{Na} + g_{Cl}$. The membrane resistance diminishes during the initial part of the action potential, then increases above its diastolic value (Fig. 12.9d). Estimations of membrane conductances indicate a fall in g_K during the "plateau" and a slow decrease in g_{Na} after the large initial increase. During the action of epinephrine, resistance increases during the plateau and this increase seems associated to a greater fall in g_K. Acetylcholine produces a progressive shortening of the action potential and a corresponding shortening of the time course of g_K and of g_{Na}, the relation between g_K and voltage appearing unchanged.

Ventricular Tissue. In the case of ventricular tissue theoretical difficulties arising from the geometry of the tissue are still greater than in the atrium. However, a first attempt was made in 1958 by Coraboeuf et al.[76] to determine membrane resistance variations appearing in guinea-pig ventricule at the beginning of the plateau. Because in this tissue the membrane potential is about the same at the top of the ascending phase of the action potential and during the early part of the plateau, it was of interest to determine if the potential was maintained at this stable value with or without any resistance change. Experiments were made with two microelectrodes separated by 50–100 μ.

Figure 12.9a shows that the amplitude of the electrotonic potential is quite small at the very top of the ascending phase (point B) but increases markedly 5–10 msec later (point A), although the membrane potential between these two points differs by less than 2 mV. During the plateau the resistance increases progressively but does not exceed (in this experiment) the diastolic value.

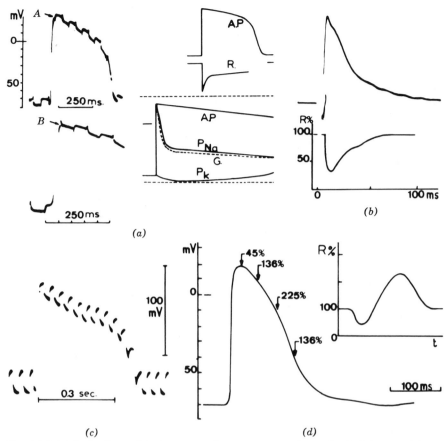

Fig. 12.9 (*a*) Left: Potential changes owing to square pulses of polarizing anodal current in isolated guinea-pig ventricle. Membrane resistance is small at top of action potential (*B*), high a few milliseconds later (*A*). Right: Hypothetical concomitant conductance changes (the later part of lower schema, corresponding to repolarization phase, has been suppressed). (Coraboeuf, Zacouto, Gargouil, and Laplaud[76]. Reprinted by permission of Gautheir-Villars.) (*b*) Membrane resistance changes during ventricular action potential of mouse. (Kuriyama, Goto, Maeno, Abe, and Ozaki[80]. Reprinted by permission of Professor Katsuki, and Igaku Shoin.) (*c*) Membrane resistance changes during action potential of sheep myocardium. (Weidmann[60]. Reprinted by permission of *Helv. Physiol. Pharmacol. Acta*.) (*d*) Calculated membrane resistance changes during action potential of guinea-pig atrium[75].

As in Purkinje fibers, the very short initial systolic decrease in resistance was considered as being related to an increase in sodium permeability. The development of the rapid increase in resistance following this initial systolic decrease was interpreted[76] as being the consequence of a possible decrease of potassium permeability during the plateau (anomalous rectification), associated with a more or less complete sodium inactivation. In fact, it was proposed (Fig. 12.9a) that the sodium conductance remains at an appreciable level all through the plateau. Similar propositions were made at about the same time, [77–78] and chiefly by Noble [79] who applied to the action potential of Purkinje fibers the Hodgkin-Huxley equations.

The fact that in the above experiments the resistance remained smaller during the plateau than during the diastole cannot be considered as a characteristic of ventricular tissue. While Kuriyama et al. [80] in the mouse ventricle and Weidman [60] in sheep myocardium indicate similar results (Fig. 12.9a and c) other investigators observe on the contrary that the ventricular resistance is higher during the plateau than during the diastole – as was already shown for Purkinje and atrial tissues. Results already obtained on turtle heart with extracellular electrodes [81–84] generally supported this view. According to other authors [85,86] the resistance at the crest of the rabbit ventricular action potential is not very different from the resting resistance – but a short time later the systolic resistance increases above the resting value.

More precisely, Johnson and Tille [87] described in rabbit right ventricular endocardial preparations two types of fibers, P and V fibers, the first having characteristics of conducting tissue and the second being the true ventricular myocardial fibers. The authors used in this investigation double-barreled microelectrodes and observed that in two-thirds of the investigated V fibers, the amplitude of the action potential remains constant for all values of hyperpolarizing current used, while in the remaining third this amplitude increases under the action of the current. The increase in amplitude of the action potential and the development of a more or less marked initial brief spike owes to the fact that the displacement of the ascending phase is less than that of the resting potential and of other parts of the response. When a spike does not develop and, consequently, when the amplitude of the response practically does not change, one observes (particularly with polarizing current of small intensity) an increase in the maximum rate of depolarization, creating a sort of "hump" on the ascending phase; all intermediaries may be observed between this hump and the spike. The fact that the hump may be limited to a part of the ascending phase shows that the decrease in resistance is itself very brief and terminates before the crest of the action potential is reached.

A few measurements have been made on guinea-pig and rat ventricular tissue maintained at 24°C also by using double-barreled microelectrodes [156]. Guinea pig and rat were chosen because of their very different cardiac activity. Figure 12.10 shows the effect of anodal polarization on a

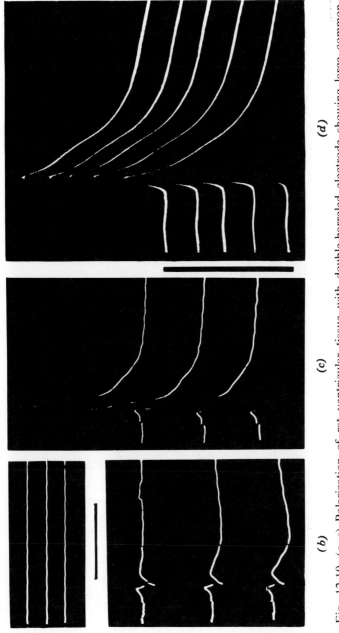

(a)

(b)

(c)

(d)

Fig. 12.10. (a–c) Polarization of rat ventricular tissue with double-barreled electrode showing large common (coupling) resistance. Top: tip of electrode is in surrounding medium; center; it is "sealed" on tissue surface; bottom; it is inside fiber. Notice the considerable artifacts when such an electrode is used. (d) Polarization similar to preceding one, but with low coupling resistance electrode. Successive potential steps correspond to polarizing currents of 1.5; 3.1; 5.0; and 7.8×10^{-7} A. Vertical bar: 100 mV; horizontal bar: for a–c, 100 msec; for d. —50 msec.

perfused rat heart. In such an experiment (shown in Fig. 12.10d), the displacement of the resting potential under the effect of a 10^{-7}-A applied current is of about 15 mV. The voltage displacement at the top of the action potential is less (12.8 mV rather than 15 mV), so that the amplitude of the action potential increases by 2.2 mV. The evolution of resistance during the plateau is rather difficult to follow; in the rat, after the very brief decrease observed during the spike, the resistance increases (frequently but not always above its resting value) from the beginning of the initial upper component of the plateau (Fig. 12.11). In the guinea-pig, after an initial decrease, the resistance increases more slowly than in the rat and may remain lower than its diastolic value during an appreciable part of the plateau (Figs. 12.12 and 12.13).

When a single microelectrode is used for polarizing and recording, the results are not fundamentally different from the above[88]; however, records were also obtained that showed a systolic increase in resistance above the diastolic value from the beginning of the plateau. This effect seemed all the more marked as the action potential was of a greater amplitude. On the other hand, it is clear that when the plateau of the cardiac ventricular action potential is lengthened under the action of a low-calcium high-sodium medium (Fig. 12.14)[89] or under the action of EDTA[88], the systolic membrane resistance increases progressively to values much higher than during diastole.

Values of the resting polarization resistance may be estimated as already indicated or by using the Frank and Fuortes method[41]. This method consists of dividing the change in amplitude of the action

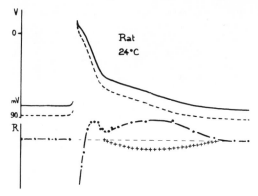

Fig. 12.11. Rat ventricular action potentials with (interrupted curve) and without (continuous) polarizing current, superimposed at crest of their spikes. Lower continuous curve (filled circles) represents corresponding relative resistance. In some experiments, a small decrease in resistance developed during the plateau (crosses).

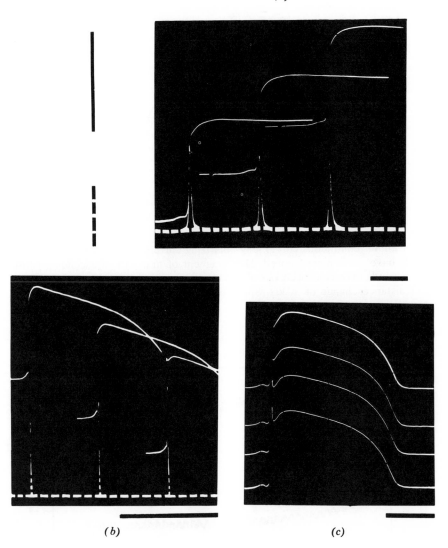

Fig. 12.12. Polarization of guinea-pig ventricular tissue with long steps of current of increasing (b and c) or decreasing (a) intensity. Each step corresponds to same values as in Fig. 12.10. In (a) no spike develops, but speed and height of ascending phase are increased by current; in last record of (b) and last three records of (c), spikes are visible. In (a) and (b), sharp vertical deflections above horizontal dotted lines are first derivative of potential. Vertical continuous bar: 100 mV; vertical interrupted bar: 100 V/sec; horizontal bars: (a) 10 msec, (b) and (c) 100 msec.

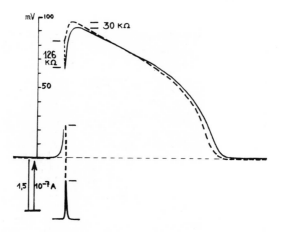

Fig. 12.13. Guinea-pig ventricular action potentials with (interrupted curve) and without (continuous curve) long polarizing current of 1.5×10^{-7}A. Diastolic levels have been superimposed. Displacement of first part of ascending phase corresponds to resistance change of 126 kΩ, displacement of upper part of plateau to resistance change of 30 kΩ. Speed of ascending phase increases from 60 to 150 V/sec.

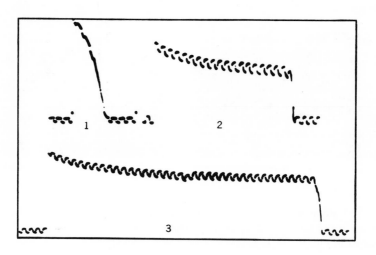

Fig. 12.14. Membrane resistance changes in course of guinea-pig ventricular action potential. (1) normal solution: (2 and 3) Na-rich, Ca and Mg-free solution. Two separate electrodes are used; the distance between them is different in 1, 2, and 3. (Garnier, Coraboeuf, and Paille[89]. Reprinted by permission of Masson.)

potential produced by the polarization by the current. It is based upon the assumption that the crest of the action potential is not displaced by the current because the membrane resistance at that time is supposedly extremely low. In such a case all the voltage change occurs because of the modification of the resting membrane resistance. It is clear that for this reason the value obtained is generally too low and must be considered as a lower limit of the true polarization resistance. Using this method, Johnson and Tille[90] found on the rabbit ventricle a mean resting polarization resistance of 45 kΩ in V fibers and of 106 kΩ in P fibers. Similar values of about 40 kΩ were obtained in rat ventricle (Table 12.1)[156]. In the guinea pig, the results are also similar, but one has to consider the fact — already observed by Johnson and Tille in the rabbit ventricle — that the amplitude of the action potential does not necessarily increase for small values of the current, or increases only very little. If the Frank and Fuortes method is used, it may be more difficult than in the rat to distinguish between "spike" and "plateau." The fact that in Coraboeuf and Vassort's observations the resistance at the beginning of the plateau remains often lower than during the diastole may lead to apparently very low values for diastolic polarization resistance if the change in the amplitude of the plateau is considered instead of the change in the amplitude of the ascending phase. Values of 14 kΩ were obtained in the first case, rather than 59 kΩ in the second. The second value is the most plausible, as the first component of the ascending phase is probably the only part of the guinea-pig ventricular action potential corresponding to the initial increase in permeability to sodium ions (see Figs. 12.12, 12.13 and 12.15). On the other hand, if we consider the diastolic displacement for a given current, mean values of 153 kΩ for rat and 187 kΩ for guinea pig were obtained by Coraboeuf and Vassort for the polarization resistance after substracting the coupling resistance (measured when the double-barreled electrode is dipped into Ringer's solution). Measured with a single electrode, the input resistance had a much higher value of 1.45 MΩ (mean of 4 determinations). In the mouse ventricle, Kuriyama et al.[80] reported for effective resistance values between 35 and 38 kΩ. However, Tarr and Sperelakis[36], using two cemented microelectrodes, obtained voltage-current relations with slopes of 6.4 MΩ for cat and 5.5 MΩ for frog in the case of substantial electrotonic interaction considered by these authors as corresponding to both electrodes being impaled in the same cell.

Johnson and Tille[86,90] did not find any detectable nonlinearity in the relation between polarizing current and displacement of the membrane potential — except for strong currents during the later part of the action potential. They concluded from this observation that the ionic

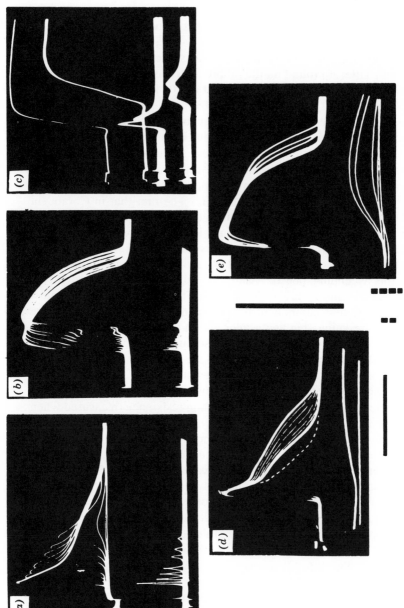

Fig. 12.15. Action of injection of TTX (upper) and Mn (lower) on ventricular action potentials of perfused rat (a, d) and guinea pig (b, c, e) hearts. Under action potentials are shown in a, b and c their first derivatives and in d and e the apex contraction. Horizontal bar; 100 msec for a and d, 200 msec for b and e, 40 msec for c. Vertical scales, continuous: 100 mV; interrupted: 2g (Coraboeuf, and Vassort[91]. Reprinted by permission of Gauthier-Villars.)

conductances are not voltage-dependent but only time-dependent. In fact this apparent linearity (also observed by other authors) may be the consequence of the geometry of the tissue, as we shall see below.

The uncertainties of interpretation affecting the ventricular resistance measurements and the need for more information concerning the observed differences between rat and guinea-pig ventricular action potentials led to an attempt—by comparing the action of supposed inhibitors of these permeabilities—to delineate the possible participation of ionic permeabilities in the development of the electrical response. We shall briefly report here the effects of tetrodotoxin (TTX) and of manganese ions[91].

The action of TTX is different in rat and guinea-pig ventricle. As shown in Fig. 12.15, TTX progressively suppresses the spike and all the initial part of the rat action potential, leaving the plateau unchanged; this suppression leads to a marked decrease in the amplitude of the response. In the guinea pig, on the other hand, the amplitude of the action potential is only slightly modified and the plateau remains unchanged in the range of doses used. However, the first component of the ascending phase, which frequently does not reach the top of the action potential in isolated guinea-pig heart, is greatly depressed by TTX. This observation suggests that the initial increase in permeability to sodium ions, responsible for the full development of the spike in rat ventricle, is limited in guinea pig to the early part of the ascending phase. Such a fact could explain earlier observations[92,93] according to which guinea-pig ventricular action potentials do not diminish as expected in low Na media.

As Hagiwara and Nakajima pointed out, Mn^{2+} has been considered as an inhibitor of calcium permeability[94]. In rat, Mn modifies very little the amplitude of the spike but it does decrease the amplitude and duration of the plateau. In guinea pig, it slows down markedly the second part of the ascending phase, thus decreasing the amplitude of the response while the first component of the ascending phase remains unaltered. It abolishes contraction in both species. This observation suggests, but does not demonstrate, a participation of calcium ions in the development of mammalian cardiac potential[91]. Similar conclusions were drawn [94,95] from experiments on frog ventricle. It will be recalled that the existence of a negative non-sodium current was observed in Purkinje fibers[64,65,96,97]. Such a current could be a Ca^{2+} current, but it seems to be independent of external calcium concentration[98]*.

*Since the writing of this chapter, it was shown that Mn^{2+} acts in cardiac tissue as an inhibitor of a "slow" channel allowing Na^+ and Ca^{2+} to penetrate into the cell (O. Rougier et al., *Compt. Rend. Acad. Sci.* **266**, 802 (1968)). With respect to calcium current, see also H. Reuter, *J. Physiol (London)*, **192**, 479 (1967).

12.3.3   Intercalated Discs

An important problem of cardiac physiology closely related to membrane resistance measurements is the problem of electrical connections between adjacent cells. Since the discovery by electronmicroscopists that the heart cannot be considered as an histological syncytium[99–101], the problem of electrical resistance of intercalated discs has been the subject of discussion.

Two opposite opinions have been supported: most physiologists [74,102–107] believe that transmission is brought about by local circuit currents. This opinion was strongly established on electrophysiological ground by Weidman[38], in the case of Purkinje fibers; he showed that if the membrane dc resistance (2,000 Ω cm^2) was similar to that of skeletal muscle, the specific resistance of myoplasm was relatively low (105 Ω cm). This could not be the case if individual cells making up a fiber were separated by membranes of resistance high enough to prevent transmission by local currents. According to Sperelakis et al. (see below), the transmission is not electrical but owes to some "junctional" phenomenon from cell to cell – probably of chemical origin. Their opinion was strengthened by histochemical observations of different enzymatic reactions localized in the region of intercalated discs[108,109], particularly of a cholinesterase activity that was inhibited by pretreatment with eserine [110].

Sperelakis et al.[111] measured in isolated frog ventricle the dc resistance between two microelectrodes, one intracellular and the other extracellular, by passing through them constant current pulses of approximately $0.5 . 10^{-9}$ A. The microelectrodes were selected for high impedance ($10-25$ MΩ) and low tip potentials. The data for the resistance of the cells were obtained from usual downward micropunctures and also from micropunctures made during withdrawal of the electrode through cells previously impaled. In the first case very high resistances were obtained (12.4 ± 0.5 MΩ); in the case of previously punctured cells – with consequently more or less injured membranes – values were much lower (3.7 ± 0.3 MΩ), although such cells exhibited stable normal resting potentials with regular action potentials. This observation led Sperelakis et al. to consider that the criteria of normal resting and action potential could be inadequate criteria for assessing good puncture.

The very high resistance measured by these workers when a penetration was made in an uninjured cell was interpreted by them as indicating that cardiac cells are functionally small units and do not form a functional syncytium. When both microelectrodes were impaled in cells and when the distance between the electrodes was great enough (0.5–1.0 mm) to make certain that two different cells had been penetrated, the average measured dc resistance was 24.8 MΩ, which is twice the value for single cells. This observation appeared as another argument against the concept of a functional cardiac syncytium. It has been shown, however, that these results can also be explained by an increase of the resistance of the microelectrodes owing to the resistivity of the cytoplasm around the tip of the microelectrode[42].

Hoshiko et al.[112] and Sperelakis and Hoshiko[113] measured impedance in cat ventricular trabeculae and papillary muscle and, for comparison, in frog sartorius muscle, using ac Wheatstone bridge at frequencies ranging between 10 and 10,000 cps. Considering that there are two parallel pathways for applied currents, intracellular pathway (ip) and extracellular pathway (ep), they found that the ratio of resistances of the two pathways rip/rep is about 20 (5.560 Ω cm/281 Ω cm) in heart and only 2.4 (454 Ω cm/188 Ω cm) in skeletal muscle. They found also that the impedance is independent of frequency in skeletal muscle but decreases at high frequencies in cardiac muscle. They interpreted this fact and the very high resistance of the cell pathway in cardiac muscle as suggesting the presence in this tissue of high resistance $-$ high capacitance transverse membranes.

Sperelakis et al.[114] also observed a reversible blocking of conduction in hypertonic solution (three times isotonic sucrose), while Sperelakis [115] showed a dissociation between conduction and direct response to longitudinal electrical fields in cat papillary muscle soaked in isotonic sucrose solutions. Impairment of conduction and prepotentials, considered as evidence in favor of junctional transmission, was observed in frog ventricular strips submitted to calcium-free solution (used to reduce mechanical activity and facilitate micropuncture[116]; or in trypsin-dispersed chick heart cells cultured in vitro[117]. From the slope of the average voltage-current curve, the mean input resistance for cultured chick heart cells was 7 MΩ (value corresponding to about 500 Ω cm² for a cell 200 μ long, 15 μ large, and 5 μ thick)[117]. In cat and frog cardiac muscle similar results (weak electrotonic interaction between contiguous cells $-$ cell resistance of 6.4 MΩ for cat and 5.5 MΩ for frog) were reported[36].

In cultured chick embryo heart cells treated with trypsin it is possible to obtain different patterns of organization: isolated single cells and clusters of cells such as strands, sheets, or multilayered preparations [117–119]. The same technique of dispersion by trypsin has been used in young rat hearts[120]. In chick embryo heart isolated single cells are usually damaged by the microelectrode and show low resting and action potentials[117,119]. Cardiac cells beating at an independent rate become synchronous upon physical contact[120], but such a synchronism does not appear if two independently beating cells are pushed together in close contact for a few minutes[70].

According to several authors[117,119], in a cluster of cells the change in membrane voltage produced by polarizing current seems to be independent of the distance to the polarizing electrode. For the first group, this independence is found (for interelectrode distance of 100 μ) even when visible cell boundaries separate the polarizing and recording electrodes; for the second group, electrical interactions, frequently absent or weak, seem—when present for great interelectrode distances (150–400 μ)—to be the consequence of the presence between the electrodes of unusually large cells. Thus Crill et al.[119] conclude that the few cells of a single cluster are in a highly effective electrical syncytium, while Lehmkuhl and Sperelakis[117] consider rather that the junctional membranes are of high resistance.

Other authors have produced experimental arguments leading to conclusions opposite to those of Sperelakis and his group. To determine whether the intercalated discs offer a high or a low resistance to the movement of ions, Weidman[60,102–105] studied the repartition of radioactivity with respect to distance in one half of a bundle of parallel fibers excised from a sheep ventricle—the other half being loaded by ^{42}K. In such an experiment a plot of radioactivity against distance gives a value of the space constant λ of the fibers. The value found averaged 1.55 mm; that is to say, about 13 times the length of a single cell, estimated at 125 μ (80 discs per centimeter of bundle length).

Figure 12.16 shows the difference between steady-state distribution of ^{42}K and ^{82}Br within a bundle loaded with K*Br*, radioactive bromide ion being used as a (nonideal) extracellular tracer. In three experiments λ_{Br} was of the order of 0.5 mm, that is, three times less than λ_K. The apparent diffusion coefficient D is related to λ and to the time constant τ for conventional uptake or loss[105] of ^{42}K by (12.6):

$$\lambda^2 = D\tau. \tag{12.6}$$

From D, and from the number of discs per centimeter, the permeability of the disc, the potassium resistance may be calculated; it was found to be 3 Ω cm^2.

This value concerns fibers in which intracellular mixing of K ions is supposed to be perfect and in which all electrical changes across the disc are carried by K. If these conditions do not hold, the electrical resistance of discs would be still lower. If, for example, it is assumed that there is no intracellular mixing owing to convection, two-thirds of the above value might be attributed to the longitudinal resistance between two discs,

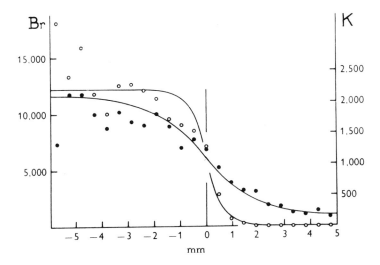

Fig. 12.16. Steady-state distribution of ^{42}K (filled circles) and ^{82}Br (open circles) within bundle of sheep myocardium exposed to active solution (left) and washed by inactive solution (right). Ordinates are counts/min. Theoretical curves are drawn with $\lambda = 1.95$ mm for ^{42}K and $\lambda = 0.45$ mm for ^{82}Br. (Weidmann[105]. Reprinted by permission of Professor Weidmann and the J. Physiol.)

only $1\,\Omega$ cm^2 being caused by the disc itself. From several reasons given[105], it seems obvious that radioactive potassium really moves longitudinally through the cells and not within the interspaces. The above results agree with previous observations by Kavaler[121] on the same type of preparation, which showed that electrotonic potentials produced by large extracellular electrodes decreased with a space constant corresponding to several times the cell length. Similar results were described by Woodbury[12] on rat auricle polarized by a microelectrode.

Tanaka and Sasaki[35], measuring the electrotonic spread in mouse ventricular myocardium, never obtained a discontinuous curve—except for a very rapid potential decay in the close vicinity of the electrode. Such a decay can be explained, as we shall see below, by an "origin" distorsion or "transverse" effect and cannot be related to a large resistance of the intercalated discs (Fig. 12.17).

It must also be noted that the coexistence of strong and weak electrotonic interactions in the same preparation of mammalian ventricular tissue (a phenomenon observed by Tarr and Sperelakis[36]) was also described by Tille[122], who explained it by positing the existence of two types of cells (P and V) and also by the fact that the true ventricular fibers (V) do not form a single freely interconnected syncytium but rather

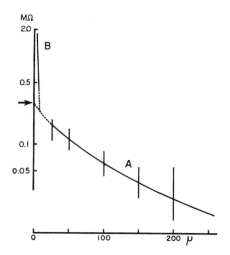

Fig. 12.17. Electrotonic spread in mouse ventricular myocardium. Curve A is obtained with two separate microelectrodes inserted on line parallel to fiber direction. Curve *B* is obtained with coaxial intracellular microelectrodes and corresponds to "origin" or "transverse" effect. Effective resistance can be obtained graphically by extrapolation of curve *A* (arrow). (Slightly modified from Tanaka and Sasaki[35]. With the permission of Professor Tanaka and the Rockefeller University Press.)

constitute small bundles or network systems with few interconnections between the systems. The existence of such a structure would be consistent with a recent concept based upon quite different arguments[123].

The opinion that the intercalated discs may possess regions of low electrical resistance is strengthened by the fact that the membrane of adjacent cells is different from the ordinary sarcolemma: the basal component is absent and one observes specialized regions (*S* regions) as well as close contact membrane regions[124], tight junctions[125], quintuple-layered membrane junction[126], or nexus[127]. Between elements of tight junctions no space exists for extracellular medium. It is interesting to note that in mouse cardiac muscle the fine structure of specialized junctions is characterized by the fact that the opposing cell surfaces are periodically interconnected by very thin opaque bridges with interspaces of about 125 Å[124,128]. In the case of certain tight junctions, Muir[129] describes between the adjacent membranes in low-calcium medium a row of dots with a periodicity of about 90 Å; however, such a periodicity does not appear in electronmicrographs of tight junction when both membranes are in perfect contact[127].

These facts may be related to observations showing that in different tissues, for example, in salivary gland cells or Malpighian tubular cells, true intercellular communications do exist—probably at the level of "septate junctions"[130]. In such regions the membranes are joined periodically by electron-opaque bridges separated by 150 Å intervals. Although transverse densities may represent overlap artifacts[131], it is not impossible, according to Loewenstein[130], that such bridges are water channels sheathed in a circumferential diffusion barrier—for example, a liquid-crystalline structure of phospholipids oriented in hexagonal arrays of cylinders filled with water[132, 133]. The fact is that septate junctions are permeable not only to ions but also to molecules as large as serum albumin with a molecular weight of 69,000. In other tissues, such as urinary bladder and liver, the cells show membrane junctions of the type termed *zona occluden*[134], which are probably functionally similar to the septate junctions. In the case of salivary glands the membrane specific resistance of septate junctions was determined by passing currents through a microelectrode and by recording the resulting membrane voltages at varying distances giving a junctional membrane resistivity lower than $10 \, \Omega \, \text{cm}^2$ and more probably between 0.1 and $1 \, \Omega \, \text{cm}^2$. The apparent similarity of structure between cardiac specialized or tight junctions and low-resistivity junctional membranes of unexcitable cells makes it probable that intercalated discs are a region of electrical continuity between adjacent cardiac cells.

Such similarity extends to the sensitivity to uncoupling agents such as calcium removal, trypsin, or hypertonic solutions; all three suppress the electrical coupling in unexcitable cells. Absence of calcium, when prolonged enough, also impairs cardiac conduction and produces physical dissociation of heart cells and tight junctions[129]. Hypertonic sucrose solutions are also able to separate (reversibly) cardiac cell membranes[127], with concomitant suppression of electrical coupling between cells. On the other hand, it may be noted that calcium is necessary to seal the membrane, both in cardiac[135] and epithelial cells[130]. In the latter type of cells it was demonstrated that extracellular calcium is able, upon coming into contact with a just-dissociated junctional area, to increase considerably the electrical resistance so that in a few minutes the cell behaves like a highly resistant single unit. Such facts could probably be of help in interpreting some results obtained by Sperelakis and his group.

12.3.4 Slope Resistance and Chord Resistance

The main interest in obtaining membrane resistance measurements would be to get information on ionic membrane permeabilities, that is, on chord conductances as they are considered in the Hodgkin-Huxley theory[139, 140]. In fact, information obtained dealt essentially with slope conductances (or resistances).

If we consider the current-potential characteristic of a nonlinear conducting system in a steady or quasi steady state and with small changes of current and potential, the ratio of these small changes represents a resistance $r = \partial E/\partial I$. If these changes are made slowly, we have the condition

$$r \to r_\infty = \partial E/\partial I, \qquad (12.7)$$

where $r \to \infty$ corresponding to an indefinitely long time. The symbol r_∞, given by the slope of the steady state characteristic, is the slope resistance (or zero frequency resistance). If the changes are made very rapidly, in a time shorter than that for the onset of non-linearity,

$$r \to r_0 = (E - E_{eq})/I \qquad (12.8)$$

where r_0 is the chord resistance or infinite frequency resistance[136]. The relationship between slope conductance (G) and chord conductance (g) can be obtained by differentiating (12.9) with respect to E, giving I as the sum of the individual ionic currents[12] — essentially sodium and potassium currents:

$$I_i = g_{Na}(E - E_{Na}) + g_K(E - E_K) \qquad (12.9)$$

$$G = \frac{\partial I_i}{\partial E} = g_{Na} + (E - E_{Na})\frac{\partial g_{Na}}{\partial E} + g_K + (E - E_K)\frac{\partial g_K}{\partial E}. \qquad (12.10)$$

This equation shows that it is possible to defined a slope conductance for each ion, for instance, sodium:

$$G_{Na} = g_{Na} + (E - E_{Na})\frac{\partial g_{Na}}{\partial E}. \qquad (12.11)$$

Accordingly, the slope conductance for a given ion would be equal to the corresponding chord conductance only if it is not voltage-dependent.

Figure 12.18, as redrawn[136], makes more visible the difference between chord and slope resistance in the case of the peak-inward sodium current. Chord resistances are always positive. Slope resistance is negative in the case of the figure. If the change in voltage is measured a given length of time after the application of the current and if the voltage-dependent conductance changes require about the same time to reach completion, it is clear that the measurements concern slope and not chord conductance. Woodbury[12] considers that this is precisely the case in experiments such as those made by Weidmann[40], in which the membrane is polarized through one microelectrode by durable current pulses (several time constants long), the resultant change in potential being measured at the end of the current pulse.

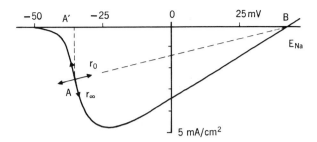

Fig. 12.18. Difference between slope resistance and chord resistance. Peak inward current. Sufficiently rapid changes of potential and current are along the zero time or infinite frequency line r_0 (chord resistance). Slow small changes are along steady characteristic giving zero frequency or slope resistance r_∞, which is negative in this example. Potential difference $A'B$ represents Na driving force of A. (Redrawn from Cole and Moore[136]. With the permission of Professor Cole and the Rockefeller University Press.)

Computations of cardiac action potentials by Noble[137] show clearly the limitations of the method of resistance measurements in throwing light upon the underlying ionic chord conductance changes. The solution of the modified Hodgkin-Huxley equations leads to action potentials closely resembling the ones recorded experimentally in the pacemaker region of Purkinje fiber — except for the speed of the ascending phase, which is slower than normal.

The time course of the computed conductance changes (Fig. 12.19a) shows a very important initial rise in g_{Na} during the spike of the action potential, which is then followed by a plateau. The value of g_{Na} during the plateau is about eight times the lowest diastolic value. Immediately after the beginning of the ascending phase, g_{K1} falls (anomalous rectification). This early effect is followed by a small and very slow rise in g_{K2} (delayed rectification), so that g_K, after a rapid initial decrease, increases slowly — remaining, however, under its diastolic value until the end of the plateau (at that time, the increase in membrane potential results in immediate increase in g_{K1}, followed by a slow fall in g_{K2}). The curve corresponding to the total membrane conductance ($g_{Na} + g_K$) has been drawn in Fig. 19a to show what sort of conductance changes should be recorded if the experimental measurements had concerned chord conductances instead of slope conductances (or resistances).

To imitate the experimental records[40], tensions produced by sinusoidal currents were superimposed on a computed action potential [137]. Figure 12.19b shows the approximative envelope of such sinusoidal potential variations and the rough corresponding slope conductance

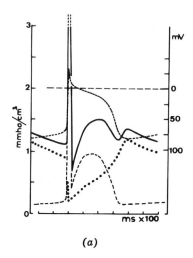

(a)

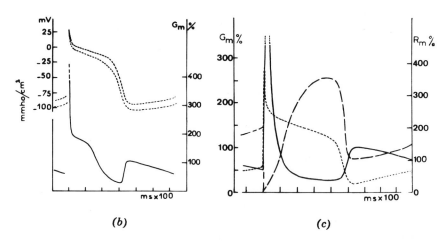

(b)   (c)

Fig. 12.19. (a) Computed action potential (dotted curve) and time course of conductance changes redrawn on linear scale (after Noble[137]. With the permission of Professor Noble and the *J. Physiol.*) Interrupted curve, g_{Na}; filled circles, g_K; continuous curve: $g_{Na} + g_K$, corresponding to total chord conductance. (b) Dotted curves: envelope of potential variations produced during computed cardiac cycle by sinusoidal currents (redrawn from Noble[137]. With the permission of Professor Noble and the *J. Physiol.*) Continuous curve: approximate corresponding impedance changes. (c) Action potential (dotted curve), relative membrane resistance (interrupted curve), and corresponding relative conductance (continuous curve); (redrawn from Weidmann[7]. With the permission of Hans Huber.)

variations. Important differences exist between this curve and the preceding one: the fall below the diastolic value appearing at the end of the spike is no longer visible and the progressive increase during the plateau is replaced by a progressive decrease. As already stated, the reason for such differences is that the factors determining the amplitude of the voltage waves are the slope and not the chord conductances. During the plateau, and as the repolarization proceeds, the Na slope conductance falls more markedly than the K slope conductance rises — which leads to a progressive increases in membrane impedance. If the duration of the current pulses or waves delivered by the polarizing microelectrode is long compared to the Na time constants, but short compared to the K time constants, adequate information can be obtained from the steady-state Na current-voltage relation and from instantaneous K current-voltage relations at different stages during the action potential[137].

The calculated conductance curve of Fig. 12.19b imitates rather closely the experimental curve drawn from Weidmann's results (Fig. 12.19c), except for one minor difference: the part of the plateau during which the conductance remains greater than during the diastole is longer in the computed response than in the experimental record.

12.3.5 Influence of Syncytial Structure on Membrane Resistance Measurements

Woodbury and Crill[74] consider that the simplest model able to account for electrotonic spread in atrium tissue is a single planar cell of practically infinite extent, bounded by two parallel membranes. In this case the decrement of potential with distance from a polarizing microelectrode would be described by a zero-order Bessel function of the second kind with imaginary argument. The rapid falling off of the curve with distance near the origin is explained by the rapid increase of the membrane area available to current penetration. This model cannot, however, be considered as a strict representation of auricular tissue. It is clear, for example, that 1 cm² of atrium contains a much greater area of membranes than 1 cm² of sheet surface, owing to the cylindrical shape of cells arranged in close successive layers.

Using the same model, Noble[138] studied the alterations introduced in the current-voltage determinations of cardiac muscle by the geometry of this tissue in the case of point polarization. In the instance of a membrane having a linear voltage-current relation the voltage change produced in the neighborhood of a polarizing electrode increases only by 30% when the membrane resistance increases by 400%. This shows that in the case of a sheet membrane, voltage measurement does not constitute a very sensitive means of delineating membrane resistance changes, especially when the voltage recording electrode is very close to the polarizing one. It has been established in the case of nerve tissue[139, 140] and Purkinje fibers[7, 40, 66] that the membrane presents a nonlinear current-voltage relation. If such a nonlinearity exists also in cardiac syncytium, electrical polarization will produce nonuniform and more limited changes in membrane resistance, and it is a priori possible that the method would fail to permit detection even of large nonlinearities in the membrane current-voltage relation. This could explain Johnson and Tille's[86,87,90] results, which showed linear polarizing current-voltage relations.

Starting from equations previously developed[79,137], Noble[138] considers the case of uniform polarization of the membrane and determines the membrane current-voltage relations described by:

$$I_{Na} = g_{Na}(E_m - E_{Na}) \qquad (12.12)$$

$$I_K = (g_{K1} + g_{K2})(E_m - E_K) \qquad (12.13)$$

where g_{Na} is numerically estimated taking into account Hodgkin and Huxley's results on squid axon and Weidmann's measurements on Purkinje fiber[141]. According to Fitzhugh[142] and to George and Johnson[143] an increase in the duration of the action potential may be explained by greatly slowing the rise in potassium permeability. However, an anomalous rectification is made necessary by the high resistance during the plateau: g_{K1} is described by a purely empirical equation and g_{K2} by the Hodgkin and Huxley[139,140] equation with a reduced maximal value and with rate constants divided by 100 in order to take into account Hutter and Noble's[61] results. Figure 12.20a shows such computed membrane current-voltage relation (at 100 and 200 msec from the beginning of the action potential). Both curves show a region of negative slope conductance.

Noble then considers the case of a point polarization (sheet membrane). He computes the variations with distance of the voltage displacement V and the concomitant variations of the intracellular current I_i and of the membrane current I_m given by the above equations ($I_m = I_{Na} + I_K$). At the point where the polarizing current I_0 is delivered, that is, at $r = 0$, $I_i = I_0$. Figure 12.20b shows the computed variations in V, I_m, and I_i at 100 msec and 200 msec for the same hyperpolarizing current supposed to be delivered by the microelectrode acting as a cathode. It may be seen that at both instants during the plateau, and for short distances around the microelectrode, the steady-state voltage deflection V decreases very rapidly and then, for larger distances, much more slowly.

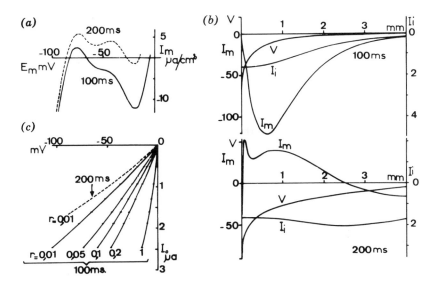

Fig. 12.20. (*a*) Computed membrane current-voltage relations (uniform polarization) at 100 and 200 msec. (*b*) Computed relations in V, I_m, and I_i at 100 and 200 msec. At $r = 0$, $I_i = I_0$ (total polarizing current). (*c*) Computed relations between total polarizing current and membrane voltage displacement for different distances r separating polarizing and recording electrodes. (Redrawn from Noble[138]. With the permission of Professor Noble and the Rockefeller University Press.)

From the figure it is obvious that the applied current-voltage relations will considerably differ from the membrane current-voltage relations, as the applied current has the same value and direction at 100 and 200 msec, while I_m changes its direction between these two instants in the vicinity of the electrode ($r < 2.5$ mm). Figure 12.20*c* shows that the relations between applied current and voltage displacement (at different distances away from the polarizing electrode) are in fact almost linear. This important theoretical study by Noble shows that one cannot hope to detect nonlinearities of the cardiac membrane by using microelectrode polarization if the cardiac syncytium may be assimilated to a sheet membrane.

Such an effect owes to a redistribution of the current. Let us suppose that the membrane resistance increases as a consequence of the polarization produced by the applied current. This increase will be large only in the immediate vicinity of the electrode, that is, in a very small area where the polarization is by far the greatest. As polarization decreases quite rapidly with distance, all the sheet around this small area will have an almost unchanged resistance; thus the greatest part of the current will cross the membrane far from the electrode. Consequently the potential difference across the membrane near the electrode will be much smaller than it would have been if the current had been divided equally among all elements of the membrane, each of them being submitted to the same increase in resistance. George[144], using Hodgkin-Huxley equations modified according to Noble also obtained nonlinear voltage-current relations at the plateau potential in the case of a cable, and linear relations when the current is injected at the center of a sheet.

Studies concerning types of cardiac syncytium models different from the sheet membrane can be found in other published reports[35,145–147].

Berkinblit and co-workers, in their mathematical study of different types of syncytia, have calculated the input resistance of their models as a function of l/λ, in which l is the distance between two adjacent nodes of branching. When l is large with respect to the space constant ($l/\lambda \geqslant 2$), the value of the input resistance becomes independent of l and depends only on the number, n, of branches emerging from each node. It tends to l/n of the characteristic resistance; any syncytium then behaves as n infinite cables emerging from one point. For smaller and smaller values of l, the input resistance decreases rapidly (Fig. 12.21) because the

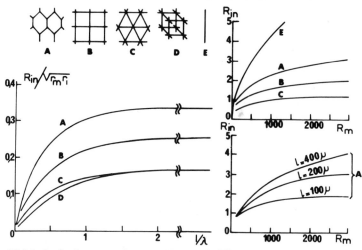

Fig. 12.21. Left: Input resistance (R_{inp}) of different syncytia, A, B, C and D (schematically drawn in upper part of Figure), as a function of the ratio $1/\lambda$. Right: upper portion shows input resistance of syncytia A, B and C and of linear cable E to membrane resistance (Ω cm^2). Lower portion shows input resistance of syncytium A to membrane resistance for different distances between nodes of branchings. (Drawn after Berlkinblit, Kovalev, Smolyaninov, and Chailakhyan [146]. With the permission of Professor Berlkinblit and Pergamon.)

current applied to a node may flow out into the external medium not only through the membranes of the elements branching off the given node but also through the membranes of the elements branching off the neighboring nodes, as they then become close enough to the polarized node to be largely accessible to the current.

An important point concerns the extension of the above considerations to a three-dimensional syncytium. As the resistance of the external medium may be neglected and as the input resistance is essentially dependent upon the number of branches emerging from a node, it seems logical to assume that the spatial arrangement of these branches does not influence the resistance. Thus the input resistances of a plane and of a three-dimensional syncytium having the same number of branches should be similar. This is, for instance, the case of a plane triangular and of a spatial cubic syncytium (six branches coming from each node in both cases). For the different syncytium models, Fig. 12.21 shows the influence of the membrane resistance on the input resistance.

Tanaka and Sasaki[35] also studied the electrical properties of lattice (square) and established the theoretical distribution of the electrotonic potential along the axis of one of its constituent fibers (diameter, 10μ; internodal distance, 100μ; λ, 0.5 mm). As already seen in the case of a sheet membrane, the electrotonic potential falls off rapidly in comparison with the exponential decay (zero-order modified Bessel function of the second kind), but for relatively large values of x the decay may be regarded as exponential and λ may be determined. With the values for fiber diameter and internodal distance indicated above, the model is, of course, much too loose compared with cardiac tissue, and the relative area of extracellular space is much too large. A better approximation is obtained if the lattice is folded so that the square meshes become very slender rhombi with adjacent sides almost touching, such a disposition being able to account for the observed directional differences of electrotonic spread in the myocardium.

In the close vicinity of the polarizing electrode, Tanaka and Sasaki described another effect that is superimposed on the "lattice" effect and corresponds to a still more rapid decay of potential (Fig. 12.17). They measured it with coaxial microelectrodes in the transverse direction. This "origin" or "transverse" effect [148, 149] may owe to the fact that around the tip of the microelectrode the electrotonic potential does not spread only longitudinally but also invades the surface of the fiber in all directions—as in the case of a sheet membrane that would be bent to form a cylinder. Because of the transverse effect, the apparent input resistance may be very high (several MΩ); but the effective resistance, obtained graphically by extrapolation of the curve corresponding to the electrotonic spread in regions beyond $50\,\mu$ from the polarizing electrode, is much lower (a few hundred kilohms).

It may be indicated here that another effect contributes to increase the "effective" cell resistance. This effect, observed when a single electrode is used instead of two, was described by Schanne, Kawata, and their co-workers[42,150,151], who obtained values around 3 MΩ with a single electrode instead of 0.2–0.4 MΩ with two separate electrodes[152, 153]. This difference may be explained by the presence of an additional resistance between the tip of the microelectrode and the cable structure of the fiber. Such a resistance, situated outside the tip of the microelectrode could correspond to the change in the convergence effect of the current toward the tip when the electrode passes from the Ringer's solution to the myoplasm.

Johnson and Sommer[147] consider a model in which a fiber is connected to n other fibers of infinite length, the polarizing microelectrode being impaled at a distance L from the branching. The relationship between the input resistance and the membrane resistance is shown in Fig. 12.22. If the membrane resistance is low, the effect of branching is negligible because λ is small in respect to L and the input resistance approaches that of an unterminated fiber of infinite length (R_c). If the membrane resistance is high, the input resistance approaches R_c/n. If L is negligible in respect to λ and if n^2 is large compared to 1, the input resistance R_{inp} may be approximately:

$$R_{\text{inp}} = \frac{\sqrt{r_i\,r_m}}{n} + r_i L = \frac{R_c}{n} + r_i L. \qquad (12.14)$$

One may notice that in this expression the second term is independent of the membrane resistance r_m. Choosing the same values as those used by Johnson and Sommer[147] (average fiber diameter of $10\,\mu$, core resistivity of $150\,\Omega$ cm, membrane resistance of $1,000\,\Omega$ cm^2, and $n = 5$), it is possible to calculate the approximate value of both components of the input resistance:

$$R_{\text{inp}} \simeq 1.6 + 0.8 \text{ M}\Omega. \qquad (12.15)$$

This shows that the participation to the input resistance of a component independent of r_m is not negligible; it attains one-third of the input resistance value. The electrotonic potential decay [Mazet, this laboratory, unpublished] is exponential beyond the point of branching but not between the microelectrode and this point; this constitutes a reflection phenomenon that is caused by the fact that the resistance of the whole of the branches is different from the characteristic resistance of the fiber.

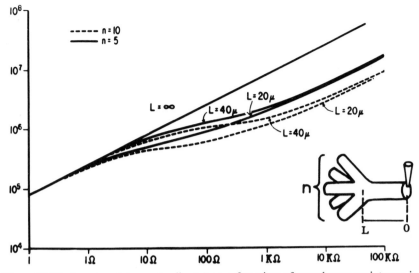

Fig. 12.22. Input resistance (ordinate) as a function of membrane resistance in Ω cm^2 (abcissa) for polyfurcated cable. (After Johnson and Sommer[147]. With the permission of Professor Johnson and the Rockefeller University Press.)

Mazet also studied a more complex model formed of fibers dividing into n similar branches at regular intervals d, the distance between the microelectrode (point A) and the first branching (point B) being also d. The resistance R_0 of one of the n branches at B (if d is small compared with λ) is approximately

$$R_0 \simeq \frac{n}{n-1} \cdot \frac{R_c d}{\lambda} = \frac{n}{n-1} r_i d. \qquad (12.16)$$

For $n = 5$ and $d/\lambda = 0.1$ the approximation is better than 1%. If the point A is supposed to be connected to branches similar to those emerging from B, the input resistance R_{inp} will be:

$$R_{inp} = \frac{R_0}{n+1} = \frac{n}{n^2-1} \cdot r_i d. \qquad (12.17)$$

It is clear that in such a model, provided that the interval between successive branchings is small compared with the space constant, the input resistance is almost completely independent from the membrane resistance. A similar conclusion was also drawn[145] in the case of a dichotomously branching open syncytium (model resembling the preceding one, but with $n = 2$). The electrotonic potential decay of the model studied by Mazet is shown in Fig. 12.23. Corresponding to every branching point, the curves present angular points between which each curve is the sum

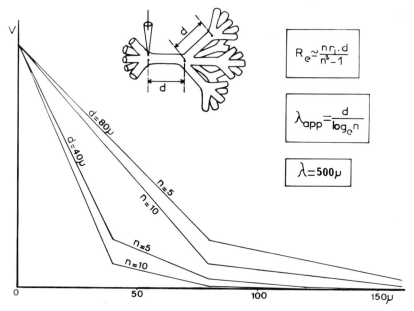

Fig. 12.23. Electrotonic potential decay of repeatedly polyfurcated syncytium. Angular points of each curve are located on exponential (with a space constant, $\lambda_{app.}$ given by indicated formula). (From Mazet, unpublished.)

of two exponentials. Angular points of a given curve are located on an exponential the space constant of which is approximately

$$\lambda_{app} = \frac{d}{\log_e n} \qquad (12.18)$$

REFERENCES

[1] G. Ling, and R. W. Gerard, *J. Cellular Comp. Physiol.*, **34**, 383 (1949).

[2] W. L. Nastuk, and A. L. Hodgkin, *J. Cellular Comp. Physiol.*, **35**, 39 (1950).

[3] E. Coraboeuf, and S. Weidmann, *Compt. Rend. Soc. Biol.*, **143**, 1329 and 1360 (1949).

[4] L. A. Woodbury, J. W. Woodbury, and H. H. Hecht, *Circulation*, **1**, 264 (1950).

[5] M. H. Draper, and S. Weidmann, *J. Physiol. (London)*, **115**, 74 (1951).

[6] Mc. C. Brooks, B. F. Hoffman, E. E. Suckling, and O. Orias, *Excitability of the Heart*, Grune and Stratton, New York, 1955.

[7] S. Wiedmann, *Elektrophysiologie der Herzmuskelfaser*, Hans Huber, Bern and Stuttgart, 1956.

[8] P. F. Cranefield, and B. F. Hoffman, *Physiol. Rev.*, **38**, 41 (1958).

[9] E. Coraboeuf, *J. Physiol. (Paris)*, **52**, 323 (1960).

[10] B. F. Hoffman, and P. F. Cranefield, *Electrophysiology of the Heart*, McGraw-Hill, New York, 1960.

[11] W. Trautwein, *Ergeb. Physiol. Biol. Chem. Exp. Pharmakol.*, **51**, 131 (1961).

[12] J. W. Woodbury, *Handbook of Physiology, Circulation I*, Am. Physiol. Soc., Washington, D.C. 1962.

[13] D. W. Kennard, in *Electronic Apparatus for Biological Research* (P. E. Donaldson, ed.), Butterworths, London, 1958.

[14] K. Frank, and M. C. Becker, in *Physical Techniques in Biological Research* (W. L. Nastuk, ed.), vol. 5A, Academic, New York, 1964.

[15] L. A. Woodbury, H. H. Hecht, and A. R. Christopherson, *Am. J. Physiol.*, **164**, 307 (1951).

[16] W. Trautwein, and K. Zink, *Pfluegers Arch. Ges. Physiol.* **256**, 68 (1952).

[17] B. F. Hoffman, and E. E. Suckling, *Am. J. Physiol.*, **170**, 357 (1952).

[18] E. M. Vaughan Williams, *J. Physiol. (London)*, **147**, 3P (1959).

[19] J. W. Woodbury, and A. J. Brady, *Science*, **123**, 100 (1956).

[20] E. Coraboeuf, and D. Breton, *Compt. Rend. Soc. Biol.*, **157**, 986 (1963).

[21] P. H. Benoit, *Compt. Rend. Soc. Biol.*, **156**, 1465 (1962).

[22] J. K. Pruett, and E. F. Woods, *J. Appl. Physiol.*, **21**, 1071 (1966).

[23] B. Renshaw, A. Forbes, and B. R. Morrison, *J. Neurophysiol.*, **3**, 74 (1940).

[24] J. S. Coombs, J. C. Eccles, and P. Fatt, *Australian J. Sci.*, **16**, 3 (1953).

[25] J. S. Coombs, J. C. Eccles, and P. Fatt, *J. Physiol. (London)*, **130**, 291 (1955).

[26] P. H. Benoit, and E. Coraboeuf, *Compt. Rend. Soc. Biol.*, **149**, 1435 (1955).

[27] E. Coraboeuf, R. Distel, and J. Boistel, *Collo. Intern. Microphysiol.*, *Centre Nat. Rech. Sci. (Paris)*, (1955).

[28] D. R. Curtis, in *Physical Techniques in Biological Research* (W. L. Nastuk, ed.). vol. 5A, Academic, New York, 1964.

[29] J. Boisseau, and J. Boistel, *J. Physiol. (Paris)*, **57**, 529 (1965).

[30] L. Elson, *J. Sci. Instru.*, **30**, 40 (1953).

[31] V. A. Vis, *Science*, **120**, 152 (1954).

[32] N. W. Carter, F. C. Rector, and D. W. Seldin, *J. Clin. Invest.*, **46**, 920 (1967).

[33] R. Henderson, *J. Appl. Physiol.*, **22**, 1179 (1967).

[34] T. Tomita, *Japan. J. Physiol.*, **6**, 327 (1956).

[35] I. Tanaka, and Y. Sasaki, *J. Gen. Physiol.*, **49**, 1089 (1966).

[36] M. Tarr, and N. Sperelakis, *Am. J. Physiol.*, **207**, 691 (1964).

[37] Y. M. Gargouil, R. Tricoche, and C. Bernard, *Compt. Rend. Acad. Sci.*, **257**, 1801 (1963).

[38] S. Weidmann, *J. Physiol. (London)*, **118**, 348 (1952).

[39] O. Schanne, Chapter 14 of this book.

[40] S. Weidmann, *J. Physiol. (London)*, **115**, 227 (1951).

[41] K. Frank, and M. G. F. Fuortes, *J. Physiol. (London)*, **134**, 451 (1956).

[42] O. Schanne, H. Kawata, B. Schäfer, and M. Lavallée, *J. Gen. Physiol.*, **49**, 897 (1966).

[43] W. Trautwein, K. A. Deck, and R. Kern, *Pfluegers Arch. Ges. Physiol.*, **278**, 13 (1963).

[44] H. H. Hecht, O. F. Hutter, and D. W. Lywood, *J. Physiol. (London)*, **170**, 5P (1964).

[45] J. Boistel, and P. Fatt, *J. Physiol. (London)*, **144**, 176 (1958).

[46] D. Noble, *Physiol. Rev.*, **46**, 1 (1966).

[47] M. Vassale, *Am. J. Physiol.*, **210**, 1335 (1966).

[48] E. Carmeliet, *J. Physiol. (London)*, **156**, 375 (1961).

[49] E. Carmeliet, *Chloride and Potassium Permeability in Cardiac Purkinje Fibers*, Presses Académiques Européennes, Brussels, 1961.

[50] O. F. Hutter, and D. Noble, *J. Physiol. (London)*, **157**, 335 (1961).

[51] K. A. Deck, R. Kern, and W. Trautwein, *Pfluegers Arch. Ges. Physiol.*, **280**, 50 (1964).

[52] K. A Deck, and W. Trautwein, *Pfluegers Arch. Ges. Physiol.*, **280**, 63 (1964).

[53] M. Lapicque, and C. Veil, *Chronaxie des Fibres de Purkinje*, Doin, Paris, 1925.

[54] J. F. Nonidez, *Am. Heart J.*, **26**, 577 (1943).

[55] E. Coraboeuf, R. Distel, and J. Boistel, *Compt. Rend. Soc. Biol.*, **147**, 1757 (1953).

[56] K. Matsuda, in *Electrical Activity of Single Cells* (Y. Katsuki, ed.), Igaku Shoin, Tokyo, 1960.

[57] B. F. Hoffman, *Circulation*, **24**, 506 (1961).

[58] A. L. Hodgkin, and A. F. Huxley, *J. Physiol. (London)*, **106**, 341 (1947).

[59] S. Weidmann, *Am. J. Physiol.*, **183**, 671 (1955).

[60] S. Weidmann, *Helv. Physiol. Pharmacol. Acta*, **19**, C35 (1961).

[61] O. F. Hutter, and D. Noble, *Nature (London)*, **188**, 495 (1960).

[62] A. E. Hall, O. F. Hutter, and D. Noble, *J. Physiol. (London)*, **166**, 225 (1963).

[63] R. E. McAllister, and D. Noble, *J. Physiol. (London)*, **186**, 632 (1966).

[64] J. Dudel, K. Peper, R. Rudel, and W. Trautwein, *Pfluegers Arch. Ges. Physiol.*, **295**, 197 (1967).

[65] J. Dudel, K. Peper, R. Rudel, and W. Trautwein, *Pfluegers Arch. Ges. Physiol.*, **295**, 213 (1967).

[66] J. J. Chang, and R. F. Schmidt, *Pfluegers Arch. Ges. Physiol.*, **272**, 127 (1960).

[67] J. Dudel, and W. Trautwein, *Arch. Exp. Path. Pharmakol.*, **232**, 393 (1958).

[68] H. Reuter, *Arch. Exp. Path. Pharmakol.*, **251**, 152 (1965).

[69] W. Trautwein, and D. G. Kassebaum, *J. Gen. Physiol.*, **45**, 317 (1961).

[70] N. Sperelakis, and D. Lehmkuhl, *J. Gen. Physiol.*, **47**, 895 (1964).

[71] W. Trautwein, S. W. Kuffler, and C. Edwards, *J. Gen. Physiol.*, **40**, 135 (1956).

[72] W. Trautwein, and J. Dudel, *Pfluegers Arch. Ges. Physiol.* **266**, 324 (1958).

[73] J. Dudel, and W. Trautwein, *Experientia*, **12**, 396 (1955).

[74] J. W. Woodbury, and W. E. Crill, in *Nervous Inhibition* (E. Florey, ed.), Pergamon, New York, 1961.

[75] H. A. Fozzard, and W. Sleator, *Am. J. Physiol.*, **212**, 945 (1967).

[76] E. Coraboeuf, F. Zacouto, Y. M. Gargouil, and J. Laplaud, *Compt. Rend. Acad. Sci.*, **246**, 2934 (1958).

[77] A. M. Shanes, *Pharmacol. Rev.*, **10**, 165 (1958 .

[78] A. J. Brady, and J. W. Woodbury, *J. Physiol. (London)*, **154**, 385 (1960).

[79] D. Noble, *Nature*, **188**, 495 (1960).

[80] H. A. Kuriyama, M. Goto, T. Maeno, Y. Abe, and S. Ozaki, in *Electrical Activity of Single Cells* (Y. Katsuki, ed.), Igaku Shoin, Tokyo, 1960.

[81] D. Rapport, and G. B. Ray, *Am. J. Physiol.*, **80**, 126 (1927).

[82] A. Rosenblueth, and E. C. del Pozo, *Am. J. Physiol.*, **139**, 514 (1943).

[83] J. A. E. Eyster, and W. E. Gilson, *Am. J. Physiol.*, **150**, 572 (1947).

[84] P. F. Cranefield, J. A. E. Eyster, and W. E. Gilson, *Am. J. Physiol.*, **167**, 450 (1951).

[85] E. A. Johnson, P. A. Robertson, and J. Tille, *Nature*, **182**, 1161 (1958).

[86] E. A. Johnson, and J. Tille, *Australian J. Exp. Biol. Med. Sci.*, **38**, 509 (1960).

[87] E. A. Johnson, and J. Tille, *J. Gen. Physiol.*, **44**, 443 (1961).

[88] P. Guilbault, J. Delahayes and M. Paillard, *Compt. Rend. Soc. Biol.*, **160**, 2031 (1966).

[89] D. Garnier, E. Coraboeuf, and A. Paille, *Compt. Rend. Soc. Biol.*, **155**, 2430 (1961).

[90] E. A. Johnson, and J. Tille, *Nature*, **192**, 663 (1961).

[91] E. Coraboeuf, and G. Vassort, *Compt. Rend. Acad. Sci.*, **264**, 1072 (1967).

[92] E. Coraboeuf, and M. Otsuka, *Compt. Rend. Acad. Sci.*, **243**, 441 (1956).

[93] J. Deleze, *Circulation Res.*, **7**, 461 (1959).

[94] S. Hagiwara, and S. Nakajima, *J. Gen. Physiol.*, **49**, 793 (1966).

[95] R. Niedergerke, and R. K. Orkand, *J. Physiol. (London)*, **184**, 291 and 312 (1966).

[96] J. Dudel, K. Peper, R. Rudel, and W. Trautwein, *Pfluegers Arch. Ges. Physiol.*, **292**, 255 (1966).

[97] H. Reuter, *Pfluegers Arch. Ges. Physiol.*, **287**, 357 (1966).

[98] J. Dudel, K. Peper, and W. Trautwein, *Pfluegers Arch. Ges. Physiol.*, **288**, 262 (1966).

[99] F. S. Sjöstrand, and E. Anderson, *Experientia*, **10**, 369 (1954).

[100] D. H. Moore, and H. Ruska, *J. Biophys Biochem. Cytol.*, **3**, 261 (1957).

[101] A. R. Muir, *J. Biophys. Biochem. Cytol.*, **3**, 193 (1957).

[102] S. Weidmann, *J. Physiol. (London)*, **153**, 32P (1960).

[103] S. Weidmann, in *Electrophysiology of the Heart* (B. Taccardi and G. Marchetti, eds.), Pergamon, New York, 1965.

[104] S. Weidmann, *Ann. N.Y. Acad. Sci.*, **137**, 540 (1966).

[105] S. Weidmann, *J. Physiol. (London)*, **187**, 323 (1966).

[106] L. Barr, and W. Berger, *Pfluegers Arch. Ges. Physiol.*, **279**, 192 (1964).

[107] W. G. Van Der Kloot, and B. Dane, *Science*, **146**, 74 (1964).

[108] G. Bourne, *Nature*, **172**, 588 (1953).

[109] D. J. Goldstein, *Anat. Record*, **134**, 217 (1959).

[110] F. Joo, and B. Csillik, *Nature*, **193**, 1192 (1962).

[111] N. Sperelakis, T. Hoshiko, and P. M. Berne, *Am. J. Physiol.*, **198**, 531 (1960).

[112] T. Hoshiko, N. Sperelakis, and R. M. Berne, *Proc. Soc. Exp. Biol. Med.*, **101**, 602 (1959).

[113] N. Sperelakis, and T. Hoshiko, *Circulation Res.*, **9**, 1280 (1961).

[114] N. Sperelakis, T. Hoshiko, R. F. Keller, and R. M. Berne, *Am. J. Physiol.*, **198**, 135 (1960).

[115] N. Sperelakis, *Circulation Res.*, **12**, 676 (1963).

[116] T. Hoshiko, and N. Sperelakis, *Am. J. Physiol.*, **201**, 873 (1961).

[117] D. Lehmkuhl, and N. Sperelakis, *Am. J. Physiol.*, **205**, 1213 (1963).

[118] R. Fange, H. Persson, and S. Thesleff, *Acta Physiol. Scand.*, **38**, 173 (1956).

[119] W. E. Crill, R. E. Rumery, and J. W. Woodbury, *Am. J. Physiol.*, **197**, 733 (1959).

[120] I. Harary, and B. Farley, *Exp. Cell Res.* **29**, 451 and 466 (1963).

[121] F. Kavaler, *Am. J. Physiol.*, **197**, 968 (1959).

[122] J. Tille, *J. Gen. Physiol.*, **50**, 189 (1966).

[123] T. Cooper in *Nervous Control of the Heart* (W. C. Randall, ed.), Williams and Wilkins, Baltimore, 1965.

[124] F. S. Sjöstrand, E. Anderson-Cedergren, and M. M. Dewey, *J. Ultrastruct. Res.*, **1**, 271 (1958).

[125] K. R. Porter, and M. A. Bonneville, in *An Introduction to the Fine Structure of Cells and Tissues*, 2nd ed., Lea and Febiger, Philadelphia, 1964.

[126] A. R. Muir, *J. Anat.*, **99**, 27 (1965).

[127] L. Barr, M. M. Dewey, and W. Berger, *J. Gen. Physiol.*, **48**, 797 (1965).

[128] F. S. Sjöstrand, and E. Andersson-Cedergren, in *The Structure and Function of Muscle*, Vol. 1, Academic, London, 1960.

[129] A. R. Muir, *J. Anat.*, **101**, 239 (1967).

[130] W. R. Loewenstein, *Ann. N. Y. Acad. Sci.*, **137**, 441 (1966).

[131] J. D. Robertson, *Ann. N. Y. Acad. Sci.*, **137**, 421 (1966).

[132] V. Luzzati, and F. Husson, *J. Cell. Biol.*, **12**, 207 (1962).

[133] V. Luzzati, F. Reiss-Husson, E. Rivas, and T. Gulik-Krzywicki, *Ann. N. Y. Acad. Sci.*, **137**, 409 (1966).

[134] M. G. Farquhar, and G. E. Palade, *J. Cell. Biol.*, **17**, 375 (1963).

[135] J. Deleze, in *Electrophysiology of the Heart* (B. Taccardi and G. Marchetti, eds.). Pergamon, New York, 1965.

[136] K. S. Cole, and J. W. Moore, *J. Gen. Physiol.*, **44**, 123 (1960).

[137] D. Noble, *J. Physiol. (London)*, **160**, 317 (1962).

[138] D. Noble, *Biophys. J.*, **2**, 381 (1962).

[139] A. L. Hodgkin, and A. F. Huxley, *J. Physiol. (London)*, **116**, 449 (1952).

[140] A. L. Hodgkin, and A. F. Huxley, *J. Physiol. (London)*, **117**, 500 (1952).

[141] S. Weidmann, *J. Physiol. (London)*, **127**, 213 (1955).

[142] R. Fitzhugh, *J. Gen. Physiol.*, **43**, 867 (1960).

[143] E. P. George, and E. A. Johnson, *Australian J. Exp. Biol. Med. Sci.*, **39**, 275 (1961).

[144] E. P. George, *Physics Letter*, **1**, 305 (1962).

[145] E. P. George, *Australian J. Exp. Biol. Med. Sci.*, **39**, 267 (1961).

[146] M. B. Berkinblit, S. A. Kovalev, V. V. Smolyaninov, and L. M. Chaila-khyan, *Biophysics*, **10**, 341 (*Biofizica*, **10**, 309) (1965).

[147] E. A. Johnson, and J. R. Sommer, *J. Cell. Biol.*, **33**, 103 (1967).

[148] A. L. Hodgkin, and W. A. H. Rushton, *Proc. Roy. Soc. (London), Ser. B.* **133**, 444 (1946).

[149] P. Fatt, and B. Katz, *J. Physiol. (London)*, **120**, 171 (1953).

[150] O. Schanne, R. Kern, and B. Schäfer, *Naturwissenschaften*, **49**, 161 (1962).

[151] H. Kawata, O. Schanne, and D. Krakat, *Pfluegers Arch. Ges. Physiol.*, **278**, 5 (1963).

[152] P. Fatt, and B. Katz, *J. Physiol. (London)*, **115**, 320 (1951).

[153] P. Jenerick, *J. Cellular Comp. Physiol.*, **42**, 427 (1953).

[154] S. Weidmann, *Ann. N. Y. Acad. Sci.*, **65**, 663 (1957).

[155] H. A. Fozzard, *J. Physiol. (London)*, **182**, 255 (1966).

[156] E. Coraboeuf, and G. Vassort, unpublished.

CHAPTER 13

Cation and Hydrogen Microelectrodes in Single Nephrons

RAJA N. KHURI

Department of Physiology, American University of Beirut

This chapter was written with the support of U.S. Public Health Service Grant 5 RO1 AM 09275 and of a medical research grant from the American University of Beirut.

13.1 INTRODUCTION

13.1.1 Micropuncture Technique in Renal Physiology

The two major techniques used in studying renal physiology are the clearance technique and the micropuncture technique. The clearance approach is indirect and consists of examining the plasma and the product of the kidney operations (the urine). It treats the kidney itself as a "black-box." The micropuncture technique is a direct approach that examines the operations of the single units of the kidney, the nephrons. Its usefulness lies in permitting evaluation of the contributions of different nephron segments, delineation of the electrochemical forces at the cellular level, and construction of models of the cellular mechanism involved.

The tubular fluid of single nephrons courses down a lumen with an inner diameter that is not too different in magnitude from the diameter of single epithelial cells (Fig. 13.1). Therefore the degree of miniaturization required for intracellular measurements is also required for in-situ intraluminal measurements in single nephrons. In order to get into the intracellular phase, a microelectrode must go through a surface cell membrane. But to lie in the intraluminal phase, a microelectrode has to go through the entire thickness of an epithelial layer. Figure 13.2 shows an ion-sensitive and a reference microelectrode positioned for an in-situ intraluminal measurement in a segment of a single nephron.

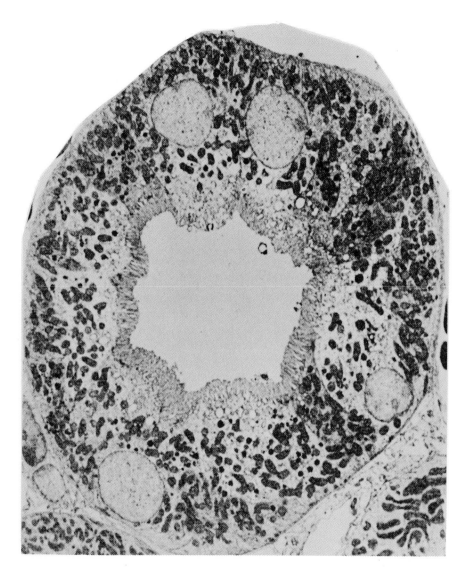

Fig. 13.1. Proximal convoluted tubule: cross-section, low-magnification electron-micrograph.

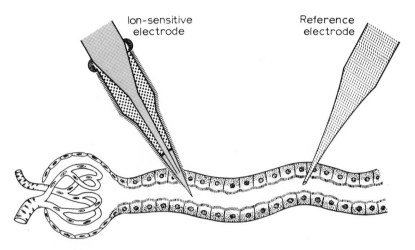

Fig. 13.2. Indicator and reference microelectrode in lumen of proximal tubule.

13.1.2   Cation Glass Microelectrodes in Renal Micropuncture Work

The most important advantage of glass microelectrodes is their ability to measure hydrogen ions and the alkali cations in vivo under conditions of normal spontaneous flow; this eliminates the potential errors inherent in the sampling and collection of tubular fluid, as the composition of the sample of tubular fluid may be a function of the rate of collection. Moreover, evaporation, gaseous exchanges, and contamination of the tubular fluid can also be avoided. Measurement of pH in vivo is particularly advantageous because it eliminates the uncertainty involved in the arbitrary choice of a CO_2 tension with which to equilibrate the microsample of collected fluid; loss of CO_2 to the ambient air is also avoided. For in-vivo measurement of pH, microelectrodes of pH glass are to be preferred to quinhydrone microelectrodes; the latter are sensitive to oxidation-reduction reactions and have significant protein errors and an alkaline error even in moderately alkaline solutions.

Even when used for cation analyses in vitro, glass microelectrodes have some distinct advantages over the other microanalytic techniques (e.g., colorimetric and photometric). First, the microsample is not consumed in the measurement and is, therefore, available for other determinations as well as replicate measurement of the same sample. Second, micropotentiometry does not require either sample dilution or assessment of sample volume, both of which provide potential sources of error. The fact that glass electrodes measure ion activities rather than concentrations is a further advantage under conditions in which ion binding or complex formation occurs. In addition it is generally recognized that physiological processes are more often functions of ionic activities than of concentration.

13.2 MICROELECTRODE CELLS AND COMPONENTS

13.2.1 Components of a Microelectrode Cell

A potentiometric cell consists of two half-cells: (a) a cation indicator glass electrode and (b) a reference electrode. The emf of a cell is the algebraic sum of the two half-cell potentials. Each half-cell incorporates an inner reference solution and an inner reference metallic electrode that is reversible to one of the ions of the inner solution bathing it. The inner reference electrodes commonly used for microelectrodes are the same as for electrodes in general: mercury-mercurous chloride (Hg–Hg_2Cl_2 or calomel) and silver-silver chloride (Ag–AgCl). They are both reversible to Cl^-.

An electrochemical cell used for pH measurement may be represented as follows:

$$Hg;Hg_2Cl_2, KCl\,(satd) \underset{E_j}{//} test\ sample \underset{E_{H+}}{/} H^+\text{-glass} \underset{E_R}{/} pH\ buffer, Hg_2Cl_2;Hg$$

where (;) represents a metal-electrolyte boundary, (,) separates two different solutes, (/) is a liquid-glass electrode interface, and (//) is a free liquid-liquid junction.

The potential of such a cell is the algebraic sum of all its boundary potentials. The boundary potentials of the left-hand calomel inner reference are balanced by the potentials of the mirror image boundaries of the right-hand calomel. The observed potential of the cell is then equal to the sum of E_{H+}, E_j, and E_R. The potential at the reference surface of the glass electrode E_R may be safely regarded as being maintained constant. E_j represents the liquid-liquid junction potential at the interface at which the test solution meets the salt bridge. If E_j at the salt bridge-test solution boundary is not different from its value at the salt bridge-buffer standard boundary, then E_j subtracts out in the calibration of the cell. This leaves E_{H+}, the potential at the glass test solution interface, as the only boundary potential of the cell that is subject to variation. This potential is a function of the activity of the ion in the solution bathing the indicator surface of the glass and to which the electrode glass is reversible.

The indicator and reference half-cells may be kept as two separate microelectrodes or may be combined into one physical unit. The single unit pH glass ultramicroelectrode of Khuri et al.[1] shown in Fig. 13.11

is an example of the latter. The combination of the glass and the reference electrodes into a single unit is of advantage in in-situ measurements, particularly in situations in which double punctures may produce leakage of fluid and increase the possibility of short-circuiting the membrane potential.

13.2.2   The Indicator Microelectrode Half-Cell

The indicator half-cell of a microelectrode unit consists of an ion-sensitive glass membrane the reference surface of which is bathed by a stable buffered reference solution which is in contact with an inner reference metallic electrode. The indicator surface of an ion-sensitive glass microelectrode is, of necessity, very small indeed. Unless this highly restricted membrane functions reversibly in the electrochemical sense, the microelectrode cell cannot be used to determine the activity of the ion in question.

The ion-sensitive glass membrane has some important physical and electrical properties. Ideal electrode response is achieved only when the membrane thickness is below a certain critical value. Diamond and Hubbard[2] found for each electrode glass a characteristic departure thickness — the maximum thickness which permits full theoretical electrode response. For Corning Code 0150 pH glass they found a limiting thickness of about 100 μ. Glass microelectrodes usually have membranes that are thinner than the limiting thickness of the electrode glass in question.

Electrode glasses generally have a chemical durability which is inferior to that of inert glasses. Optical glasses, on the other hand, have a very high chemical durability. The chemical durability has a direct bearing on such problems as the aging, storage, and maintenance of glass microelectrodes. In the author's experience, capillaries of the K^+-sensitive glass KAS_{20-5} seem to benefit from a period of aging of one to two weeks prior to the completion of the microelectrode construction. This was not the case with either the Na^+-sensitive glass NAS_{11-18} or the H^+-sensitive glass Corning Code 0150. The indicator glass surface may be stored either in distilled water or in a closed humid atmosphere, as the rate of dissolution of electrode glasses is minimized in pure water. The use of strong acids and alkalis and other corrosive agents must be avoided.

Electrode glasses are hygroscopic. Electrode function may be lost on drying and recovered on rehydration. The hydrated gelatinous surface layer seems to be essential for the generation of a boundary potential at the glass-solution interface. With glass microelectrodes the hydration of the indicator surface may be achieved either by direct (but limited) immersion in distilled water or by storage in an atmosphere of high relative humidity. The latter method is based on the observation of van Itterbeek and Vereycken[3] that glass surfaces are capable of adsorbing water vapor quite readily at room temperature. Figure 13.3 shows glass microelectrodes stored in a closed jar just above the surface of the water.

Fig. 13.3. Closed jar for microelectrode storage.

The thermal properties of electrode glasses assume importance in microelectrode construction because repeated exposure of the glass to heat may be essential in the various steps involved in the construction of microelectrodes. When the electrode glass is sealed to an inert glass to achieve a glass-to-glass seal for insulation purposes, the two glasses must have quite similar thermal characteristics — that is, similar softening points and similar thermal expansion coefficients.

The internal electrical resistance of the glass membrane is relatively high. It is a direct function of the thickness of the sensitive portion of the electrode glass membrane and an inverse function of its area. For ultramicro Na^+, K^+, and H^+ glass electrodes the internal resistance ranges from 10^7 to $10^{11} \ \Omega$. This high internal resistance magnifies the problem of insulation as a step in the construction of glass microelectrodes. (A more extensive discussion of the physical and electrical properties of glass microelectrodes is available[4].)

The two configurations of indicator glass microelectrodes used in the author's studies of ion analysis in single nephrons are shown diagrammatically in Fig. 13.4. The first is the sealed micropipette-type of glass electrodes such as those shown in Figs. 13.6 and 13.10. This type has been made to achieve the highest degree of miniaturization and is particularly

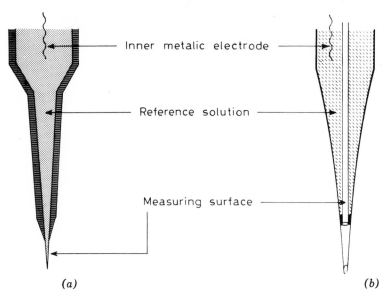

Inner metalic electrode

Reference solution

Measuring surface

(a)   (b)

Fig. 13.4. Two configurations of indicator glass microelectrodes. (a) Sealed micropipette type; (b) internal-capillary, suction type.

suited for in-situ ion analysis. Its spearhead configuration renders it mechanically suited for micropuncture work. The second type is the internal capillary, flow-through, or suction type with the inner surface acting as the measuring surface. An indicator capillary microelectrode half-cell is shown in Fig. 13.12 and in physical combination with the reference half-cell (making a single unit) in Fig. 13.11. This type is particularly suited for monitoring ionic activity of luminal fluid within fine tubular structures.

13.2.3   The Reference Microelectrode Half-Cell

The reference microelectrode half-cell consists of an inner metallic electrode and an inner reference solution. The latter usually establishes contact with the test solution through the intermediary of a salt bridge. KCl, saturated or $3 M$, is the most commonly used bridge electrolyte because it minimizes the liquid junction potential.

The conventional micropipette electrodes filled with concentrated KCl are often used as reference half-cells for the indicator glass microelectrode. In order to avoid the incorporation of a membrane potential into the total potential difference of the cell, the reference electrode should establish electrolytic contact with the same phase or compartment as does the measuring glass electrode.

The tip potential is the junctional potential developed across the tip of the Ling and Gerard micropipette electrode, which often serves as the reference microelectrode. Micropipettes with tip diameters of less than $1\,\mu$ may exhibit high tip potentials of some 20–30 mV. This may owe to obstruction of the tip by dirt or organic matter following tissue penetration. This changing tip potential necessitates recalibration of the microelectrode cell after each tissue puncture. In some studies the author mounts the reference micropipette into a holder (Fig. 13.5) after the design of Grundfest et al.[5]. The bridge electrolyte of the micropipette connects with the same solution filling a channel inside the holder; the latter connects freely with a calomel electrode as an internal reference element. This type of reference microelectrode holder allows the injection of KCl solution through the micropipette tip following each tissue insertion. With this method a stable potential may be maintained at the tip, a fresh junction being reformed by the KCl injection. The injection of KCl may also aid in localizing the tip of the reference microelectrode within the tissue.

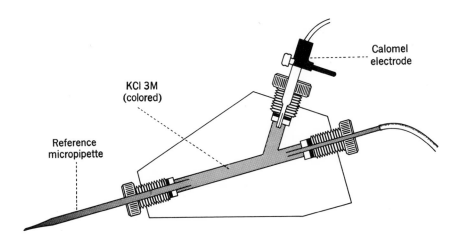

Fig. 13.5. Reference microelectrode holder with injecting mechanism (Khuri [29]. Reprinted by permission of *The Review of Scientific Instruments*).

13.3 SODIUM AND POTASSIUM IN SINGLE NEPHRONS

An accurate measurement of Na^+ and K^+ in single nephrons is needed to describe the mechanism of transport of these ions. Na^+ can be measured accurately by flame photometry even on a small volume of tubular fluid of less than 0.1 μl. In contrast, K^+ determination by flame photometry on this microscale was still uncertain, and this was reflected in the conflicting results for proximal tubular fluid K^+ concentration reported prior to the development by Malnic, Klose, and Giebisch[6] of the dual-channel "ultramicro" flame photometer.

13.3.1 Cation Glass Microelectrodes

Microelectrodes were built from glasses developed by Eisenman[7] to measure cation activity directly. Concentric cation glass microelectrodes with external insulation were used by Khuri et al.[8, 9] to measure Na^+ and K^+ in single proximal tubules of the *Necturus* and rat kidney. These devices belong to the sealed-micropipette type shown in Fig. 13.6. The first step in the construction is the pulling of the cation glass capillary tubing into micropipettes by means of a puller of the type of Alexander and Nastuk[10]. (The reader is referred to the comprehensive review by Frank and Becker[11] on micropipette pulling by machines.) The potassium glass KAS_{20-5}, like the pH glass Corning Code 0150, can be pulled into micropipettes at moderate coil temperatures. In contrast, the sodium glass NAS_{11-18} could be pulled only at very high coil temperatures. The tip of the cation micropipette was sealed by plunging it into the center of a highly heated wire loop element of a de Fonbrune Microforge. With potassium glass (KAS_{20-5}) and pH glass Code 0150 pulled at relatively low coil temperatures and with minimal time delay,

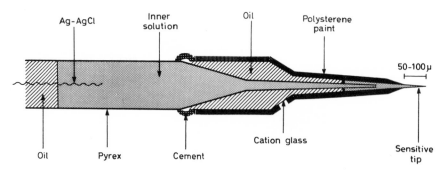

Fig. 13.6. Concentric cation glass microelectrode with external insulation (Khuri, Goldstein, Maude, Edmonds, and Solomon[8]. Reprinted by permission of the *Am. J. Physiol.*).

a significant fraction of the micropipettes formed have a spontaneously "sealed" tip. Two criteria were used to assess the patency of the tip: one was taking up water by capillary; the other consisted of injecting through the tip of a filled cation glass segment.

The sealed micropipette of cation glass had their stems cut close to the shoulder and were lined up horizontally on a ridge of Plasticine. Filler micropipettes of ordinary glass were prepared, with shanks longer than the entire cation glass segments. They were filled with the appropriate inner reference solution, mounted on micromanipulators, and connected to an injection device. Under binocular microscopy, the filler micropipette was threaded inside the cation pipette segment as far as the tip and reference electrolyte was injected to fill the tip and a small segment of the shank. To achieve some internal insulation, the rest of the cation glass segment was filled with mineral oil. Then a micropipette filled with the same reference electrolyte was introduced to establish contact with the electrolyte-filled tip of the cation glass segment. The stems of the two concentric glass micropipettes were fused with household cement. Some oil was added to the open end of the stem to prevent the evaporation of the inner reference solution. These electrodes were subsequently hydrated by immersing the tips in distilled water for a couple of days before their external insulation.

The microelectrodes were externally insulated by painting with polystyrene coil dope (G–C Electronic Co.) to within 50 to 100 μ of the tip. The insulation was performed under binocular microscopy with the aid of micromanipulators. Following insulation, a check is made to insure that the only sensitive portion of the electrode is the uninsulated tip. This step consists of immersing the uninsulated tip of the cation microelectrode into oil while the insulated shank and the tip of the reference electrode are bathed by an electrolyte to which the cation glass is sensitive. A very high voltage reading indicates adequate insulation. Otherwise, additional layers of polystyrene are applied to achieve adequate insulation. Following insulation, the cation glass microelectrodes were stored in closed jars (Fig. 13.3) containing distilled water, with the electrode tips above the water surface. Prolonged immersion in water was found to mar the insulation.

A Cary vibrating reed electrometer was used for measurements of potentials. Voltages, measured by null balance with a bucking potentiometer, were displayed on a recorder. All equipment was contained in a copper-screen cage to exclude spurious potentials, each piece being connected to the same grounded copper bar.

Cation microelectrodes were selected on the basis of having a voltage response, that is, a slope of greater than 50 mV per tenfold change in

Na^+ or K^+ concentration as well as favorable selectivity to the other cations present. The selectivity of the microelectrode to the other cations was determined by comparing the electrical potentials displayed by the cell in $0.1\text{-}M$ solutions of NaCl, KCl, and HCl.

13.3.2 Sodium Measurements in Single Nephrons

It has been shown by several investigators that proximal tubular fluid of the amphibian and mammalian kidney under normal conditions has a sodium concentration that does not change as the fluid moves along the proximal tubule and is not different from that of plasma[12–15]. In all these studies the sodium concentration was determined by flame photometry. A potentiometric analysis of Na^+ activity was in order because we were comparing a virtually protein-free tubular fluid with plasma.

The sodium-measuring cell consisted of a Na^+-indicator microelectrode made of the Na^+-sensitive glass NAS_{11-18} and a reference half-cell incorporating RbCl ($3\ M$) as a bridge electrolyte. The sodium cell used for the in-vitro analysis of proximal tubular fluid is shown in Fig. 13.7. RbCl was used rather than the usual KCl because the sensitivity of the glass NAS_{11-18} to Rb^+ is 100 times lower than to K^+. The ionic mobility of Cl^- is even closer to the mobility of Rb^+ than to that of K^+.

The following equation describes the potential of the sodium cell:

$$E_{Na} = E_{Na}^{\circ} + RT/F \ln\left[(Na^+) + k_{Na-H}(H^+) + k_{Na-K}(K^+)\right], \quad (13.1)$$

where R, T, and F are the gas constants per mole, the temperature in degrees Kelvin, and the Faraday constant, respectively; E_{Na} is the the measured potential; E_{Na}° is the standard potential of the sodium electrode; Na^+ is the sodium ion activity; and k_{Na-H} and k_{Na-K} are the selectivity constants of glass to H^+ and K^+ ions relative to the Na^+ ion. Microelectrodes of NAS_{11-18} glass gave a $k_{Na-H} \approx 10$ and a k_{Na-K} of 0.01–0.001. Therefore, in neutral body fluids of low K^+ concentration, the sodium glass electrode is essentially sensitive to the Na^+ ion only and (13.1) reduces to:

$$E_{Na} = E_{Na}^{\circ} + RT/F \ln (Na^+). \quad (13.2)$$

Proximal tubular fluid was collected at normal flow rates from *Necturus* and rat kidneys, the volume of the sample of fluid being about $0.1\ \mu l$ and $0.03\ \mu l$, respectively. The collected sample is immediately delivered as a droplet into a lucite well filled with oil (Fig. 13.7).

Table 13.1 gives the results obtained with the sodium cell on samples of serum, glomerular fluid, and proximal tubular fluid collected from five

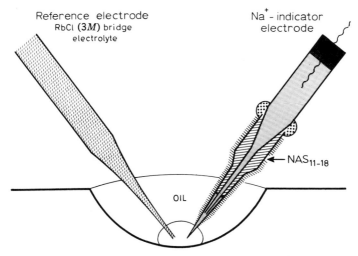

Fig. 13.7. Na$^+$ microelectrode with 3 M RbCl cell for Na$^+$ measurement.

Necturii. Several samples were also determined by means of a modified Beckman DU flame photometer. The mean potentiometric value of 89.1 mM for serum Na is in good agreement with the value of 89.7 mM obtained with the flame photometer. This supports the assumption that the activity coefficient in serum is the same as that in simple salt solution and suggests that no significant amount of Na$^+$ is bound to the plasma proteins. Sodium concentration in both glomerular and proximal tubular fluid was very close to that of serum, the ratio of [Na] glomerulus/[Na] serum being 1.00 ± 0.02 and that of [Na] tubule/[Na] glomerulus being 0.99 ± 0.01. Neither ratio is significantly different from 1.00.

Figure 13.2 shows two microelectrode half-cells in situ. When intraluminal sodium was measured in situ, the Na$^+$-sensitive microelectrode was built of the NAS$_{11-18}$ glass. The open tip of the reference electrode filled with RbCl (3 M) was placed downstream from the sodium glass microelectrode. Particular care was taken in positioning both microelectrodes within the lumen of Bowman's capsule, the proximal tubule, and superficial renal blood vessels. The [Na]glomerulus/[Na]blood ratio is 0.96 ± 0.04 and the [Na]tubule/[Na] glomerulus ratio is 1.04 ± 0.06 (see Table 13.2). Neither ratio is significantly different from unity, thus confirming the constancy of Na$^+$ concentration as the fluid flows along the proximal tubule.

Na$^+$ concentration in the plasma of 31 rats averaged 141 ± 1 mM when measured by flame photometry, as compared with a mean value of 145 ± 1 mM obtained by potentiometry on 52 samples of proximal tubular

Table 13.1 Na Concentration in Serum, Glomerular Fluid, and Tubular Fluid Measured In Vitro[8]

Necturus No.	[Na]serum(mM) Potentiometric	[Na]serum(mM) Photometric	[Na]glomerulus(mM) Potentiometric	[Na]glomerulus(mM) Photometric	[Na]tubule(mM) Potentiometric	[Na]tubule(mM) Photometric	$\dfrac{\text{[Na]glomerulus}}{\text{[Na]serum}}$ *	$\dfrac{\text{[Na]tubule}}{\text{[Na]glomerulus}}$ *
1			85.8	94.2				
2	87.6	90.2	93.0	98.0	91.8		1.06	0.99
3	89.5	86.7	86.7	89.6	87.1	87.8	0.97	1.00
4	87.2	84.8	87.1	92.1	84.3	88.6	1.00	0.97
5	92.0	97.0	88.2	96.9	89.3	92.1	0.96	1.01
Mean	89.1	89.7	88.2	94.2	88.1	89.5	1.00	0.99
S.E.	1.1	2.7	1.3	1.5	1.6	1.4	0.02	0.01

*Potentiometric.

284

Table 13.2   Na Concentration in Blood, Glomerular Fluid, and Tubular Fluid Measured in Situ[8]

Necturus No.	[Na] blood (mM)	[Na] glomerulus (mM)	[Na] tubule (mM)	$\dfrac{[Na]glomerulus}{[Na]blood}$	$\dfrac{[Na]tubule}{[Na]glomerulus}$
6	92.5	95.1	88.9	1.03	0.93
7	105.5	100.3	110.8	0.95	1.10
8	97.7	85.4	93.9	0.91	1.10
9			93.3		
Mean	98.6	93.6	96.7	0.96	1.04
S.E.	3.8	4.3	4.8	0.04	0.06

fluid. When a 5% correction for the proteins is applied for the plasma, the two values agree closely[9].

13.3.3   Potassium Measurements in Single Nephrons

Studies of K^+ concentration in the proximal tubule have led to conflicting results. In the amphibian kidney Bott et al.[16] have found the K^+ concentration in the proximal tubule to be significantly lower or, more recently, about the same[12] as that of plasma. On the other hand, Oken et al.[17] found a K^+ concentration of fluid collected from the end of the proximal tubule of *Necturus* to be 1.6 times greater than that in the glomerulus. In all the above reported instances, K^+ concentration was determined by flame photometry. To help resolve this controversy Khuri et al.[8, 9] used K^+-sensitive glass electrodes to measure K^+ in proximal tubular fluid (see Table 13.3).

The complete potentiometric determination of potassium involves two separate steps. The first step utilizes the cell shown in Fig. 13.8, which consists of a K^+-indicator electrode against a Na^+-glass reference half-cell. This cell effectively measures a complex K/Na concentration ratio. The second step consists of determining the Na^+ concentration on the same sample by means of the Na^+ microelectrode-RbCl (3 M) reference electrode system shown in Fig. 13.7.

The K^+-sensitive glasses were found to be sensitive to the several cations whose salts could be used as salt-bridge components of the reference half-cell. Thus we resorted to the use of a Na^+-glass electrode as a reference to the K^+-electrode, that is, an electrode system consisting of two glass half-cells. Khuri and Merrill[18] used a Na^+-glass electrode

Table 13.3   K Concentration of Glomerular Fluid and Tubular Fluid Measured In Vitro[8]

Necturus No.	[K] glomerulus (mM)	[K] tubule (mM)	$\dfrac{[K]\text{tubule}}{[K]\text{glomerulus}}$
18	4.0	7.8	1.9
19	2.2	4.1	1.9
20	1.4	2.8	2.0
21	1.6	3.2	2.0
22	2.4	3.4	1.4
Mean	2.3	4.3	1.8
S.E.	0.4	0.9	0.1

as a reference to an H^+-glass electrode in determining blood pH. Such electrode systems consisting of two glass half-cells have no liquid-liquid junction. There is an advantage in substituting a Na^+-glass microelectrode reference for the open-tip conventional micropipette reference filled with salt-bridge electrolyte. Salt leakage from the latter into the microvolume of test sample may alter the solution's ionic strength and affect the measuring electrode.

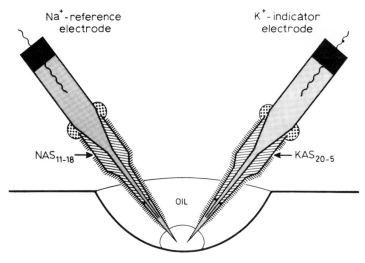

Fig. 13.8. K^+ microelectrode with Na^+ microelectrode cell for K/Na measurement.

The K–Na electrode system is a cell without a liquid-liquid junction. This cell may be represented by the following electrochemical scheme:

Ag;AgCl, KCl (0.1 M)/K-glass/test sample/Na-glass/NaCl (0.1 M),

AgCl;Ag,

writing the electrode equation for each half-cell as

$$E_K = E_K^0 + RT/F \ln[(K^+) + k_{K-Na}(Na^+)] \qquad (13.3)$$

$$E_{Na} = E_{Na}^0 + RT/F \ln(Na^+). \qquad (13.4)$$

The potential of the K–Na electrode system is obtained by subtracting the half-cell equations; this yields

$$E_{K-Na} = E_{K-Na}^0 + RT/F \ln \frac{(K^+) + k_{K-Na}(Na^+)}{(Na^+)}. \qquad (13.5)$$

It is important to note that the K–Na glass electrode cell described by (13.5) measures a concentration ratio; the activity coefficient cancels out in the ratio. In view of the relatively low selectivity of the available K$^+$-sensitive glasses, a k_{K-Na} of 0.1, the Na$^+$ contribution to the potential of the K–Na cell is large and necessitates a separate measurement of Na$^+$. Thus the use of a Na-glass as a reference half-cell did not in itself increase the steps in the potentiometric determination of K$^+$. The pair of Na and K-glass electrodes should have matched slopes which do not differ from each other by more than 2 mV/decade. The K–Na microelectrode system was calibrated in standards that contained 5 mM tris buffer, had a pH of 7.4, and contained a constant NaCl concentration and a variable KCl concentration. The measured potential was plotted against a logarithmic scale of the complex ionic ratio of (13.5). The unknowns were read off the calibration plot by linear interpolation.

K$^+$ concentration was measured in glomerular fluid and fluid collected from the distal end of the proximal tubule of *Necturus*. A mean [K]tubule/[K]glomerulus of 1.8 ± 0.1 was obtained, a value which differs significantly (p < 0.01) from unity. This is in agreement with the ratio of 1.6 ± 0.1 previously obtained by Oken et al.[17], who used flame photometry. This K$^+$ concentration ratio in *Necturus* has been subsequently confirmed by Watson et al.[19], who obtained a ratio of 1.76 by using an ultramicro filter photometer. The K$^+$ concentration ratio found is consistent with (but is not a definite proof of) passive K$^+$ transport in the *Necturus* proximal tubule.

Potassium concentration was measured with the electrodes in 52 samples of rat proximal tubular fluid. A mean ratio [K]tubule/[K]plasma of 0.70 ± 0.03 was obtained. This ratio is not in agreement with the values reported by Bott[20], Wirz et al.[21], and Litchfield et al.[22]. Watson et al.[19] obtained a tubular fluid/plasma K^+ concentration ratio of 1.34 in the rat. However, our potentiometrically determined K^+ concentration ratio is in agreement with the values reported by Bloomer et al.[23] March et al.[24], and Malnic et al.[6].

In 24 of the 52 tubules, complete casts were obtained, which permitted a quantitative identification of the site of puncture. Figure 13.9 represents a plot of the tubular fluid/plasma K^+ concentration ratio against the site of puncture in the proximal tubule. No significant correlation was obtained. Thus no evidence was provided for the tendency of K^+ concentration to fall as the fluid flows down the proximal tubule. A tubular fluid/plasma K^+ concentration ratio of 0.70 is consistent with a potassium-active reabsorptive mechanism located at the luminal membrane of the proximal tubule of the rat.

13.4   pH IN SINGLE NEPHRONS

Several uncertainties are involved in the in-vitro determination of pH on an ultramicrovolume of fluid; among them are the proper thermal equilibration, evaporation, and CO_2 loss. The in-vivo measurement of pH is advantageous because it eliminates the uncertainty involved in the arbitrary choice of a CO_2 tension with which to equilibrate the microsamples. It also avoids the loss of CO_2 to the ambient air. The measurement is carried out at the body temperature of the animal. For in-vivo measurement of pH, microelectrodes of pH glass are superior to quinhydrone

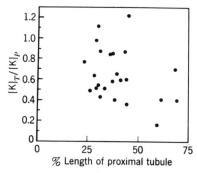

Fig. 13.9. Proximal tubule fluid: plasma potassium concentration ratio as a function of sampling site. (Khuri, Flanigan, and Oken[9]. Reprinted by permission of the *J. Appl. Physiol.*).

microelectrodes because, as we noted earlier, the latter are sensitive to oxidation-reduction reactions, have significant protein errors, and an alkaline error even in moderately alkaline solutions.

13.4.1   pH Glass Microelectrodes

The two types of pH glass microelectrodes used to measure pH in single nephrons in vivo are represented diagrammatically in Fig. 13.4.

Micropipette (Sealed) Type. Figure 13.10 shows the "sealed" micropipette type of pH glass electrode. The sensitive area may be localized to a very small segment near the tip, thus increasing the spatial resolution of the measuring micro-electrode. The fine-point tip and the taper of the shank of the micro-pipette render it highly suitable for micropuncture work.

Corning Code 0150 pH glass in the form of capillary tubing of 1 mm o.d. was broken into 2-in. segments and one end was sealed by a microflame. The capillary segments were filled with a buffer of pH 4, colored with methyl red. The open end of the capillary was then sealed. The filled capillary segment was threaded through the wire-loop of an Alexander and Nastuk type of pipette puller. The filled pH capillary was then pulled into two micropipettes at a relatively low coil temperature. About 10% of micropipettes pulled at such a setting have spontaneously "sealed" tips. Two criteria were used to test whether the tip was sealed: the taking up of colorless distilled water upon dipping the tip, and injecting colored filling solution through the tip by means of pressure applied on the stem side. These micropipettes had tip diameters of 1 μ or less.

Micropipettes with air bubbles in the shank were stored for several days in the vertical position, with their tips pointing downward in closed humid jars (Fig. 13.3). The fitting of an inner concentric micropipette into the pH micropipette segment follows the same procedure as that described for the cation glass microelectrodes described in section 13.3.1.

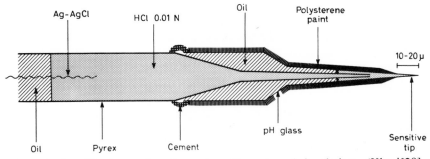

Fig. 13.10. pH glass microelectrode with external insulation. (Khuri[29]. Reprinted by permission of *The Review of Scientific Instruments*).

The uninsulated microelectrode was then allowed to age. Before external insulation, the tips were soaked in distilled water for a couple of days. The external insulation was performed by dipping the micropipettes vertically into polystyrene paint of proper consistency. This was followed promptly by deinsulation of $10-20\,\mu$ of the tip by turning the tip upward and then carefully introducing it through the interface of a droplet of polystyrene solvent. The testing for the adequacy of the insulation and the storage of the complete pH microelectrodes follows the same lines as for cation microelectrodes (described in section 13.3.1). These pH microelectrodes have an internal resistance of about $10^8\,\Omega$.

The reference half-cell for the pH microelectrode consists of a micropipette of an ordinary glass that has a beveled tip of $3-5\,\mu$. This was filled with colored KCl (3 M) and mounted on the injecting type of holder (Fig. 13.5).

Carter[25] built micropipette-type pH electrodes and insulated them externally by a selective silver staining technique. This method consisted of externally applying the silver salt on a pH capillary and heating the coated capillary to a point close to its annealing temperature. As a result, enough silver ions penetrate the glass surface to desensitize it completely. The externally insulated pH capillary is then mounted in a pipette puller. This yields insulated micropipettes except for pH-sensitive tips, perhaps caused by the pulling out of an unstained layer of glass from underneath the stain-insulated surface.

Rector et al.[26] introduced a technique of *glaze insulation* in the construction of micropipette-type pH electrodes. Segments of pH capillaries of Corning Code 0150 glass were insulated externally by painting with a coat of Pemco No. TR-514-A glaze. The painted capillary tubes were dried in air and then heated in an oven at 600°C for 10 min to fuse the glaze to the glass surface. The glazed capillaries were pulled into micropipettes. This method yielded pH microelectrodes that were completely glaze-insulated externally — except for sensitive tips of $1-2\,\mu$ in diameter and $15-20\,\mu$ in length. Rector et al. suggest that the glass of the sensitive tip is pulled from underneath the glaze-insulated superficial layers.

Using the glaze insulation technique, Rector et al. combined the pH microelectrode with the reference micropipette in a double-barreled configuration[26]. A glazed pH capillary was cemented to a Corning Code 0120 lead glass and the double-barreled capillary was mounted in a pipette puller. The capillary was twisted at the softening point and then pulled. A single double-barreled tip resulted, with an open reference side and a closed pH side. Rector et al. used the two types of pH microelectrodes described to measure the intraluminal pH in single nephrons of the rat kidney.

Internal Capillary Type. The internal-capillary, suction-type glass micro-electrodes shown in Figs. 13.11 and 13.12 have some distinct advantages, particularly for pH determination. The test solution is sucked internally — a location which prevents evaporation and gaseous exchange. Several factors contribute to the excellent electrode function of the internal capillary configuration. It possesses a large area-to-volume ratio. Its asymmetry potential is small. Hamilton and Hubbard[27] point out that in the process of drawing a capillary, the outer surface, because it is subjected to higher direct flame temperatures, may experience greater volatilization of alkalies. Therefore, the inner surface of a fine capillary may very well possess electrode function superior to that of the outer surface.

Indicator microelectrodes of the internal-capillary suction type were combined with the reference half-cell to form a single unit. A description of the steps involved in the construction of the single-unit pH glass ultramicroelectrode (Fig. 13.11) follows.

An ordinary glass capillary is pulled into a micropipette and beveled on a rotating grinding stone to a tip diameter of $5-10\ \mu$. A Corning Code 0150 pH capillary is drawn out manually in two stages on a microflame. The second drawing results in a very thin capillary segment with an o.d. of $30-60\ \mu$. Segments of these fine capillaries of $2-3$ cm in length and having relatively thin and smooth walls are selected. The pH microcapillary segment is then waxed in the stem of a larger capillary of ordinary glass; the enclosed end is left patent.

A capillary of ordinary glass is pulled into two segments. It is then filled with agar 2% gel containing KCl (3 M). A careful check is made for any discontinuities in the agar. The mounted pH capillary and the agar-

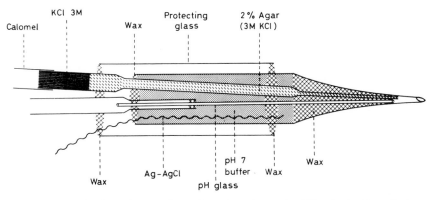

Fig. 13.11. Single unit pH glass ultramicroelectrode (Khuri, Agulian, Oelert, and Harik[1]. Reprinted by permission of Springer-Verlag).

KCl reference capillary are waxed together at two points close to their free ends. Under microscopic control the united pair of capillaries is threaded inside the micropuncture pipette. With the use of controlled heat, the wax is melted down to fill the shank of the micropipette to the tip of the pH capillary. The tip of the agar-KCl reference capillary is made to extend about 20 μ beyond the tip of the pH capillary. The stem of the micropuncture pipette is filled with a pH 7 buffer. An Ag–AgCl wire is inserted into the buffer as an internal reference element of the indicator pH capillary. The open end of the stem of the micropipette is sealed with wax. Three elements emerge from the sealed stem of the single unit electrode: (a) the stem of reference capillary, which is filled with aqueous KCl (3 M) and connects to a calomel electrode; (b) the Ag–AgCl wire of the outer reference solution of the pH glass capillary; and (c) the open stem of the pH glass capillary holder, which is fitted into polyethylene tube.

The single-unit microelectrode shown in Fig. 13.11 measures the pH of a sample of less than 0.05 μl in volume. The microelectrode was found useful for making in-vivo pH measurements of the luminal fluid of microtubular structures such as kidney tubules. The tubular fluid or the pH buffer standards are withdrawn under anaerobic conditions into the pH capillary by a combination of capillarity and very light suction. The physical combination of the glass indicator half-cell and the reference half-cell into a single unit eliminates the need for the double penetration of the tissue by two separate half-cells. When a transmembrane potential exists, both the measuring and the reference half-cells must be localized within a single phase on one side of the membrane in question; otherwise the measured potential includes a membrane potential. In cases in which there is no potential difference across a membrane separating two phases, the membrane may be impaled with the measuring microelectrode alone and the reference half-cell may be placed on the outside surface. Under such circumstances there is no advantage in combining the two half-cells into a single unit.

Figure 13.12 shows an internal capillary suction-type pH glass microelectrode with a glass-to-glass seal. In this case the indicator half-cell was not combined with the reference electrode. The use of a glass-to-glass seal has proved to be superior to sealing by wax, as it results in a higher yield of microelectrodes with ideal electrode function over a long span.

Figure 13.12 also shows microelectrode of the internal capillary type that employs a glass-to-glass seal. The Corning 0150 pH glass electrode capillary is fused at one end to the interior of the shank of the inactive glass (Corning 0120) and fused at the other end to a capillary holder of the same inactive glass. The inactive glass and the electrode glass fused readily because of their similar thermal characteristics. Microcapillaries of the sodium electrode glass NAS_{11-18} fused readily with Corning 1720 while those of the potassium glass KAS_{20-5} fused with Corning 0080. Table 13.4 gives some of the thermal properties of pH, sodium, and potassium electrode glasses in association with the selected inactive glasses with which they fuse readily.

A segment of Corning 0120 capillary with an o.d. of 1 mm is pulled into micropipettes by means of a horizontal pipette puller. The micropipettes are then beveled on a rotating grinding metal to a tip diameter of 8–12 μ. The tip is then cleaned with distilled water and acetone. Corning 0150 pH glass capillaries with an o.d. of 1 mm are pulled manually on a micro-flame. Microcapillaries with an i.d. as small as 30 μ may be produced. Special care is taken to minimize the flame exposure of the glass. Segments of these fine microcapillaries of about 1 cm in length and having relatively thin and smooth walls are selected.

The end of the pH microcapillary segment (i.d. 30 μ and 8–10 mm in length) is introduced into the terminal end of a capillary of Corning 0120 inactive glass. By means of a micromanipulator the concentric joint is centered in the wire loop of a pipette puller and sealed with controlled heat. The microcapillary, fused to its holder, is then freed. The beveled

Table 13.4   Some Thermal Properties of the Electrode and "Inert" Glass Pairs That Account for Their Readiness to Fuse Together* [28]

Glass	Soft Point (°C)	Thermal Expansion (10^{-7} in/in/°C)
pH electrode		
Corning 0150	655	110
Corning 0120	630	89
Na$^+$ electrode		
NAS_{11-18}	> 970	53
Corning 1720	915	42
K$^+$ electrode		
KAS_{20-5}	796	102
Corning 0080	695	92

*Tabulated values are quoted from Dr. N.C. Hebert, Corning Glass Works, Corning, N.Y.

micropipette of Corning 0120 glass is then mounted on a micromanipulator and introduced through the wire-loop of the pipette puller. The tip of the micropipette is extended out of the wire loop. Under microscopic control, the pH glass microcapillary is threaded as far as possible down the interior of the shank of the micropipette. The tip of the pH glass capillary is fused to the micropipette with controlled heat.

The micropuncture pipette is filled with a pH 7 buffer as an inner reference solution. An Ag–AgCl wire is inserted into the inner solution as an internal reference element of the measuring pH glass electrode microcapillary. The stem of the micropipette is then permanently sealed with wax, leaving an emerging silver wire and the holder of the electrode capillary. The latter is fitted with polyethylene tubing which connects to a syringe. A segment of thick glass is fitted around the stem of the microelectrode for mechanical protection.

The pH glass electrode microcapillary (Fig. 13.12) has a capacity of about 0.01 μl. For in-vivo measurements, the reference half-cell is placed in the extracellular space. For in-vitro measurements on small samples, the reference half-cell incorporates an ordinary micropipette (tip diameter 1 μ) filled with 3 M KCl. Following the complete filling of the pH microcapillary, the potential attains a stable value within 40 sec.

13.4.2   pH Measurements in Single Nephrons (In Vivo)

In single nephrons of the rat kidney, pH was measured in vivo with both the sealed micropipette and the internal capillary types of glass microelectrode. The risk of breaking the pH microelectrodes during the impalement of the tubular epithelium was circumvented by introducing the microelectrodes into the lumina of nephrons through preformed holes. The pH microelectrode cell was calibrated in three buffer standards maintained on the same thermostatic table on which the rat was exposed for renal micropuncture. A Cary vibrating reed electrometer was used for potential measurements. The potential was displayed on a recorder and was read after a steady value was attained. All equipment was placed inside a Faraday cage and grounded to a common copper bar.

Intraluminal pH measurements[29] in the proximal tubules were made with the sealed micropipette type of pH electrode (Fig. 13.10). The entire sensitive area of the pH microelectrode and the tip of the reference micropipette (Fig. 13.5) were localized within the lumen of a single proximal tubule (Fig. 13.2). At each measurement a fresh injection of KCl was made to free the tip of any organic matter and to localize the tip of the reference capillary in the intraluminal position. The in-vivo intraluminal pH of 20 proximal tubules of 8 nondiuretic rats yielded a mean value of 7.22 ± 0.03. This mean intraluminal pH is significantly

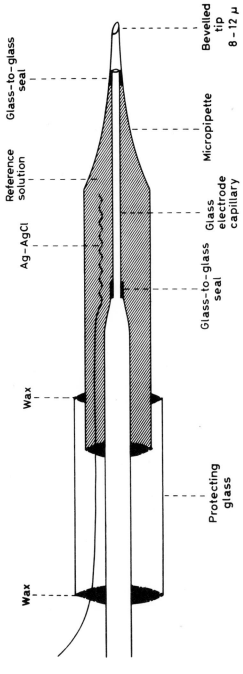

Glass-to-glass seal

Bevelled tip 8 – 12 μ

Reference solution

Micropipette

Glass electrode capillary

Ag–AgCl

Glass-to-glass seal

Wax

Protecting glass

Wax

Fig. 13.12. pH ultramicroelectrode, internal capillary type, glass-to-glass seal. (Khuri, Agulian, and Harik[28]. Reprinted by permission of Springer-Verlag.)

different (p < 0.01) from the mean arterial blood pH of 7.43 ± 0.01 obtained simultaneously in the same rats.

Using the internal-capillary, suction-type microelectrodes (Figs. 13.11 and 13.12), pH was measured[30] simultaneously in the proximal tubule and in arterial blood in vivo. The arterial blood pH was measured by cannulating the femoral artery with an enlarged version of the internal capillary electrode. When one intermittently unclamps the artery proximally, arterial blood flows into the pH capillary.

Forty proximal tubular pH measurements were made on 11 nondiuretic rats. These measurements yielded a mean proximal tubular pH of 7.18 ± 0.02 (s.e.), a value which is significantly different (p < 0.001) from the mean arterial blood pH 7.41 ± 0.01 obtained simultaneously.

The 40 proximal tubular pH measurements were performed at different points along the length of different proximal tubules. The prospective identification of beginning, middle, and late proximal convoluted tubule was made possible by the intermittent intravenous infusion of 8% lissamine green solution. The exact site of puncture for pH measurement in the tubule was determined retrospectively by Latex injection into the tubule and the microdissection and measurement of the resulting tubular cast. A plot of the measured tubular pH against the percent length along the proximal convoluted tubule displayed an inverse relationship, that is, the pH fell as the tubular fluid moved down the length of the proximal convoluted tubule. A calculated correlation coefficient $r = -0.92$ was obtained, indicating a significant inverse relationship. At the most proximal sites of the tubule, pH values as high as 7.40 were obtained, while the pH fell down to values as low as 6.80 in the most distal parts of the proximal convoluted tubule.

Rector et al.[26] measured randomly the intraluminal pH of proximal tubules of the rat kidney by using pH glass microelectrodes; they reported a mean value of 6.82 ± 0.13 (s.d.). Our mean value for proximal tubular pH of 7.18 ± 0.02 (s.e.) is higher than the reported mean value of Rector et al. Several precautions were taken to avoid factors that may falsely elevate the measured pH of the tubular fluid. To avoid decreasing the transit time of flow of tubular fluid, the pH microelectrodes used had a small capacity of 0.01 μl or less and the collection time of the fluid sample was made to exceed 1 min.

13.4.3   Bicarbonate Measurements in Single Nephrons (In Vitro)

Quinhydrone microelectrodes prepared after the design of Pierce and Montgomery[31] have been used by several investigators[32–34] to measure the concentration of bicarbonate ion in tubular fluid. In all these studies the proximal tubular fluid/plasma bicarbonate concentration ratio was significantly less then unity under normal conditions.

Khuri et al.[30] have utilized the pH glass microelectrode (Fig. 13.12) to measure the bicarbonate concentration of proximal tubular fluid. Samples were collected from different parts of the proximal tubule and placed in a Lucite cup under a thick layer of continuously circulating mineral oil maintained at 37°C and equilibrated with a gas mixture having 6% CO_2. Sufficient time was allowed for the samples to reach the CO_2 tension of the covering oil. The pH microelectrode (Fig. 13.12) and a reference micropipette filled with KCl 3 M and having a tip diameter of 1 μ or less were brought into contact with the tubular fluid sample under the equilibrated oil. A portion of the sample was sucked into the pH capillary electrode and the potential recorded. This method yielded an in-vitro pH value for the sample of tubular fluid when compared with the readings in calibrating buffer standards. Simultaneously the pCO_2 of the equilibrated oil was determined by using a Severinghaus CO_2 electrode. Then the bicarbonate concentration of the tubular fluid was calculated from

$$pH = 6.1 + \log \frac{[HCO_3^-]}{0.0301 \, pCO_2} \qquad (13.6)$$

where $[HCO_3^-]$ designates bicarbonate concentration and pCO_2 represents carbon dioxide tension in millimeters of mercury.

A mean proximal tubular fluid/plasma bicarbonate concentration ratio of 0.51 ± 0.02 (s.e.) was obtained on 28 collections from rats receiving an infusion containing NaCl 125 mM and $NaHCO_3$ 25 mM. When infusions of progressively increasing concentration of bicarbonate were administered, the proximal tubular fluid/plasma bicarbonate concentration ratio rose to 0.76 for the infusate with 75 mM bicarbonate and did not exceed that value during infusions of 150 mM $NaHCO_3$. This observation that the bicarbonate ratio remains below unity is in conflict with the data of others[34] and implies that the proximal tubular epithelium is preferentially more permeable to bicarbonate than to chloride ion.

REFERENCES

[1] R. N. Khuri, S. K. Agulian, H. Oelert, and R. I. Harik, *Pflueg. Arch. Ges. Physiol.*, **294**, 291 (1967).

[2] J. J. Diamond, and D. Hubbard, *J. Res. Nat. Bur. Std.*, **47**, 443 (1951).

[3] A. Van Itterbeek, and W. Vereycken, *Z. Physik. Chem.*, **B48**, 131 (1941).

[4] R. N. Khuri, in *Glass Electrodes for Hydrogen and Other Cations: Principles and Practice* (G. Eisenman, ed.), Dekker, New York, 1967.

[5] H. Grundfest, C. Y. Kas, and M. Altamiro, *J. Gen. Physiol.*, **38**, 245 (1954).

[6] G. Malnic, M. Klose, and G. Giebisch, *Am. J. Physiol.*, **206**, 674 (1964).

[7] G. Eisenman, *Biophys. J.*, **2**, 259 (1962).

[8] R. N. Khuri, D. A. Goldstein, D. L. Maude, C. Edmonds, and A. K. Solomon, *Am. J. Physiol.*, **204**, 743 (1963).

[9] R. N. Khuri, W. J. Flanigan, and D. E. Oken, *J. Appl. Physiol.*, **21**, 1568 (1966).

[10] J. T. Alexander, and W. L. Nastuk, *Rev. Sci. Inst.*, **24**, 528 (1953).

[11] K. Frank, and M. C. Becker, in *Physical Techniques in Biological Research*, vol. V, Part A (W. L. Nastuk, ed.), Academic, New York, 1964.

[12] P. A. Bott, *Am. J. Physiol.*, **203**, 662 (1962).

[13] D. E. Oken, G. Whittembury, E. E. Windhager, and A. K. Solomon, *Am. J. Physiol.*, **204**, 372 (1963).

[14] B. Schmidt-Nielsen, K. J. Ullrich, R. O'Dell, G. Pehling, C. W. Gottschalk, W. E. Lassiter, and M. Mylle, *Excerpta Med. Sect.*, **29**, 72 (1960).

[15] E. E. Windhager and G. Giebisch, *Am. J. Physiol.*, **200**, 581 (1961).

[16] P. A. Bott, *Am. J. Med. Sci.*, **227**, 102 (1954).

[17] D. E. Oken, and A. K. Solomon, *Am. J. Physiol.*, **204**, 377 (1963).

[18] R. N. Khuri and C. R. Merril, *Phys. Med. Biol.*, **9**, 541 (1964).

[19] J. F. Watson, J. R. Clapp, and R. W. Berliner, *J. Clin. Invest.*, **43**, 595 (1964).

[20] P. A. Bott, in *Proceedings at the Eighth Annual Conference on the Nephrotic Syndrome* (J. Metcoff, ed.), New York, 1957.

[21] H. Wirz and P. A. Bott, *Proc. Soc. Exp. Biol. Med.*, **87**, 405 (1954).

[22] J. B. Litchfield and P. A. Bott, *Am. J. Physiol.*, **203**, 667 (1962).

[23] H. A. Bloomer, F. C. Rector, and D. W. Seldin, *J. Clin. Invest.*, **42**, 277 (1963).

[24] D. G. Marsh, J. K. Ullrich, and G. Rumrich, *Pflueg. Arch. Ges. Physiol.*, **277**, 107 (1963).

[25] N. W. Carter, U.S. Patent 3129160 (1964).

[26] F. C. Rector, N. W. Carter, and D. W. Seldin, *J. Clin. Invest.*, **44**, 278 (1965).

[27] E. H. Hamilton, and D. Hubbard, *J. Res. Nat. Bur. Std.*, **27**, 27 (1941).

[28] R. N. Khuri, S. K. Agulian, and R. I. Harik, *Pflueg. Arch. Ges. Physiol.*, **301**, 182 (1968).

[29] R. N. Khuri, *Rev. Sci. Inst.*, **39**, 730 (1968).

[30] R. N. Khuri, R. I. Harik, and S. K. Agulian, *Pflueg. Arch. Ges. Physiol.*, in press (1968).

[31] J. A. Pierce, and H. Montgomery, *J. Biol. Chem.*, **110**, 763 (1935).

[32] C. W. Gottschalk, W. E. Lassiter, and M. Mylle, *Am. J. Physiol.*, **198**, 581 (1960).

[33] J. R. Clapp, J. F. Watson, and R. W. Berliner, *Am. J. Physiol.*, **205**, 693 (1963).

[34] F. C. Rector, H. A. Bloomer, and D. W. Seldin, *J. Clin. Invest.*, **43**, 1976 (1964).

CHAPTER 14

Measurement of Cytoplasmic Resistivity
by Means of the Glass Microelectrode

OTTO SCHANNE

Département de Biophysique, Faculté de Médecine, Université de
Sherbrooke

This work has been supported by grants from the Medical Research Council of
Canada, the Muscular Dystrophy Association of Canada, and the Canadian
Heart Foundation.
Professor Schanne is a Scholar of the Medical Research Council of Canada.

14.1 CYTOPLASMIC RESISTIVITY

Modern electrobiology finds itself in a strange situation: our methods
require highly sophisticated electronic circuits and the most recent products of scientific instrumentation, but our theoretical concepts date from
the second half of the nineteenth century.

A very striking example of this discrepancy in methods and theoretical
approach occurs in electrophysiological research on isolated cells. We
explain the biological potentials with the theory of Nernst[1] and we treat
the spread of the electrotonic potential with the analogy developed by
Hermann[2] and Cremer[3] of a cable and a longitudinal biological cell.
This theory of the core conductor implies that a nerve or a muscle cell
consists of an isolating sheath and a central conducting medium. Later,
this concept was generalized for the electrobiological treatment of cells
independent of their geometry: the cell consists of a membrane having the
properties of a poor conductor which controls the bioelectric potentials;
the cytoplasm is an aqueous solution containing the intracellular ions and
is therefore, electrically speaking, a rather good conductor.

Since it became possible to obtain quantitative information about the
specific resistance of the cytoplasm — and the main part of this chapter is

devoted to how this was done – the specific resistivity of the cytoplasm has been found to be higher than that of the solution surrounding the cell. Approximately speaking, the value of the cytoplasmic resistivity is at least twice as high as the resistivity of the extracellular medium (Table 14.1).

The discrepancy between the resistivity of the outside solution and the cytoplasm cannot be explained by the difference in conductance of cations inside and outside the cell. If one assumes in the case of the sartorius muscle a concentration of 140 mM sodium chloride outside and 140 mM potassium chloride inside the cell, a respective resistivity of 80 and 65 Ω cm is obtained. These values are only correct when chloride is used as an anion on both sides of the membrane. However, this is not true for the frog sartorius. If one assumes the potassium as di-H-orthophosphate inside the frog sartorius, one obtains a cytoplasmic resistivity of 198 Ω cm, which could qualitatively explain the experimental results. Analytical data, however, show that the phosphate concentration inside the muscle is insufficient to neutralize all the potassium. On the other hand, it is possible to argue that the viscosity of the cytoplasm restricts the ionic mobility. This phenomenon could also explain the increased cytoplasmic resistivity. Taft and Malm[7] measured the conductivity of potassium chloride solutions in the presence of gelatine and found a slight increase in resistivity – owing to the presence of gelatine. Katz[5] argued that the presence of myofibrils, acting as insulators inside the cell, would tend to increase the cytoplasmic resistivity.

These examples illustrate that the interpretation of measurements of the cytoplasmic resistivity has been anything but uniform. Recent results obtained with microelectrodes make the problem appear even more complicated.

Table 14.1   Resistivity of Medium and Cytoplasm of Different Tissues

	ρ cyto (Ω cm)	ρ medium (Ω cm)	source
Nonmedullated nerve (*Carcinus maenas*)	90	24 (sea water)	[4]
Skeletal muscle (sartorius)	218–248	73 (Ringer)	[5]
Cardiac muscle (Purkinje fiber of kid)	105	51 (Tyrode)	[6]

14.2   DETERMINATION OF CYTOPLASMIC RESISTIVITY

Every determination of cytoplasmic resistivity is influenced by the fundamental problem of any measurement of biological impedance: the quantity measured is of a pure electrical nature with the dimension of ohm while the quantity of interest is a resistivity with the dimension of ohm-centimeter. To obtain the value of the resistivity, it is necessary to have an electrical equivalent circuit to interpret the resistance value measured. Usually impedance can be measured with reasonable accuracy, but it is very difficult to find criteria for the correctness of the electrical model. In general, the main difficulty with the equivalent circuits is that of taking into account the properties of electrodes and the geometry of the biological system. For longitudinal cells, the analysis of the measured impedance values to determine the cytoplasmic resistivity is more complicated with a method using external electrodes (conductivity cells or two external electrodes) than with methods using microelectrodes.

14.2.1   Methods Using the Conductivity Cell

Because this chapter mainly concerns the determination of the cytoplasmic resistivity with microelectrodes, only a short review of the method using the conductivity cell will be given. More detailed information can be obtained from review articles [8–10]. The methods used for these measurements employ modifications of devices for measuring the conductivity of electrolyte solutions. The main circuit is a Wheatstone bridge with an ac generator and a suitable null-indicator as shown in Fig. 14.1. One of the impedances Z_1–Z_4 is the conductivity cell containing

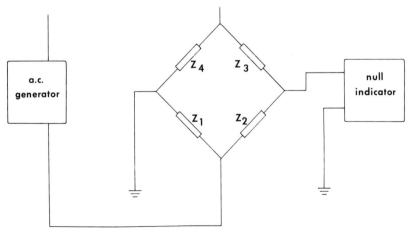

Fig. 14.1. Basic circuit for determination of biological impedances with ac bridge.

the biological object. The type of conductivity cell used depends largely upon the nature of the tissue under investigation. The nature of the impedances Z_1–Z_4 depends also on the impedance of the biological object and the frequency range used. Technical details about construction of the bridge and the conductivity cell can be found in an article by Schwan[10].

Except in the case of extremely large cells, this method results in a mean value of the electrical parameters of a certain cell population. Nevertheless, experiments using a conductivity bridge have produced the first useful information about the basic electrical parameters of biological cells. Even in the case of the frog sartorius, which has been thoroughly investigated with microelectrodes, the technique using the conductivity bridge can be an effective alternative to the microelectrode method under certain conditions[9].

14.2.2   Method Using Two Microelectrodes

This method is used more often to determine the membrane resistance of a cell by measuring its input resistance than to measure the cytoplasmic resistivity. For reasons of geometry, it has been applied successfully only to longitudinal cells such as skeletal muscle fibers, nonmyelinated nerve fibers and Purkinje fibers.

Theory. The theoretical basis for measurements on cells with core conductor structure was adapted for modern measurements by Hodgkin and Rushton[11]. They used external electrodes for their experiments, but the theoretical background for the microelectrode method can be easily derived from Hodgkin and Rushton's analysis by neglecting the influence of the extracellular medium in the equations. Figure 14.2 shows a diagram of the experimental situation, M being a membrane and $x = 0$ being a point far from the ends of the cell where the current I_0 is injected. The current will flow through the longitudinal resistance (the cytoplasm) and through the membrane back to the reference electrode in the outside

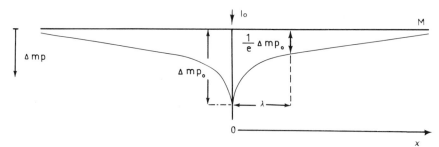

Fig.   14.2. Change of membrane potential (Δ mp) along membrane M when current is injected at $x = 0$. (For explanation, see text.)

solution. The current flow through these resistances results in a voltage drop across the membrane, and this drop is a function of the distance between the current-injecting and the potential-recording electrode. The change in the membrane potential varies in this case exponentially with the distance x. The distance where the initial change of potential has decreased by a factor of $1/e$ is called the length, space, or attenuation constant λ. The formulas controlling the potential distribution along the membrane are shown in Table 14.2. Equation 14.1 describes in a general way the distribution of an electrotonic potential as a function of the distance from the current electrode and the time t. In the simple case of a linear cable model with the potential not varying with time, (14.2) describes the potential distribution along the fiber. Equation 14.3 is the solution of (14.2) for a penetration in the middle of the fiber. The factor $\frac{1}{2}$ occurs because an equal amount of current flows in the positive and negative direction of x. In other words, with the current electrode in the middle of the fiber there are two equal input resistances in parallel. If the distance between the electrodes is zero. (14.4) holds. Rearranging (14.4), one obtains (14.5), which shows the input resistance as a function of the specific resistance of the core conductor R_i (cytoplasmic resistivity), the membrane resistance for a square centimeter of membrane R_m, and the diameter of the fiber d. Equation 14.6 represents the definition of the length constant as a function of R_m and R_i. In an actual experiment the input resistance, the length constant, and the fiber diameter can be measured under favorable conditions. Therefore the two unknowns R_i and R_m are determined by (14.5) and (14.6). A combination of (14.5) and (14.6) results in (14.7) and (14.8), which give the value of the membrane resistance and of the cytoplasmic resistivity as a function of the input resistance, the length constant, and the fiber diameter.

Among those variables, d must be determined by direct measurement with a microscope. Very rarely can the whole diameter of a fiber be seen, therefore measurements of d are subject to experimental errors. A comparison of (14.7) and (14.8) shows that the cytoplasmic resistivity depends on the square of d whereas R_m is only proportional to d with an exponent of one when the input resistance and the length constant are measured. Accordingly, in using this method one can expect a greater accuracy for the membrane resistance than for the cytoplasmic resistivity.

Circuits. The standard circuit for measurements of the input resistance and the length constant with two microelectrodes is shown in Fig. 14.3. With minor modifications, this circuit has been used for most of the determinations of cytoplasmic resistivity with two microelectrodes [6, 12, 13]. In this circuit, the current is injected by the microelectrode C,

Table 14.2   The Distribution of the Electrotonic Potential along a Longitudinal Cell

$$\lambda^2\nabla^2 V - \tau_m\frac{\delta V}{\delta t} - V = 0 \tag{14.1}$$

$$\lambda^2\frac{d^2V}{dx^2} - V = 0, \tag{14.2}$$

$$V = \tfrac{1}{2}I_0\sqrt{r_m r_i}\,e^{x/\lambda} \tag{14.3}$$

$$\frac{V_0}{I_0} = \tfrac{1}{2}\sqrt{r_m r_i} \tag{14.4}$$

$$R_{\text{inp}} = \frac{V_0}{I_0} = \tfrac{1}{2}\sqrt{r_m r_i} = \sqrt{\frac{R_m R_j}{\pi^2 d^3}} \tag{14.5}$$

$$\lambda = \sqrt{\frac{R_m}{R_i}\frac{d}{4}} \tag{14.6}$$

$$R_m = 2\pi d\lambda R_{\text{inp}} \tag{14.7}$$

$$R_i = \frac{\pi d^2}{2}\cdot\frac{R_{\text{inp}}}{\lambda} \tag{14.8}$$

$\lambda = \sqrt{\frac{r_m}{r_i}}$ = length constant   (cm)

$r_m = \frac{R_m}{d\pi}$ = membrane resistance per unit length   (Ωcm)

R_m = resistance of 1 cm² of membrane   (Ωcm²)

$r_i = \frac{4R_i}{d^2\pi}$ = longitudinal resistance per unit length   (Ω/cm)

R_i = specific resistance of material in core   (Ωcm)

R_{inp} = input resistance   (Ω)

d = fiber diameter   (cm)

$\tau_m = c_m r_m$ = time constant of membrane   (sec)

$c_m = \pi d C_m$ = membrane capacity per unit length   (F/cm)

C_m = capacity of 1 cm² of membrane   (F/cm²)

∇^2 = Laplacian

V = electrotonic potential at distance x (V)

V_0 = potential at $x = 0$ (V)

I_0 = current injected into membrane at $x = 0$ (amp)

x = distance between current-injecting and potential-recording microelectrode   (cm) and the current pulse is supplied by a square wave generator G through a resistor R. This resistor has a double function: it prevents current from being drawn from the fiber during the time the generator is connected to the circuit, and it makes the current in the circuit less sensitive to changes in the electrode resistance. V_2 is a voltmeter with electrometer input to measure the current I. P is the second microelectrode for the recording of the change in the membrane potential; V_1 is also a voltmeter with an electrometer input. Usually one oscilloscope channel in a double-beam oscilloscope is used for V_1 and V_2, together with cathode follower input stages.

There are some modifications possible in the circuit of Fig. 14.3. One can find a low resistance of several hundred kΩ between the generator and the ground. In this case the current is monitored as voltage drop across this resistance[12,14].

The resistance R in Fig. 14.3 usually has values between 20 and 100 MΩ, but some researchers omit it completely[13]. This is possible when low-resistance electrodes are used for injection of the current because such electrodes have a lesser tendency to change their resistance as a function of the current. Moreover, the muscle of the crayfish on which Henček and Zachar[13] worked has an input resistance of about 40 kΩ; that is, the voltage drop across the membrane using a microelectrode of 10 MΩ will be less than 1 mV because of the current drawn from the cell. If one is working with structures having a higher input resistance—such as the frog sartorius or the Purkinje fiber—omitting the resistance R can change the membrane potential by a few millivolts.

Evaluation of the method. As has been indicated in the *Theory* section above, a mathematical model is necessary for the application of the method. At present, the only practical model is the cable model applicable to longitudinal cells such as the algae *Nitella*, the giant axons, the Purkinje fiber, and some muscle fibers. Models for syncytial structures exist[15–17], but they have not been perfected to a degree which permits a quantitative

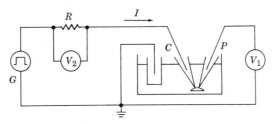

Fig. 14.3. Standard circuit for measuring input resistance and length constant. (For explanation, see text.)

evaluation of the cytoplasmic resistivity under actual experimental conditions.

There are also technical restrictions to the method for measuring the cytoplasmic resistivity with two microelectrodes: the fibers have to be on the surface of the preparation, otherwise it is not possible to penetrate them with two microelectrodes under microscopic control. Moreover, measurements of fiber diameters can be taken only on superficial fibers.

For the determination of the length constant, several penetrations in the same fiber are necessary; however, this can reduce the membrane potential by several millivolts [18]. It is also possible that the fiber is damaged by the repeated penetrations, resulting in a reduced value for the length constant.

In spite of these restrictions the method is useful to obtain the values of the cytoplasmic resistivity in longitudinal cells. It permits results from single cells, whereas the methods using extracellular electrodes or the conductivity cell give only mean values of this parameter in a certain cell population. However, the evaluation of the cytoplasmic resistivity using two microelectrodes is a time-consuming process; consequently, most of the published values were obtained in normal Ringer or Tyrode solution [6, 13]. Researchers using this method have been generally more interested in the membrane resistance than in the cytoplasmic resistivity, which was considered to be a constant. Jenerick [19] assumed the cytoplasmic resistivity of the frog sartorius muscle to be 250 Ω cm and invariant with changes in the composition of the outside solution; at that time no information about a change in the cytoplasmic resistivity was available. In 1965 Maiskii [12] measured in the frog sartorius the length constant and the input resistance in different potassium concentrations. He found a nonsignificant change in 20-mM KCl and a statistically significant decrease of the cytoplasmic resistivity in 100-mM isotonic potassium chloride solution. This was the first indication that the cytoplasmic resistivity cannot always be considered as a constant.

To sum up, one can say that the method of measuring the cytoplasmic resistivity with two microelectrodes is superior to the methods using external electrodes because it permits measurements on single cells. But the method is also limited by the availability of an electrical model for the cells under investigation, the time required for the evaluation of the results, and by technical difficulties arising from the technique of penetration with two microelectrodes.

14.2.3   Measurement of Cytoplasmic Resistivity Using One Microelectrode

If one is interested in the routine measurements of the cytoplasmic resistivity, one has to eliminate the disadvantages inherent in the methods discussed in the preceding sections; that is, one has to find a method which can fulfill the following conditions: The method has to be applicable to all cells that can be penetrated by microelectrodes (independence of a mathematical model), the technique should be as easy as possible, and one should be able to have the result of the measurement during or immediately after the measurement is made. The first two conditions can be fulfilled by using the change of the resistance of a microelectrode as an indicator for the cytoplasmic resistivity. The third condition is then reduced to a problem of instrumentation.

Theory. The first indication that there are changes of the resistance of a microelectrode as a function of the resistivity of the solution surrounding the tip of the microelectrode came from a series of resistance measurements which resulted in unusually high input resistances of biological cells [20–22]. It is now possible to show that experimenters who found high values for the input resistance used methods in which the electrode resistance entered as a constant in the calculation of the input resistance. The resistance of the microelectrode is usually measured when the microelectrode is immersed in the extracellular medium. Most of the high input resistance values can be explained using the following assumptions: the microelectrode resistance depends upon the resistivity of the solution outside the microelectrode, and the resistivity of the cytoplasm is higher than the resistivity of the surrounding solution. The second assumption is verified by experiments in which the methods produce results independent of the microelectrode resistance (Table 14.1). In the literature there is information that the microelectrode resistance depends on the concentration and therefore on the resistivity of the solution surrounding the tip of the microelectrode [23–27]; thus the first assumption is also supported by experimental evidence.

For the frog sartorius, for example, with methods independent of the microelectrode resistance, an input resistance of about 400 kΩ has been found. The values measured with methods dependent upon the resistance of the microelectrode are about 3 MΩ [28].

The difference between these two resistance values is too large to be attributed to an experimental error; therefore we have assumed that the resistance of the microelectrode changes according to the difference in resistivity of the Ringer solution and the resistivity of the myoplasm of the frog skeletal muscle [5]. To test this hypothesis it was necessary to have precise information about the change of the microelectrode resistance as a function of the concentration and composition of the solution in which the microelectrode is immersed.

Change of Microelectrode Resistance as a Function of the Resistivity of the Medium Surrounding the Microelectrode Tip. There is general agreement in the literature that microelectrode resistance is a function of the resistivity of the medium outside the tip of the microelectrode. Amatniek[23] proposed a formula for the microelectrode resistance which is independent of a diffusion mechanism. This formula predicts a linear relationship between the microelectrode resistance and the resistivity of the outside solution. The relation derived by Lanthier and Schanne[27] takes into account the diffusion owing to the concentration gradient at the microelectrode tip. Figure 14.4 shows the resistance of a microelectrode made of Supremax glass No. 2955* filled with 3 M KCl. From Fig. 14.4 it is evident that there is no linear relationship between the microelectrode resistance and the resistivity of the solution outside the microelectrode. On the contrary, it can be shown that the expression derived for a diffusion mechanism at the microelectrode tip agrees well with experimental data[27]. It should be noted here that all the theoretical data for micro-

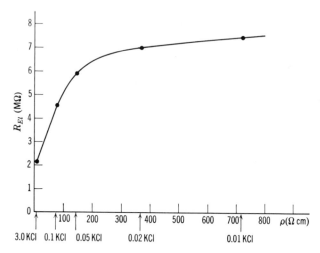

Fig. 14.4. Resistance of microelectrode R_{El} immersed in KCl solutions of different resistivities ρ. For the measurements, KCl solutions of different concentrations were used as indicated by the arrows.

*The glass Supremax No. 2955 is a brand made by Schott and Gen., Mainz, Germany. So far as we know, however, the company has discontinued the production of this glass. The supremax No. 2955 is comparable to the Pyrex Code 1720 produced by Corning Glass Works, Corning, N.Y. As compared with the Pyrex Code 7740 glass, the Supremax glass has a higher softening point. However, microelectrodes made of Pyrex Code 7740 are comparable to those made of Supremax glass.

electrodes given in this section are applicable only to electrodes made of a glass with a negligible fixed charge density (such as Pyrex Code 7740 or Supremax 2955) and filled with highly concentrated electrolytes. For a more general theory of microelectrode properties, the reader is referred to the work of Lavallée and Szabo[29].

For some applications, it is better to use a logarithmic scale on the abscissa. In the range from 20 to $1000 \, \Omega$ cm an approximately linear function is obtained, regardless of whether the resistivity or the concentration is used as the independent variable. Figure 14.5 shows the resistance of two microelectrodes as a function of the resistivity of the outside medium plotted against an abscissa with a logarithmic division. However, when the concentration is used as an independent variable, the slope of the function $R_E = f(\log C)$ is negative[25, 26].

It has to be admitted that the discussion of the change of the micro-electrode resistance has been rather simplified here. There are some other parameters which can influence the function $R_{El} = f(\rho)$. First there is the pH. For the standard electrodes made of Pyrex Code 7740 or Supremax 2955 glass, the effect of the pH is negligible as long as the micro-electrode resistance does not exceed 50 MΩ in 3 M KCl[30]. The second parameter which can influence the microelectrode resistance versus resistivity characteristic is the fact that the microelectrodes have a different sensitivity to cations. Figure 14.6 depicts the characteristic of a microelectrode made of Pyrex glass Code 7740 in NaCl and KCl solutions. In general, for a given resistivity the resistance in a NaCl solution is higher than in a KCl solution. The same effect was also observed by Kern[25] for electrodes made of Supremax glass. In Figure 14.6 each of

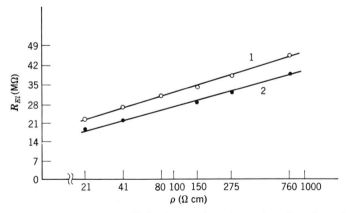

Fig. 14.5. The function $R_{El} = f(\rho)$ of two microelectrodes plotted on abscissa with logarithmic division.

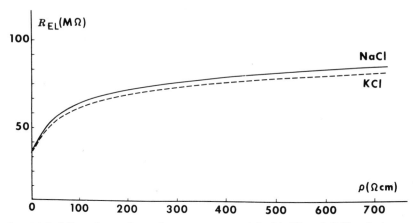

Fig. 14.6. Microelectrode resistance measured in NaCl and KCl solutions of different resistivities.

the traces are two super-imposed curves recorded at a time interval of 2 min. to check the stability of the microelectrode. In contrast to Figs. 14.4 and 14.5, which are curves constructed point by point, this characteristic was made with an automatic device which allows tracing of a resistance characteristic of a microelectrode in about 2 min [31]. This sensitivity to different ions has been observed consistently with Supremax glass and with most of our Pyrex Code 7740 electrodes filled with 3 M KCl.

Another difficulty which can be encountered is the stability of the microelectrode, that is, the time the electrode needs to reach a stable resistance value in a solution of a given resistivity. In general, electrodes with a high resistance are less stable than electrodes with low resistances. It seems that impurities at the microelectrode tip and the process of boiling enhance the instability of the microelectrode. Filling the electrodes under vacuum increases their stability [31], so that even electrodes with resistances higher than 100 MΩ give a reproducible resistance characteristic after several hours.

Principle of the Method. Figure 14.7 shows the method of measuring the cytoplasmic resistivity with one microelectrode. In principle it is only a resistivity measurement inside the cell with a correction for the input resistance of the cell. First, it is necessary to calibrate the microelectrode in solutions of different resistivities, and the calibration of the microelectrode has to be checked after the penetration. Usually the calibration is made in KCl solutions to eliminate errors caused by a different sensitivity of the electrode to Na and to K. A measurement of the microelectrode resistance in the medium and in a KCl solution of the same resistivity as the medium permits an estimation of the ion sensitivity

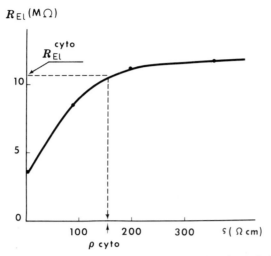

Fig. 14.7. Principle of method for measuring cytoplasmic resistivity with one microelectrode. R_1 is resistance change of microelectrode when penetrating from outside medium into cell; R_{inp}, input resistance of cell under investigation; R_{El}^{cyto}, resistance change of microelectrode caused by difference in resistivity between extracellular medium and cytoplasm. (For explanation, see text.)

of the calibration curve $[R_{El} = f(\rho)]$ to different ions. If one obtains identical values, one can assume — because the medium contains mainly sodium — that there is no strong ion specificity of the electrode. However, this check is not essential for the functioning of the method; the calibration curve was obtained with KCl solutions, therefore the same cation was used for the calibration as is inside the cell.

After penetrating the cell, the resistance of the microelectrode inside the cell is measured (R_1). This resistance represents the change in the microelectrode resistance that reflects a resistivity change between the medium and the cytoplasm (provided the ion specificity to K and Na can be ignored), R_{El}^{cyto}, in series with the input resistance of the cell R_{inp}. This input resistance has to be measured separately and then subtracted from R_1 according to the following equation:

$$R_{El}^{cyto} = R_1 - R_{inp}. \tag{14.9}$$

In longitudinal cells the input resistance can be measured with two microelectrodes. In smaller cells it is preferable to use double-barreled electrodes. This causes technical problems because of the compensation of the coupling resistance between the barrels, but in this case only one penetration is required. Figure 14.8 shows a simple circuit which was used for input resistance measurements in atrial fibers and in liver cells

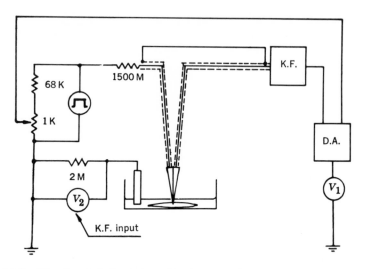

Fig. 14.8. Circuit used for measuring input resistance with double-barreled microelectrodes. *DA* is differential amplifier; *KA* cathode follower; □, square wave generator; V_1, voltage recording channel; V_2, current recording channel. (For explanation, see text.)

of the rat[32]. This circuit corresponds basically to the one in Fig. 14.3 with a provision to compensate the common resistance of the double-barreled microelectrode. When the double-barreled electrode is in the extracellular medium, a current pulse applied through the left barrel will cause a voltage drop across the common resistance of the two barrels. Therefore a part of the output voltage of the stimulator is fed into one input of a differential amplifier while the other input receives directly the voltage drop across the common resistance of the electrode. The 1-kΩ potentiometer is then adjusted so that no deflection on V_1 is visible during a current pulse.

The method of determining the cytoplasmic resistivity and correcting for the input resistance allows the application of low-resistance micro-electrodes, and it allows at the same time the measurement of the mem-brane potential, the microelectrodes used having low tip potentials. This method is preferable in all cases in which information about the membrane resistance is required in connection with measurements of the cytoplasmic resistivity.

If one wants only to measure the cytoplasmic resistivity, one can use microelectrodes with high resistances (filled with $1\,M$ KCl or $0.1\,M$ KCl), which are more sensitive to changes in the resistivity of the outside solution than are low-resistance electrodes. The resistance changes one finds with these electrodes after penetrating a cell are usually around $10\,M\Omega$ or more so that the value of the input resistance can be ignored. Because of the high resistances, one can expect that these electrodes have high tip potentials that lower the values of the recorded membrane potentials[33]. If this is true, the method does not give any information about the real value of the membrane potential.

Circuits. All the work done on instrumentation has aimed at shortening the time for the evaluation of a single measurement. To calculate the results of the first experiments, about one week was necessary for the evaluation. The calibration curve had to be composed of single points. Each of these points represented a resistance and had to be calculated from a voltage-divider circuit[34]. The circuits, developed for the method of measuring the cytoplasmic resistivity, are of three types: a measuring head for the cathode follower to facilitate the resistance measurements, analog circuits to do the computation, and a fully automatic device to trace the calibration curve of the microelectrode.

Measuring Head. In 1964 Schanne, Kern, and Kawata developed a measuring head for the Bak cathode follower which was widely used in the measurements of the cytoplasmic resistivity[34, 35]. Figure 14.9 shows the principle of this unit. The free end of the resistor R_N is connected to a generator. R_N in series with the microelectrode resistance constitutes a voltage divider. Actually R_N is a set of resistors to measure resistances between 1 and $60\,M\Omega$ with reasonable accuracy. R_L are

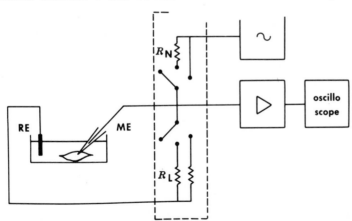

Fig. 14.9. Principle of circuit of measuring head for microelectrode resistances. *ME* is microelectrode; *RE*, reference electrode; $\sim$, sine wave generator; $\triangleright$, cathode follower; R_N and R_L are precision resistors used to measure electrode resistance (R_N) and to short-circuit or substitute microelectrode (R_L).

resistances which make it possible to short-circuit the microelectrode, or they can be used as substitutes for the microelectrode in tests of the electrical circuit. Usually we used low-frequency ac (5–30 cps) for the measurements because this allowed us to eliminate complications from the dc component of the membrane potential.

Analog Circuits. Using the voltage-divider circuit shown in Fig. 14.9, the evaluation of the microelectrode resistance can be made according to the following:

$$R_{El} = \frac{V_{El}}{V_0 - V_{El}} R_N, \tag{14.10}$$

where R_{El} is the microelectrode resistance, R_N is a resistance in series with the microelectrode (see Fig. 14.9), V_0 is the voltage at the output of the sine wave generator, and V_{El} is the voltage drop across the micro-electrode. Equation 14.10 shows that the microelectrode resistance — as measured with the circuit of Fig. 14.9 — can be obtained from a quotient with a difference in the denominator, multiplied by a constant. Thus the value of the microelectrode resistance can be calculated with a modified multiplier divider (Fig. 14.10)[38]. The circuit works with operational amplifier modules and uses for the division the principle of logarithmic conversion. In the diagram there are three main parts: to the left is a rectifier circuit for the input ac signal, because the logarithmic modules accept only positive signals; the middle part is the difference stage; and to the right is the multiplier divider circuit. Figure 14.11 shows an actual measurement made by means of this circuit. The calibration curve was made point by point with an X–Y plotter as read-out instrument. In the measurement shown in Fig. 14.11 the correction for the input resistance, measured with two separate microelectrodes, was used.

Because of the rectifier circuit and the principle of logarithmic conversion used for the circuit of Fig. 14.10, the response time of the system is rather low (about 500 msec for a change of 10 MΩ). A more advanced circuit was used for a system employing a digital voltmeter as read-out [37].

Automatic Recording of the Microelectrode Resistance. An automatic recording system for the microelectrode resistance has been designed, with which it is possible to obtain the calibration curve of the micro-electrode on graph paper within 2 min[31, 38]. The advantage of the method used is that the automatic recorder traces the calibration curve continuously, whereas all the former methods used a plotting process point by point. The applicability of this instrument is not restricted to measurements of the cytoplasmic resistivity. Because the time of

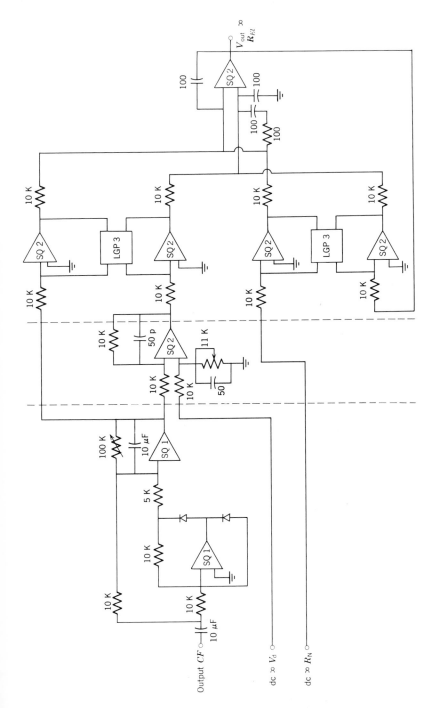

Fig. 14.10. Analog circuit for calculating microelectrode resistance.

315

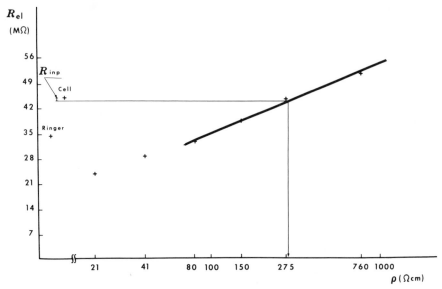

Fig. 14.11. Determination of cytoplasmic resistivity in frog sartorius by means of analog circuit. Evaluation of cytoplasmic resistivity was made according to (14.9). In left part of diagram difference between points called "Cell" and "Ringer" corresponds to R_1 in (14.9). Distance ⌶ called R_{inp} in the independently measured input resistance of cell. Level of horizontal line to calibration curve corresponds to R_{El}^{cyto} in (14.9).

evaluation of the resistance measurements and for tracing the $R_{El} = f(\rho)$ characteristic is greatly reduced, the method is very useful for theoretical work on microelectrode properties such as sensitivity to ions and dependence of microelectrode resistance upon the current density at the microelectrode tip.

Evaluation of the Method. Determining the cytoplasmic resistivity by means of change of the microelectrode resistance allows measurements which are independent of a mathematical model for the investigated cell. Because only one microelectrode is used for the penetration, the technique is not more difficult than any other penetration technique using single or double-barreled microelectrodes; further, because of the recently developed instruments, the time required to evaluate a single measurement is shorter than that required to analyze the phenomenology of an action potential recorded with standard techniques. Thus all the conditions outlined at the beginning of Section 14.2.3 are fulfilled by this method. The fact that the method is independent of a mathematical model

for a cell allows measurements of the cytoplasmic resistivity on cells
which were hitherto inaccessible for individual measurements.

14.3   RESULTS

Although the emphasis of this study lies on methods, it is expedient to
give a short review of results obtained with recent measurements of the
cytoplasmic resistivity.

14.3.1   The Cytoplasmic Resistivity of Small Cells

Because the method for the measurement of the cytoplasmic resistivity
as described in Section 14.2.3 is independent of the geometry of the cell
under investigation, the value of the cytoplasmic resistivity from small
cells or from larger cells for which the current distribution is unknown can
be determined. Using this method, we obtained a value of 120 Ω cm for
the cytoplasmic resistivity of rat atrial cells[39]. This value is close to
the one found by Weidmann in Purkinje fibers of the kid[6]. At present,
no agreement has been reached over a model for the theoretical treatment
of the current distribution in syncytial structures[17]. Thus it is impossible to obtain the value of the cytoplasmic resistivity with one of the
methods described in Section 14.2.2.

The question arises as to the size of the smallest cell in which the
cytoplasmic resistivity can be measured using one microelectrode. Preliminary measurements have shown that the cytoplasmic resistivity is also
higher in the cytoplasm than in the medium in rat liver cells [own experiments, unpublished] and in cells in tissue culture[40]. The current distribution during a resistivity measurement in the immediate neighborhood
of the microelectrode tip is not known, but it seems safe to assume that
a reliable measurement of the cytoplasmic resistivity is possible when the
microelectrode tip is surrounded by a sphere of cytoplasm with a radius
of 3-5 μ. For this reason it is advisable to make the measurements of the
cytoplasmic resistivity in the center of the cell and not close to the
membrane. As a rule of thumb, one can say that the cytoplasmic resistivity
can be determined in a cell in which the membrane potential can be
reliably measured — provided the cell is close to the surface of the preparation. Cells in deeper layers of a preparation tend sometimes to show
high values of R_1 (14.9). This is probably an erroneous result caused by
the additional resistance of the cell layers separating the tip of a microelectrode from the external medium where the reference electrode is
immersed.

14.3.2 The Problem of Whether the Cytoplasmic Resistivity is a Constant

Measurements of the cytoplasmic resistivity obtained in normal Ringer solution (2.5 mM K) with different methods give the same value for the cytoplasmic resistivity of about 250 Ω cm in the frog sartorius[5, 41]. However, as already mentioned, measurements of the electrical membrane constants with two microelectrodes in higher potassium concentration resulted in lower values for the cytoplasmic resistivity — a result confirmed by measurements of the same quantity using one electrode. An approximate calculation showed that the cytoplasmic resistivity of a sartorius muscle changed from 246 Ω cm in a Ringer solution containing 2.5 mM K[41] to about 100 Ω cm in a Ringer solution containing 100 mM potassium[28]. The change in cytoplasmic resistivity seems to occur gradually; Maiskii found a nonsignificant decrease of the cytoplasmic resistivity in 20 mM K, whereas a statistically significant decrease of the cytoplasmic resistivity was found in a Ringer solution containing 100 mM K[12]. Our own measurements confirm qualitatively the results obtained by Maiskii. Figure 14.12 shows the result of a simultaneous measurement of the input resistance (R_{inp}) and the change of the microelectrode resistance when the microelectrode tip penetrated from the outside medium into the intra-

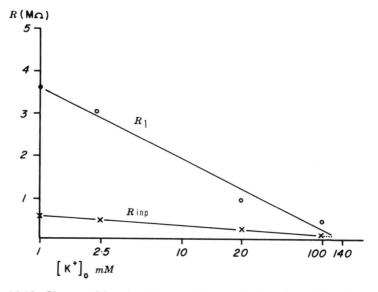

Fig. 14.12. Change of input resistance (R_{inp}) and microelectrode resistance in series with input resistance (R_1) in frog sartorius muscle immersed in media of different K concentrations. R_{inp} are actually measured values. R_1 is difference of apparent microelectrode resistance measured in extracellular medium and inside the cell. (Modified from [28].)

cellular medium of a frog sartorius muscle—R_1 in (14.9)—in different potassium concentrations. In this case the resistance R_1 represents the input resistance in series with the resistance change of the microelectrode caused by a resistivity change of the medium surrounding the microelectrode—see (14.9) and [28]. The difference between R_1 and R_{inp} can be attributed to a change in cytoplasmic resistivity inside the muscle (R_{El}^{cyto}) resulting from the change in the composition of the Ringer solution. The fact that the difference $R_1 - R_{inp}$ decreases with increasing potassium concentrations in the Ringer solution indicates that the cytoplasmic resistivity decreases gradually with increasing potassium concentration in the medium. On the other hand, Fatt, working with a method using a conductivity cell, did not find a significant change in the cytoplasmic resistivity in potassium concentrations up to 32.5 mM[9]. However, Fatt worked under conditions in which chloride and calcium were omitted from the solution. Because of these differences in the experimental design, his results cannot be compared with the results reported here.

The cause of the decrease of the cytoplasmic resistivity in high concentration of potassium is not known at present, but several working hypothesis are possible; they are presented below.

Our measurements were made in Ringer solutions in which the product of $[K]_0[Na]_0$ was kept constant and therefore the Donnan equilibrium across the fibers was not maintained after the solution was changed from normal potassium concentration to high potassium concentrations. The measurements were taken after the muscle fiber was in the test solution for 1 h, and the reduced cytoplasmic resistivity could be caused by a higher intracellular ion concentration—most probably chloride, because the gradient for potassium was kept constant across the membrane, as can be concluded from the values of the resting potentials[28].

Another possibility is that the membranes of the T system have a lowered resistance in high potassium concentrations. The resistance change of these membranes could influence the apparent core conductor resistance measured with the method employing two microelectrodes. For a measurement of the cytoplasmic resistivity with one microelectrode, however, the immediate vicinity of the microelectrode tip has to be influenced by these membranes (see Section 14.3.1) to obtain an influence on the measured cytoplasmic resistivity.

A third hypothesis would be that the potassium ions bound inside the cell at normal potassium concentrations of the extracellular medium are liberated at higher extracellular potassium concentrations and were thus increasing the concentration of free potassium inside the cell[42]. Such a phenomenon could also cause a decrease in cytoplasmic resistivity.

The working hypotheses mentioned in the previous paragraphs have to be evaluated by additional experimental work. However, the fact that there is a decrease in the cytoplasmic resistivity when the composition of the outside medium is changed[41] suggests control measurements of the cytoplasmic resistivity whenever the measurement of the input resistance is used to calculate the membrane resistance. Accordingly, to use the values of the cytoplasmic resistivity published elsewhere in the literature to evaluate the membrane resistance from input resistance measurements is safe only when the cytoplasmic resistivity is determined under the same conditions as the input resistance.

REFERENCES

[1] W. Nernst, *Z. Phys. Chem.*, **4**, 129 (1889).

[2] L. Hermann, in *Handbuch der Physiologie*, vol. 2, Vogel, Leipzig, 1879.

[3] M. Cremer, *Z. Biol.*, **37**, 550 (1899).

[4] A. L. Hodgkin, *J. Physiol. (London)*, **106**, 305 (1947).

[5] B. Katz, *Proc. Roy. Soc., Ser. B*, **135**, 506 (1948).

[6] S. Weidmann, *J. Physiol. (London)*, **118**, 348 (1952).

[7] R. Taft and L. E. Malm, *J. Phys. Chem.*, **43**, 499 (1939).

[8] K. S. Cole and H. J. Curtis, in *Electric Physiology in Medical Physics*, vol. 2 (O. Glasser, ed.), Year Book Publishers, Chicago, 1950.

[9] P. Fatt, *Proc. Roy. Soc., Ser. B*, **159**, 606 (1964).

[10] H. P. Schwan, in *Physical Techniques in Biological Research*, vol. 6 (W. Nastuk, ed.), Academic, New York, 1963.

[11] A. L. Hodgkin and W. A. H. Rushton, *Proc. Roy. Soc., Ser. B*, **133**, 444 (1946).

[12] V. A. Maiskii, *Biofizika*, **8**, 649 (1965).

[13] M. Henček and Z. Zachar, *Physiol. Bohemoslov.*, **14**, 297 (1965).

[14] P. Fatt and B. Katz, *J. Physiol. (London)*, **115**, 320 (1951).

[15] J. W. Woodbury and W. E. Crill, in *Nervous Inhibition* (E. Florey, ed.), Pergamon, New York, 1961.

[16] D. Noble, *Biophys. J.*, **2**, 381 (1962).

[17] E. Coraboeuf, *Chapter 12 in this book.*

[18] W. L. Nastuk and A. L. Hodgkin, *J. Cellular Comp. Physiol.*, **35**, 39 (1950).

[19] H. P. Jenerick, *J. Cellular Comp. Physiol.*, **42**, 427 (1953).

[20] T. Araki and T. Otani, *J. Neurophysiol.*, **18**, 472 (1955).

[21] N. Sperelakis, T. Hoshiko and R. M. Berne, *Am. J. Physiol.*, 198, 531 (1960).

[22] O. Schanne, R. Kern and B. Schäfer, *Naturwissenchaften*, **49**, 161 (1962).

[23] E. Amatniek, *IRE Trans. Med. Electron. (PGME)*, **10**, 3 (1958).

[24] R. Kern, O. Schanne and K. Krakat, *Excerpta Med.* International Congress Series, **48**, 755 (1962).

[25] R. Kern, thesis, University of Heidelberg (1966).

[26] G. Szabo, thesis, Université de Montréal (1966).

[27] R. Lanthier and O. Schanne, *Naturwissenschaften*, **53**, 430 (1966).
[28] O. Schanne, H. Kawata, B. Schäfer and M. Lavallée, *J. Gen. Physiol.*, **49**, 897 (1966).
[29] M. Lavallée and G. Szabo, *Chapter 6 of this book.*
[30] G. Küchler, *Pflueg. Arch. Ges. Physiol.*, **280**, 210 (1964).
[31] O. Schanne, M. Lavallée, R. Laprade and S. Gagné, *IEEE Proc.*, **56**, 1072 (1968).
[32] O. Schanne and E. Coraboeuf, *Nature*, **210**, 1390 (1966).
[33] R. H. Adrian, *J. Physiol. (London)*, **133**, 631 (1956).
[34] O. Schanne, R. Kern and H. Kawata, *Z. Biol.*, **114**, 371 (1964).
[35] A. F. Bak, *Electroencephalog. Clin. Neurophysiol.*, **10**, 745 (1958).
[36] O. Schanne and C. Ménard, free communication WB1 read at the 11th Meeting of the Biophysical Society, Houston, Texas (1967).
[37] R. Laprade and C. Ménard, free communication at the Intracellular Glass Microelectrodes Conference, May 23–25, 1967, Montréal, Qué., Canada.
[38] S. Gagné and O. Schanne, free communication at the Intracellular Glass Microelectrodes Conference, May 23–25, 1967, Montréal, Qué., Canada.
[39] O. Schanne, L. Thomas and E. Ceretti, *Federation Proc.*, **25**, 2515 (1966).
[40] O. Schanne, R. Kern and B. Schäfer, *Pflueg. Arch. Ges. Physiol.*, **274**, 3 (1961).
[41] O. Schanne and E. Ceretti, free communication at the 23rd International Congress of Physiological Sciences, Tokyo (1965).
[42] S. E. Simon, in *Membrane Transport and Metabolism* (A. Kleinzeller and A. Kotyk, eds.), Academic, London, 1961.

CHAPTER 15

Measurement of Activity of Hydrogen, Potassium, and Sodium Ions in Striated Muscle Fibers and Nerve Cells

P. G. KOSTYUK, Z. A. SOROKINA and YU. D. KHOLODOVA

Institute of Physiology, Kiev

15.1 INTRODUCTION

Glass electrodes are now being used more and more frequently for measurement of ionic activity in solutions. Such electrodes have several advantages, especially for biological systems: their function is free from interference with redox systems, colloids, toxic factors, etc., and they are very convenient for examination of small quantities of fluids.

Glass electrodes were first used in biological measurements for the determination of pH in minced muscles. It was shown that for a resting muscle pH varies in the region from 7.04 to 7.30[1–3]. In fatigued muscle this value falls (depending on the season) to 6.26–6.60. Maison, Orth, and Lemmer[4] used glass electrodes to detect pH changes in the medium surrounding the muscle, and Dubuisson[5, 6] used them for pH measurements on muscle surface during contraction and relaxation. Disteche[7] improved the method of surface pH measurements, using electrodes with a sensitive membrane about 4 mm in diameter; the electrode was fixed to the surface of a hanging muscle and could follow its movements. Measurements of pH with extracellular glass microelectrodes were also made on frozen mouse brain tissue by Vladimirova and Holms (see Epstein[8]).

A glass electrode for intracellular pH measurements was first prepared by Caldwell[9, 10]. The external diameter of this electrode was about 100–150 μ, which limited its use to very large crab muscle fibers (*Carcinus maenas* and *Maia squinado*) and squid giant axons (*Loligo forbesii*). The electrode was in the form of a glass tubing sealed at the end, filled with 0.1 N HCl and covered on its surface (except for the tip, about 500 μ long) with shellac. It could not penetrate the cell membrane; therefore a hole was made in the membrane beforehand through which both the selective and the reference electrode were inserted into the fiber. The slope of the electrode function was 55–58 mV/log a_H, and the time of equilibration reached 2–3 sec. In all fibers investigated by Caldwell the value of intracellular pH was about 7.

Two years later the experiments of Caldwell were repeated by Spiropoulos[11] on giant axons of the squid *Loligo pealii*. According to his data the value of intracellular pH in axoplasm was $7·35 \pm 0·1$.

Glass electrodes of this type could not be widely used in biological investigations, mainly because their large diameters prevented penetration into cells without considerable destruction of intracellular structures. Such electrodes, of course, could not be used for measurements in muscle fibers and nerve cells of higher animals, which made it necessary to improve the technique of preparing glass electrodes for cellular measurements so as to render them comparable in size with usual open-tip microelectrodes for membrane potential measurements. Such electrodes were made first for the intracellular measurements of hydrogen-ion activities [12–15]. Micropipettes were produced with the help of a conventional microelectrode puller. The glass with hydrogen function (McInes or ES–I $NaCaS_{22-6}$) has a very low melting point; thus the tip of the micropipette was usually already sealed after pulling. The external diameter of the hydrogen-sensitive tip of such electrodes varied between 0.5 and 1 μ; they were used for intracellular measurements of hydrogen-ion activities in frog sartorius muscle fibers and giant nerve cells of certain molluscs. Such measurements gave important data concerning the mechanism of hydrogen-ion distribution between the cell and external medium.

Later, Lavallée[16] and Lavallée and Szabo[17] proposed to use as cation-sensitive microelectrodes micropipettes with open tips and outer diameter below 0.5 μ. They suggested that it is possible to consider the opening at the tip of the microelectrode as a pore in a glass membrane filled with inside solution. This suggestion is supported by electron-microscopic findings that pore diameter in artificial membranes can reach the value of 0.15–0.1 μ. The steepness of the electrode function of such "unsealed" pH microelectrodes made from Corning glass $NaCaS_{22-6}$ varied from 6.8 to 50 mV/log a_H. Measurements on contracting atrial fibers of rat immersed in Ringer-Krebs solution of pH 7.4 under temperature 30°C gave the value of internal pH 6.9 ± 0.06.

Glass electrodes are now used also for biological measurements of activity of sodium and potassium ions. The presence of metal function in glass electrodes had been established in 1923. A strict proof of the possibility of transition from a hydrogen function to a complete sodium function was given by Schultz[18, 19]. Also important was the systematic study of relative sensitivity of glass electrodes carried out by Eisenman et al[20–22], which showed that the glass selectivity depends on the chemical composition of the glass. The introduction of Al^{3+} increases the sensitivity of glass toward alkaline ions. The relative sensitivity toward sodium or potassium depends on the amount of Al^{3+} in glass. Cation-sensitive electrodes for biological measurements were used first by Friedman et al.[23, 24] for determination of sodium-ion activity in the blood of a dog in vivo and in vitro. Sugioka and Portnoy and Thomas and Gurgian (see Eisenman[22]) prepared sodium as well as potassium electrodes and measured the activity of these ions in the circulating blood during long periods of time. Ungar[25] used cation-selective electrodes for evaluating the status of sodium and potassium ions in brain extracts.

Ling[26] has found that if Corning glass $NaCaS_{22-6}$ is covered with a thin colloid film it becomes a more prominent potassium function than the colloid film itself. This principle was used by Gotoh et al.[27] to prepare potassium-selective electrodes for continuing recording of ion activity in blood and brain. Such electrodes were completely insensitive to pH changes from 6 to 9.

Intracellular measurements of sodium-and potassium-ion activities were made first by Hinke[28, 29]. Hinke's electrodes had tip diameters of 20–30 μ, lengths of 70–90 μ, and were covered with glass insulation. Because of large diameters, such electrodes could be used for measurement only on large objects such as crab and lobster muscle fibers (*Carcinus meanas and Homarus vulgaris*) that have diameters from 200 to 500 μ.

Lev and Buzhinsky[30] improved the technique of preparation of cation-selective microelectrodes. They produced potassium electrodes with outer-tip diameters from 0.6 to 1.5 μ so that the electrodes could be used for measurements on such classic objects as frog striated muscle fibers. If potassium-selective electrodes with different values of relative selectivity were used, activity of both sodium and potassium ions could be calculated. Using such electrodes, Lev[31, 32] estimated the value of activity coefficients for sodium and potassium ions in muscle fibers.

Even electrodes with tip diameter of about 1 μ are not quite suitable for accurate measurements in most cells. They certainly depolarize the surface membrane and produce substantial damage of intracellular structures, which can result in changes of the status of intracellular electrolytes. Therefore it is rather important to decrease the external diameter of the tip of such electrodes as far as possible. Sorokina[33] succeeded in preparing potassium-selective electrodes (from a four-components potassium glass) and sodium electrodes (from NAS_{11-18} glass) with tip diameters below 0.5 μ. Such electrodes could be used for activity measurements in cells of different tissues; data have already been obtained on frog striated muscle fibers and on giant nerve cells of molluscs.

15.2 METHOD

15.2.1 Theory of the Glass Electrode

According to Nikolsky and Schultz[34–38] the potential of a glass electrode depends on the difference of chemical potentials of H ions and M ions in the boundary layer of glass and outside solution. The value of this potential is expressed by the equation:

$$\phi = \phi_0 + \frac{RT}{F}\ln(a_H + Ka_M), \tag{15.1}$$

where K is the constant of electrochemical equilibrium for substitution of ions in the surface layer of glass by ions from the external solution. This constant is a highly important characteristic of glass because it determines the specificity of glass electrode function. The smaller the value of K, the larger the region where the electrode behaves as an ideal hydrogen electrode without switching to a metal function.

If the solution has two types of metal cations (M_1 and M_2) and the activity of hydrogen ions is relatively low (pH 5.6) then

$$\phi = \phi_0 + \frac{RT}{F}\ln(a_{M1} + K_{M_1M_2}a_{M2}), \tag{15.2}$$

where

$$K_{M_1M_2} = \frac{K_{M_2H}}{K_{M_1H}}. \tag{15.3}$$

In this simple theory[34] it was assumed that the activities of hydrogen ions and metal ions in glass are equal to their concentrations:

$$a_H = c_H \quad \text{and} \quad a_M = c_M, \tag{15.4}$$

so that activity coefficients of ions do not depend on the degree to which those ions are replaced in glass or by other ions. In a later development of the theory the possible heterogeneity in the binding of hydrogen and metal ions in glass was taken into account. This involved a more complicated expression for the potential of a glass electrode[36].

The expression of Nikolsky is identical to the expression proposed by Eisenman:

$$E = E_0 + \frac{RT}{F}\ln[(A^+)^{1/n} + K_n^{1/n}(B^+)^{1/n}]^n, \tag{15.5}$$

where K is an empirical constant for the given glass with the same physical meaning as the constant in Nikolsky equation and n is also an empirical constant equal to 1 in solutions containing sodium and hydrogen ions and 3.5 in solutions with hydrogen and potassium ions.

The cationic properties of microelectrodes are characterized by the slope of the relation between $E(\phi)$ and pM. This relation determines the region of electrode function. A deviation from the main cationic function is significant when there is a deflection from the linear relation for more than 2 mV (this is equal to 0.05 units of pM).

Microelectrodes from hydrogen-sensitive glass $NaCaS_{22-6}$ have no sodium or potassium function in the pH range from 1 to 10 if they are immersed in solutions with physiological concentrations of NaCl or KCl at room temperature.

The mean value of the selectivity constant for pH microelectrodes is 1.2×10^{-7} (from 10 measurements); therefore, we can practically neglect the product Ka_{Na} for hydrogen electrodes.

The selectivity constant for sodium-selective electrodes from NAS_{11-18} glass ($K_{Na/K}$) for pH range from 6 to 8 varies from 0.01 to 0.015. Potassium electrodes have the lowest selectivity, $K_{K/Na}$ being 0.11–0.23.

It is clear that during electrometric determination of ionic activity, the measured potential difference contains two items; one of them is the diffusion potential at the junction between the medium and the reference electrode. This potential cannot be calculated with sufficient accuracy because such calculation requires the knowledge of individual ion activities. An approximate evaluation of the value of the diffusion potential can be made by means of Planck's or Henderson's equations, because in such conditions we know the ionic strength of solutions that produce a liquid junction. Such calculation indicates that the diffusion potential can change in limits from -1.8 to -4.0 mV; this can produce an error in activity determinations up to ± 0.06 units of pM.

Changes in diffusion potential that occur when the electrodes are inserted into the cell cannot be calculated because the ionic mobility and concentrations in the cell protoplasm are not known. The situation is even more complicated if we take into account the fact that the protoplasm is a water solution of polyelectrolytes and therefore the activity coefficients of main cations in the protoplasm and in usual solutions can be different (see below).

We can still assume, however, that the error introduced into activity measurements by a diffusion potential in such a case is insignificant. This can be to some extent proved in experiments when we use a reversible electrode (pH microelectrode) as a reference electrode for sodium and potassium activity measurements. The difference in potential measurements obtained by use of a salt bridge and a pH microelectrode was 3.8 mV (mean value from 10 measurements). Thus the error introduced into activity determination by the diffusion potential cannot be more than 0.05 units of pM.

15.2.2 Construction of Ultrafine Cation-Selective Microelectrodes

Ultrafine cation-selective microelectrodes were made from the following glasses: hydrogen glass $NaCaS_{22-6}$ (ES-I or MacInes), sodium glass NAS_{11-18} (Eisenman), and potassium glass $KABS_{20-9-5}$ manufactured by the Department of Physical Chemistry of Leningrad University. The first step was the preparation of micropipettes in a usual microelectrode puller (Fig. 15.1). The glasses used differ considerably in their melting properties; therefore the pulling regimen is different for different kinds of electrodes.

The hydrogen glass is soft-melting; therefore, with suitable heating temperature and pulling force, it is possible to obtain micropipettes that are already almost sealed at the tip. Such sealing can be controlled to some extent by immersion of the tip in water and control of the absence of capillary absorbtion (which can be easily checked under light microscopy).

Micropipettes from sodium and potassium glass must be specially sealed under microscopic control by a heated platinum loop. The distance from the loop and the necessary heating must be found by several trials so as to preserve a maximal fine tip and achieve complete sealing. Use of the same procedure for hydrogen glass is difficult because the tip usually fuses into a drop.

Sealed micropipettes are filled by gentle boiling in methyl alcohol under reduced pressure and then transferred into distilled water for 2 h. After this procedure, electrodes are immersed in 0.1 N solutions of HCl, KCl, or NaCl for one to three days. The connection of the selective electrode to the electrometer is made by an Ag–AgCl electrode or by an amalgamized platinum wire.

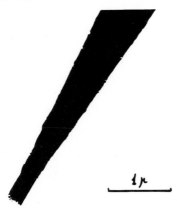

Fig. 15.1. Electronmicroscopic picture of tip of micropipette made from $NaCaS_{22-6}$ glass.

The insulation of cation-selective microelectrodes can be achieved by several methods. They can be covered by a shellac layer from a saturated alcohol solution so that only the tip (5–10 μ) remains free (Fig. 15.2a). Such electrodes can be used only for measurements in superficially located cells. The effectiveness of insulation must be specially checked (by resistance measurements during slow immersion of the electrode in a solution).

Much more convenient to use are selective microelectrodes with glass insulation. A sealed and filled micropipette is introduced by a micromanipulator into another pipette of larger diameter made from neutral glass. The space between two pipettes at the tip and at the wide end is filled with wax or dental cement. It is rather important that the shape of the inner pipette fit the shape of the outer one – the latter must be more conical. After some training it is possible to manufacture microelectrodes with the tip diameter of the insulating part about 1 μ and the length of the cation-selective part of a few microns, with tip diameter below 0.5 μ.

During preparation of hydrogen and potassium electrodes another method of insulation can also be used. A tubing from cation-sensitive glass is placed into another tubing of larger diameter made from neutral glass. Both tubings are fixed together with dental cement and pulled to

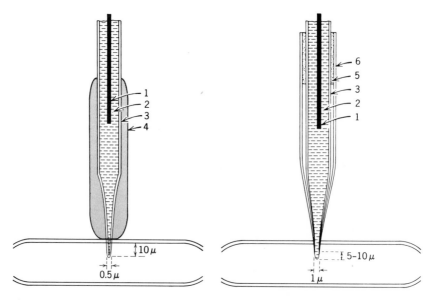

Fig. 15.2. Different types of cation-selective microelectrodes. (1) Ag–AgCl electrode; (2) electrolyte solution; (3) microelectrode from selective glass; (4) shellac insulation; (5) dental cement; (6) insulating glass.

a micropipette in an electrode puller. When heating temperature and pulling force are chosen, microelectrodes can be produced which are insulated at the whole length—except for the tip—by glass insulation. Both glasses are melted together at the tip, and the inner pipette is usually also sealed (Fig. 15.2b). The tip diameter of such electrodes can be easily made to measure below 0.5 μ. Because of some variation in the thickness of the walls of glass tubings and their diameters, the length of the insulated part is variable; complete insulation of the tip occurs only seldom. Selection of electrodes is necessary for the length of the cation-selective tip; a length of 10–20 μ is good for activity measurements on conventional neurophysiological objects.

Selective microelectrodes with glass insulation can be used for measurements on superficially as well as on deeply located cells accessible for usual microelectrode investigations.

Hydrogen and potassium electrodes with a $E/\log a_M$ slope above 45 mV/log a_M have a comparatively low resistance (from 50 to 500 MΩ). Obviously the glass membrane at the tip of such electrodes is very thin. Electrodes with lower resistance usually are also much less sensitive. Sodium electrodes have a much higher resistance (10^{10}–10^{12} Ω) and require special input electrometers.

Great attention must be paid to the standard potentials of cation-sensitive microelectrodes. In electrodes from hydrogen and potassium glass, such potentials are usually low (a few millivolts). However, in sodium electrodes made from NAS$_{11-18}$ glass, standard potentials can be quite large. For exactness of measurements, the absolute value of this potential is actually less important than its stability. Usual changes are within the range 0.1–1 mV/h. Less stable electrodes are rejected. In most electrodes potential values are stable in pM region from 0 to 2; fluctuations are observed only in solutions with lower activities, so that the slope of the electrode function changes. Therefore, in long-lasting experiments, repeated calibrations of electrodes are necessary.

An important question in considering measurements with intracellular cation-selective microelectrodes is the type and location of a reference electrode. The reference electrode can be placed outside or inside the cell (Fig. 15.3). In the first position the recorded potential difference includes the potential of the selective electrode, the potential of the reference electrode, and the transmembrane potential. Therefore a special measurement of the transmembrane potential with open-tip microelectrode is necessary—and then its value must be subtracted from the recorded value. The measurement of the transmembrane potential is also subject to error because the tip potential of the microelectrode can change when the device is transferred from outside solution to inside the

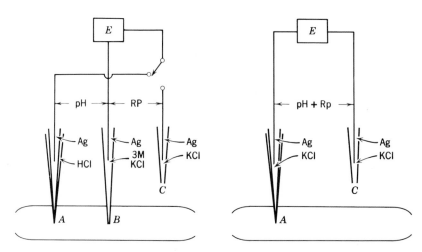

Fig. 15.3. Different types of electrode positioning during measurements of intracellular pH. (*A*) Hydrogen selective microelectrode; (*B*) intracellular reference electrode; (*C*) extracellular electrode; (*E*) electrometer input.

cell. Such changes can be considerable if the tip potential is high; thus it is important to choose electrodes with minimal tip potentials.

In the second position of the reference electrode, the transmembrane potential is omitted and the activity is calculated immediately from the potential difference between the selective and the reference electrode. The electrode connected to an Ag–AgCl or calomel half-cell can be used in this case for transmembrane potential measurements, together with another outside electrode.

The insertion of the selective electrode and reference electrode into the same cell is sometimes quite difficult. The best way is to have made a previous fixation of the tips of both electrodes at a distance of several microns and at the same level. Then both electrodes are moved together toward the cell with one micromanipulator.

All cation-selective microelectrodes must be previously calibrated in standard solutions (Fig. 15.4) and calibration curves must be prepared. Because of the possibility of changes in electrode function, a repeated calibration after actual measurement is advisable (Fig. 15.5).

15.3   ACTIVITY OF HYDROGEN IONS IN MUSCLE FIBERS AND NERVE CELLS

The determination of activity of hydrogen ions in different cells gives very close results. In striated muscle fibers of a frog the mean pH value

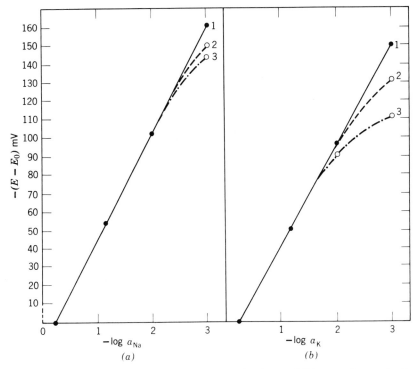

Fig. 15.4. Calibration curves for (*a*) sodium-selective and (*b*) potassium-selective microelectrode. Calibration solutions: (*a*) (1) NaCl + KCl 0.001 *M*; (2) NaCl + KCl 0.01 *M*; and (3) NaCl + KCl 0.1 *M*. (*b*) (1) KCl + NaCl 0.001 *M*; (2) KCl + NaCl 0.01 *M*; and (3) KCl + NaCl 0.05 *M*.

is 7.15 ± 0.06. Measurements of fibers in muscle with preserved circulation and innervation brought practically the same value, 7.12 ± 0.09 (see Table 15.1). In striated muscle fibers of a rat the value is 7.34. The mean value of intracellular pH in giant neurons of gastropodal mollucs (*Helix* and *Planorbis*) in summer and autumn is 7.26 ± 0.10.

The transmembrane potential varied in all these cells from -40 mV to -90 mV. A parallel measurement of transmembrane potential and intracellular pH did not reveal any relation between them. Even during big changes of the transmembrane potential in muscle fibers, the intracellular pH remained practically constant. Corresponding examples are shown in Fig. 15.6; the first of them was obtained by continuous recording of the resting potential and pH from the same muscle fiber with both electrodes remaining inside all the time; the second was made by several single recordings from many fibers. Seasonal changes of internal pH in

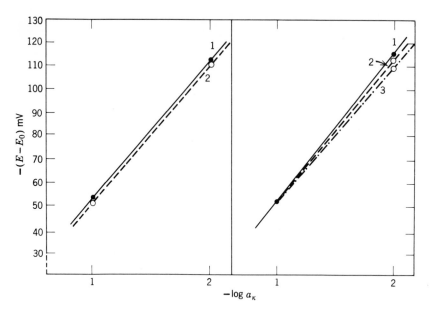

Fig. 15.5. Different types of changes of calibration curves of two potassium-selective electrodes with time.

Table 15.1   Values of Resting Potential (in mV) and Intracellular pH*
of Frog Sartorius Muscle
(Mean values ±S.D.) Taken at 30-min Intervals

Preparation	30 min	60 min	90 min	120 min
Muscle in situ	97.2 ± 2.4	96.4 ± 1.8	96.7 ± 1.3	96.4 ± 1.3
	(7.12 ± 0.07)	(7.12 ± 0.07)	(7.12 ± 0.07)	(7.12 ± 0.09)
Muscle in bicarbonate Ringer saline	90.4 ± 2.9	89.7 ± 2.0	89.0 ± 2.6	87.3 ± 2.4
	(7.12 ± 0.07)	(7.12 ± 0.07)	(7.12 ± 0.09)	(7.15 ± 0.06)
Muscle in phosphate Ringer saline	88.7 ± 2.8	88.2 ± 2.6	87.3 ± 3.0	86.4 ± 2.3
	(7.12 ± 0.07)	(7.12 ± 0.07)	(7.15 ± 0.06)	(7.15 ± 0.08)

*Figures in parentheses.

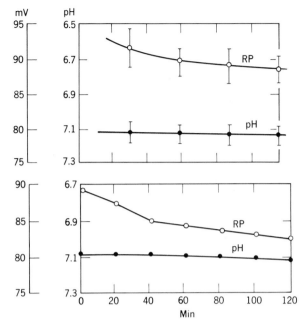

Fig. 15.6. Changes in resting potential (mV) and internal pH of frog sartorius muscle fibers from beginning of experiment. Top: Data from single measurements in different fibers (each point is mean value of 50 measurements; vertical bar indicates standard error). Bottom: Data obtained during continuous location of electrode in single muscle fiber. Abscissae are numbered in minutes.

molluscs' nerve cells are shown in Fig. 15.7. The pH value is almost constant despite large variations of membrane potential and external pH.

The measured values of intracellular pH and resting potential were used for solving the question of whether the relationship of hydrogen-ion activity inside and outside the cell is in correspondence with Donnan equilibrium. In the case of a Donnan equilibrium there must be a following relation:

$$K^+_{inp}/K^+_0 = Cl^-_0/Cl^-_{inp} = H^+_{inp}/H^+_0 = HCO^-_{3\,0}/HCO^-_{3\,inp}. \tag{15.6}$$

It can be also expressed as

$$H^+_{inp}/H^+_0 = \exp(EF/RT), \tag{15.7}$$

where E is the resting potential.

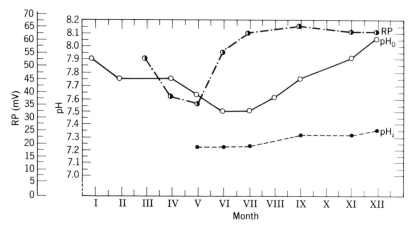

Fig. 15.7. Seasonal changes of resting potential and intra- and extracellular pH in nerve cells of the mollusc *Helix pomatia*. (Redrawn from [14].)

For the nerve cells with mean values of resting potential of 59.7 mV (*Helix*) and 55.6 mV (*Planorbis*) and outside activity of hydrogen ions $1.55 \cdot 10^{-8} M$ (pH 7.81) and $2.82 \cdot 10^{-8} M$ (pH 7.55), the intracellular activity of hydrogen ions would be $1.62 \cdot 10^{-7}$ and $2.57 \cdot 10^{-7} M$ (pH 6.79 and 6.59).

In frog muscle fibers with mean resting potential about 90 mV and pH outside 7.36–7.40, the internal pH must be 5.8–5.9.

Thus it is clear that in physiological conditions the distribution of hydrogen ions does not follow the Donnan equilibrium. Obviously the resting potential is determined by the electro-chemical gradient of other ions; although the membrane permeability to such ions can be less than to hydrogen ions, the concentration of ions outside and inside the cell is much higher.

In all investigated cells the resting potential depends approximately linearly on the external concentration of potassium ions. Decrease in outside potassium brings a small increase in resting potential; internal pH in such conditions remains practically unchanged. During increase in outside potassium the resting potential decreases; but the changes of internal pH are negligible compared with the changes in resting potential (Fig. 15.8). In solution with 100 mM potassium, the internal pH changes only 7.12–7.20 (i.e., not more than for 0.1 pH unit [13]).

Even large changes in external pH with different buffer systems had little effect in changing the internal pH. The latter changed slowly (in 1 to 2 h) in the same direction as the pH of the external solution. After a few hours the internal pH became stabilized at a certain level still far

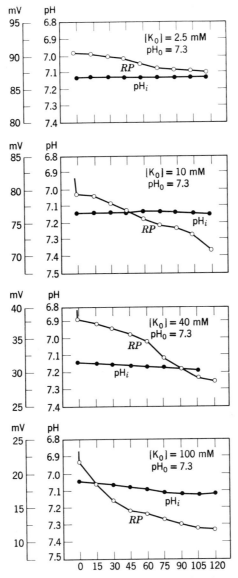

Fig. 15.8. Effect of varying outside potassium concentrations on resting potential and intracellular pH in frog sartorius muscle fibers. Outside solution was prepared by equimolar replacement of sodium by potassium. Abscissae show time in minutes after muscle is immersed into testing solution[12] (Kostyuk, and Sorokina[12]. Reprinted by permission of the Czechoslovak Academy of Sciences.).

335

from the level that could be predicted by the Donnan equilibrium (Fig. 15.9). The only influence which could effectively and rapidly change the internal pH was saturation of the external solution with CO_2. The difference between the influence of CO_2 and that of the buffer solution was striking—in all investigated cells the saturation of the external solution with 100% CO_2 brought a change of internal pH from 0.5 to 1.00 pH unit. The new pH level is reached in frog striated muscle fibers in 30–60 sec. [12]. The changes in internal pH are completely reversible; the value returns to the initial level with the same velocity after removal of CO_2.

The differences in the behavior of the resting potential and internal pH are also quite clear when different metabolic inhibitors are affecting cells —0.1 mM DNP decreases the resting potential of frog muscle fiber to 50% after two h; 0.2 mM DNP produces the same effect in 1–1.5 h. However, the internal pH remains practically unchanged even 5–6 h after the application of DNP[12].

The poisoning of muscle fiber with DNP has practically no effect on the velocity and amplitude of changes in internal pH during saturation of the external solution with CO_2 (although the restoration time becomes somewhat increased).

In explaining the above-mentioned data concerning the intracellular pH, we have to agree with several assumptions already made by Caldwell [9, 10].

1. The protoplasm of muscle and nerve cells (and, probably, of other cells) has a large buffer capacity compared with the amount of hydrogen and hydroxyl ions that can pass through the cell surface membrane.

2. At the same time, the surface membrane of the cell is easily permeable to CO_2; this high permeability makes CO_2 the single agent effective in changing the internal pH. Therefore, the CO_2-bicarbonate buffer system is the main system that determines the value of internal pH. Obviously the bicarbonate ions which are formed in the protoplasm cannot pass easily through the surface membrane; thus the inner bicarbonate system cannot equilibrate with the outer one, and the value of internal pH remains different from the external one even during saturation of the external solution with CO_2.

3. There must be a mechanism of active transport of hydrogen ions from the cell against the existing electrochemical gradient at the surface membrane. The value of internal pH is different by approximately one pH unit from the value predicted by passive distribution of hydrogen and hydroxyl ions across the membrane; this is true for isolated tissue as well as for tissues in situ. Under such conditions there is enough time for transfer through the membrane of the amount of hydrogen ions necessary

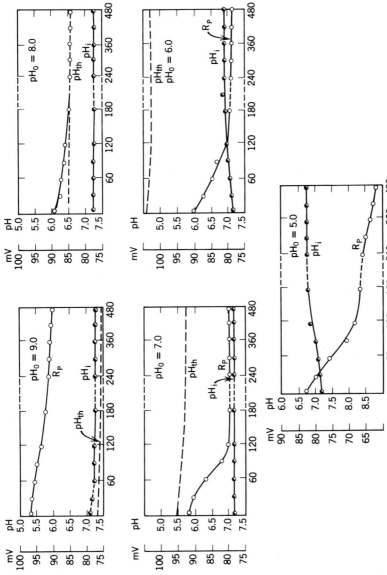

Fig. 15.9. Effect of varying external pH on resting potential and intracellular pH of frog sartorius muscle fibers. Abscissae show time in minutes after immersion of muscle into solution. pH_{th} is the value of internal pH expected if hydrogen ions were distributed across the membrane according to Donnan equilibrium (Kostyuk and Sorokina[12]. Reprinted by permission of the Czechoslovak Academy of Sciences.).

337

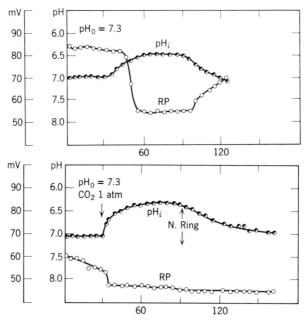

Fig. 15.10. Effect of CO_2 and 0.1 mM DNP on resting potential and intracellular pH of frog sartorius muscle fibers. Top: first arrow, CO_2 (1 atm) was bubbled through the saline; at second arrow, saturated external solution was replaced by normal Ringer. Bottom: same procedure was repeated on muscle poisoned with 0.1 mM DNP[12]. Abscissae list time in seconds.

for saturation of the inner buffer system, but such saturation never takes place. Probably the mechanism of active transport of hydrogen ions is similar to the mechanism of active transport of other cations; but no more detailed suggestions can be made. In several specialized types of cells (for instance, in cells of gastric mucosa) the active transport of hydrogen ions is well known.

The existence of such effective mechanisms for stabilizing the value of pH inside the cell is very interesting; probably the stability of internal pH is highly important for normal performance of intracellular processes — especially for the normal course of enzymatic reactions.

15.4 ACTIVITY OF SODIUM AND POTASSIUM IONS IN MUSCLE FIBERS AND NERVE CELLS

Measurements of activity of sodium and potassium ions in muscle fibers and nerve cells gave the following results. Mean value of a_K in muscles fibers of a frog sartorius muscle in situ is $0.096 \pm 0.002\,M$ and a_{Na} is $0.007 \pm 0.0001\,M$ (Table 15.2). If one takes the analytical concentration of such ions in muscle fibers and neglects possible errors produced by the diffusion potential, the apparent activity coefficient (γ) for the corresponding ions can be determined as 0.73 and 0.39 [33].

Measurements on isolated muscles have also been made several hours after preparation and immersion into Ringer's solution. The isolation resulted in a small decrease in both intracellular potassium concentration and a_K. The mean value of γ_K remained unchanged. The sodium concentration and a_{Na} increased at the same time, the mean value of γ_{Na} also remaining unchanged. Variability of a_K and a_{Na} value in different fibers was greater in isolated muscles than in muscles in situ (Table 15.2).

In giant neurons of molluscs the mean value of a_K was $0.073 \pm 0.0004\,M$ (*Helix*) and $0.039 \pm 0.0002\,M$ (*Planorbis*); a_{Na} was $0.013 \pm 0.00001\,M$ and $0.008 \pm 0.00002\,M$, respectively. Analytical concentrations of potassium and sodium ions in neurons (Table 15.3) were $0.093-0.031\,M$ (*Helix*) and $0.053-0.014\,M$ (*Planorbis*). Thus the apparent activity coefficients calculated under the same conditions are 0.78–0.73 for potassium and 0.53 for sodium [39, 40].

As one can see, both in muscle fibers and in nerve cells there is a considerable difference between activity coefficients for potassium and sodium ions. Our findings are in agreement with results obtained by Hinke [29] and by Lev [31, 32].

The presence of experimental data on intracellular activities of main inorganic cations presents several questions about theoretical interpretation. The main feature in all cells studied until now is a decrease in activity of certain cations compared to the analytical concentration, and an obvious decrease in their activity coefficient. Generally speaking there can be two reasons for such decrease in measured activity: a decrease in concentration of the dissolved substance (owing to formation of some unsoluble product or isolation of a certain amount of ions of water in some special compartments unaccessible to the measuring glass membrane) and a decrease in energy of dissolved particles (because of interaction among them).

Irregular distribution of potassium and sodium inside the cell is obvious, although the excess of one ion over another in different cell fractions is not very high [41, 42]. The tubular system in muscle fibers, for instance, could be the site of sodium retention – but it has been pointed out in this chapter that in the nerve cell with quite different intracellular structure the decrease in sodium activity is about the same as in muscle fibers.

One must also consider the presence of bound water, which can exclude

Table 15.2 Values of Resting Potentials and Activity of Potassium in Frog Muscle Fibers

Isolated Muscles

Expt. No.	MP (mV)	a_K	γ_K	C_K (M/Kg f.w.) calc.	a_K calc.	γ_K calc.
1	86.1	0.099	0.73			
2	82.7	0.091	0.68			
3	85.3	0.095	0.70			
4	81.1	0.095	0.70			
5	81.1	0.093	0.69			
6	75.6	0.090	0.67			
7	86.5	0.099	0.73			
8	80.6	0.091	0.68			
9	84.0	0.095	0.70			
10	88.1	0.099	0.73			
11	85.1	0.095	0.72			
MEAN	83.3	0.094	0.70	0.1328	0.102	0.754
±S.D.		±0.001		±0.0003		

Nonisolated Muscles

Expt. No.	MP (mV)	a_K	γ_K	C_K (M/Kg f.w.)	a_K calc.	γ_K calc.
1	85.2	0.095	0.71			
2	85.1	0.095	0.71			
3	86.1	0.098	0.72			
4	87.9	0.095	0.71			
5	87.6	0.098	0.72			
6	84.8	0.097	0.71			
7	85.2	0.095	0.71			
MEAN	85.9	0.096	0.71	0.1355	0.101	0.752
±S.D.		±0.002		±0.0002		

Table 15.2 (contd.) **Values of Resting Potentials and Activity of Sodium in Frog Muscle Fibers**

			Isolated Muscles									Nonisolated Muscles			
Expt. No.	N. of Meas.	MP (mV)	a_{Na}	γ_{Na}	C_{Na} (M/Kg f.w.)	a_{Na} calc.	γ_{Na} calc.	Expt. No.	N. of Meas.	MP (mV)	a_{Na}	γ_{Na}	C_{Na} (M/Kg f.w.)	a_{Na} calc.	γ_{Na} calc.
1	21	82.6	0.0105	0.44				1	10	84.9	0.0078	0.41			
2	20	84.1	0.0083	0.34				2	8	85.4	0.0075	0.40			
3	20	83.9	0.0091	0.38				3	11	86.9	0.0070	0.37			
4	20	83.1	0.0091	0.38				4	8	87.0	0.0070	0.37			
5	19	86.4	0.0078	0.32				5	3	85.5	0.0075	0.40			
6	12	85.3	0.0086	0.36											
7	20	81.7	0.0107	0.45											
MEAN ±S.D.		83.8 ±0.1	0.0091 ±0.0002	0.39	0.0238 ±0.0002	0.020	0.869	MEAN ±S.D.		85.9 ±0.1	0.0073 ±0.0001	0.39	0.0188 ±0.0002	0.016	0.875

Table 15.3 Values of Resting Potentials and Activity of Potassium in Giant Nerve Cells of Snails

Helix pomatia

Expt. No.	N. of Meas.	MP (mV)	a_K	γ_K	C_K (M/Kgf.w.)
1	10	53.6	0.073	0.78	
2	5	54.6	0.073	0.78	
3	12	52.4	0.072	0.77	
4	10	60.7	0.074	0.79	
5	5	50.9	0.073	0.78	
6	12	58.6	0.074	0.79	
7	10	54.3	0.072	0.77	
8	5	52.4	0.073	0.78	
MEAN		54.7	0.073	0.78	0.0933
±s.d.		±0.8	±0.0004		±0.0013

Planorbis corneus

Expt. No.	N. of Meas.	MP (mV)	a_K	γ_K	C_K (M/Kgf.w.)
1	10	51·3	0.039	0.73	
2	15	50.7	0.039	0.73	
3	25	55.8	0.038	0.72	
4	5	53.3	0.039	0.73	
5	10	56.7	0.038	0.72	
6	10	53.4	0.039	0.73	
7	10	55.2	0.039	0.73	
8	5	50.1	0.039	0.73	
MEAN		53.3	0.039	0.73	0.0534
±s.d.		±0.6	±0.0002	0.73	±0.0010

Table 15.3 *(contd.)* **Values of Resting Potentials and Activity of Sodium in Giant Nerve Cells of Snails**

		Helix pomatia						Planorbis corneus			
Expt. No.	N. of Meas.	MP (mV)	a_{Na}	γ_{Na}	C_{Na} (M/Kg f.w.)	Expt. No.	N. of Meas.	MP (mV)	a_{Na}	γ_{Na}	C_{Na} (M/Kg f.w.)
1	3	65.2	0.013	0.42		1	3	51.7	0.007	0.42	
2	8	50.4	0.012	0.40		2	2	50.8	0.008	0.48	
3	2	58.7	0.012	0.40		3	3	49.4	0.008	0.48	
4	3	53.4	0.012	0.40		4	3	55.6	0.008	0.48	
5	3	52.6	0.013	0.42		5	3	53.6	0.008	0.48	
6	2	55.8	0.013	0.42		6	4	55.0	0.007	0.42	
7	3	52.0	0.013	0.42							
MEAN		54.2	0.013	0.42	0.03106	MEAN		52.7	0.008	0.46	0.0138
± S.D.		±0.6	±0.00001		±0.00109	MEAN		±0.8	±0.00002		±0.0094

the metal ions and therefore cannot be taken into account in calculations [43].

However, it is important to consider also a second possibility—the decrease in energy of ions in protoplasm. In the existing practice of measurement, obtained activity coefficients are compared with the activity coefficients of the sodium and potassium chloride solutions and on the basis of this comparison a conclusion is made about the state of ions in the protoplasm of the cell. However, it is essential to consider that for polyelectrolytes the individual activity coefficients of ions differ considerably from those in aqueous chloride solutions of similar concentration. When comparing the activity coefficients obtained in solutions of electrolytes of concentration about $0.15-0.2$ M (which corresponds to the cell concentrations) with the activity coefficients of aqueous solution of the same ions, we already obtain a definite difference in activity coefficients (about $0.03-0.04$ units). When comparing activity coefficients of the ions in polyelectrolyte solutions with their activity coefficients in aqueous solution of similar concentration, the difference is much more significant. This is confirmed by our measurements of activity coefficients of the ions in acrylic polyelectrolytes with the carboxyl groups characteristic for cellular polyelectrolytes. Such polyelectrolytes were: copolymer of acrylamide and acrylic acid obtained while introducing into the molecule ions at concentrations similar to cellular ones ($0.015-0.05$ M); hydrolyzed polyacrylamide with a different degree of hydrolysis; and salts of hydrolyzed polyacrylnitril with different kinds of links. A series of experiments was carried out on polymeric polystyrensulfuric monofunctional resins of Dauex type (ionic group of strong acidic type—SO_3H) with different contents of the crosslinking agent divinylbensol (16 and 5%). The swelling of the resins was 130 and 190% and the capacity was $5.2-5.3$ meq/g. The ionic strength of initial solutions and ratio of ion concentrations for each ionic strength varied respectively.

These measurements showed a change of activity coefficients of electrolytes in the same mixture before copolymerization (i.e., in monomer solution) and after the copolymerization was carried out in a system of sodium hydrosulfide-potassium persulfate-triethanolamine. The dependence of the obtained coefficients upon the density of charges was noted, as were data concerning the nature of the change of the activity coefficient with the increase of the polymer concentration.

From the series of data given in Table 15.4 it is obvious that the activity coefficient of sodium in monomer is slightly different from the activity coefficient in solutions of similar concentrations, while after polymerization (in copolymer) the activity coefficient diminishes. It should be noted that the activity coefficient (counterions association) is not dependent on

Table 15.4   The Activity Coefficient of Sodium Ion in the Mixture of Acrylamide and Acrylic Acid before and after Copolymerization

Molar Ratio of Components Acrylamide:Acrylic Acid	Concentration (M/liter)	Activity in Monomer (M/liter)	Activity in Polymer (M/liter)	γ_{Na} in Monomer	γ_{Na} in Polymer
0.95:0.05	0.02496	0.02272	0.02008	0.91	0.80
0.70:0.30	0.10096	0.08679	0.06694	0.86	0.67
0.70:0.30	0.15096	0.10572	0.08400	0.74	0.55

molecular weight of polyelectrolyte, the change of which was achieved by conditions of polymerization. Evidently the influence of the charges of nearest groups in the chain of polyelectrolyte reaches the asymptotic value at the low degree of polymerization.

In Table 15.5 values of the activity coefficients in samples of hydrolyzed polyacrylamide with different content of sodium-ion in the chain (depending on the degree of hydrolysis) are shown. The degree of hydrolysis was calculated according to the content of hydrogen in the polymer. The activity coefficient of Na (concentration, 0.1 M/liter) in the solution of nonionic polyacrylamid was equal to its value in NaCl solution. The introduction of Na into the chain on polymer reduced the activity coefficient.

The change of activity coefficient of Na-ion is shown in Fig. 15.11. It is evident that with the increase of the polymer concentration with the Na-ion in the chain, the activity of Na-ion becomes a smaller and smaller value of its analytical concentration.

Table 15.5   The Activity Coefficient of Sodium Ion in the Samples of Hydrolyzed Polyacrylamide

Polyacrylamide Hydrolysis (%)	Nitrogen (%)	Experimental concentration (M/liter)	Theoretical concentration (M/liter)	Activity (M/liter)	γ_{Na}
0	19.7	0.1000	—	0.0870	0.87
17	16.4	0.0237	0.0112	0.0214	0.89
54	9.1	0.0394	0.0360	0.0257	0.65
70	6.0	0.0460	0.0480	0.0320	0.65

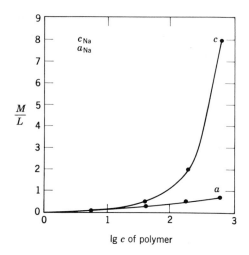

Fig. 15.11. Dependence of activity (a) and concentration (c) of sodium ions on concentration of polymer.

It is important to stress that low values of the activity coefficient of ions are observed not only in slightly acidic carboxyl groups but also in strongly acidic, well dissociated sulfuric groups — the ions of which are freely exchangeable and take part in different exchange processes.

The diminution of activity coefficients in polyelectrolytes is probably caused by electrostatic inactivation in an intensive field of closely placed charges. This inactivation can also occur because of the fact that the energy of bonds of absorbed ions is determined not only by the strength of the ionic bonds but also by the strength of covalent bonds of ions with the inactive groups of polyelectrolyte or of insoluble ionic complex and, possibly, by Van der Vaals forces of intermolecular interaction.

All these phenomena can evidently take place in the cellular protoplasm and in this respect comparison of the values of activity coefficients of ions in the cell with the values obtained in chloride solutions is incorrect for solving the question of the status of ions. The diminution of the calculated activity coefficient for certain ions inside the cell can be a result of the decrease in their energy of the type characteristic for polyelectrolytes (without diminution of the exchangeability of such ions); at the same time, this does not exclude the possibility of chemical binding of a certain amount of ions or their isolation in special cell compartments unaccessible for the measuring electrode. Obviously other methods are required for solving this question, among them determination of the concentration and activity of sodium and potassium in different intracellular structures.

REFERENCES

[1] L. Michaelis, and A. Kramsztyk, *Biochem. Z.*, **62**, 180 (1914).
[2] F. Lipman, and O. Meyerhof, *Biochem. Z.*, **227**, 84 (1930).
[3] K. Furusawa, and P. M. T. Kerridge, *J. Mar. Biol. Assoc. UK*, **14**, 657 (1927).
[4] Gl. L. Maison, O. S. Orth and K. E. Lemmer, *Am. J. Physiol.*, **121**, 311 (1938).
[5] M. Dubuisson, *Pflueg. Arch. Ges. Physiol.*, **239**, 314 (1937).
[6] M. Dubuisson, *Proc. Roy. Soc.*, **B, 137**, 63 (1950).
[7] A. Disteche, *Nature*, **187**, 119 (1960).
[8] Ya. A. Epstein, *Biokhimiya*, **13**, 434 (1948).
[9] P. C. Caldwell, *J. Physiol.*, **126**, 169 (1954).
[10] P. C. Caldwell, *J. Physiol.*, **142**, 22 (1958).
[11] C. S. Spiropoulos, *J. Neurochem.*, **5**, 185 (1960).
[12] P. G. Kostyuk, and Z. A. Sorokina, in *Membrane Transport and Metabolism* (A. Kleinzeller and A. Kotyk, eds.). Academic, London, 1961.
[13] Z. A. Sorokina, *Tsitologiya*, **3**, 48 (1961).
[14] Z. A. Sorokina, *Zh. Evol. Biokhim. Fiziol.*, **1**, 343 (1965).
[15] P. G. Kostyuk, in *Problems of Biophysics*, Moscow, 1964.
[16] M. Lavallée, *Circulation Res.*, **15**, 185 (1964).
[17] M. Lavallée, and G. Szabo, *Digest 6th Inter. Conf. Med. Electron. Biol. Eng.* Tokyo (1965).
[18] M. M. Shultz, *Uch. Zap. Leningr. Gos. Univ. Ser. Khim. Nauk.* **13**, 80 (1953).
[19] M. M. Shultz, *Vestn. Leningr. Univ., Ser. Fiz. i Khim.*, **4**, 174 (1963).
[20] G. Eisenman, D. O. Rudin and J. U. Casby, *Science*, **126**, 831 (1957).
[21] G. Eisenman, *Biophys. J.*, **2**, 259 (1962).
[22] G. Eisenman, in *Problems of Biophysics*, Moscow, 1964.
[23] S. M. Friedman, J. D. Jamison, J. A. M. Hinke and C. L. Friedman, *Proc. Soc. Biol. Med.*, **99**, 727 (1958).
[24] S. M. Friedman, and C. L. Friedman, *Anat. Record* **138**, 129 (1960).
[25] D. Ungar, *Tsitologiya*, **1**, 627 (1959).
[26] G. N. Ling, *J. Gen. Physiol.*, **43**, 149 (1960).
[27] F. Gotoh, J. Tazaki, K. Hamaguchi and J. S. Meyer, *J. Neurochem.*, **9**, 81 (1962).
[28] J. A. M. Hinke, *Nature*, **184**, 1257 (1959).
[29] J. A. M. Hinke, *J. Physiol.*, **156**, 314 (1961).
[30] A. A. Lev. and E. P. Buzhinski, *Tsitologiya*, **3**, 614 (1961).
[31] A. A. Lev, *Nature (London)*, **201**, 1132 (1964).
[32] A. A. Lev, *Biofizika*, **9**, 686 (1964).
[33] Z. A. Sorokina, *Byul. Eksperim. Biol. i. Med.*, **12**, 17 (1964).
[34] B. P. Nikolsky, *Zh. Fiz. Khim.* **10**, 495 (1937).
[35] B. P. Nikolsky, and M. M. Schultz, *Zh. Fiz. Khim.* **34**, 1327 (1962).
[36] B. P. Nikolsky, and M. M. Schultz, *Vestn. Leningrad Univ., Ser. Fiz. Khim.*, **n4**, 73 (1963).
[37] M. M. Shultz, and T. M. Ovchinnikova, *Vestn. Leningrad Univ., Ser. Fiz. Khim.*, **2**, 129 (1954).

[38] M. M. Shultz, A. I. Parfenov, I. V. Peshekhonova and A. A. Belyustin, *Vestn. Leningrad Univ., Ser. Fiz. Khim.*, **4**, 98 (1963).
[39] Z. A. Sorokina, *Fisiol. Zh. Kiev.* **12**, 776 (1966).
[40] Z. A. Sorokina, *III Jenaer Symposium* (1966).
[41] H. B. Steinbach, *Am. J. Physiol.*, **163**, 236 (1950).
[42] H. Naora, M. Jzawa, V. G. Allfrey and A. E. Mirsky, *Proc. N.Y. Acad. Sci.*, **48**, **5**,853 (1962).
[43] J. A. M. Hinke, Chapter 16 in this book.

CHAPTER 16

Glass Microelectrodes in the Study of Binding and Compartmentalization of Intracellular Ions

J. A. M. HINKE

The Department of Anatomy, The University of British Columbia

I am indebted to Dr. W. A. Webber for assistance with electronmicroscopy and the amino-acid analysis; to Mrs. I. Ingraham for much of the analytic work; and to Mr. D. Gayton and Mr. S. G. A. McLaughlin for carrying out some of the experiments. This work was supported by the British Columbia Heart Foundation and the Medical Research Council of Canada.

16.1 INTRODUCTION

Knowing the degree of structural complexity that characterizes the cell interior, one can no longer rationally assume that the cellular water and electrolytes are distributed homogeneously. A monovalent ion may be concentrated in a given fluid-filled compartment or it may be bound out of solution by association with complex organic polyelectrolytes; in either case, however, a heterogeneous distribution of the ion results.

One approach to the study of cellular ion distribution is the measurement of free concentration in the cytoplasm. Such a measurement can be made by impaling the cell with ion-selective microelectrodes. At the present time, microelectrodes can be constructed which are selective to H^+, Na^+, K^+, and Cl^- ions[1].

In our hands reliable and durable microelectrodes have been constructed with "working diameters" (where the insulation ends) as small as 5–10 μ. Unfortunately, such a size is still too large for many cells, but it is more than satisfactory for such cells as the squid axon[2, 3], the crab [2], lobster[4], and barnacle muscle[5,6], and even the frog muscle[7–9]. These fibers are sufficiently large to tolerate microelectrode impalement without serious structural or chemical alterations, provided only one impalement is made and the broken membrane is allowed to seal around the microelectrode.

Rather than participate in the development of small microelectrodes (see the chapters by M. Lavallée and A. A. Lev), we have elected to use the relatively large closed-tipped microelectrodes[1]. Our studies are confined to the single muscle fiber of the giant barnacle, *Balanus nubilus*[5]. The average dimensions of a single fiber from this crustacean are 1,500–2,000 μ in diameter and 30–40 mm long. Such large fibers tolerate impalement by microelectrodes up to 30 μ in diameter without showing serious changes in membrane potential and ionic composition.

In this chapter, experiments are described which combine the measurements of Na^+, K^+, and Cl^- activities in the myoplasm with the chemical analysis of total Na^+, K^+, and Cl^- in the fiber during isotonic, hypertonic, and hypotonic conditions. Other experiments are presented which define the size of the extracellular space of the single fiber. From this information, we show that a large fraction of fiber Na^+ and a significant fraction of fiber water can be found neither in the myoplasm nor in the extracellular space. These findings support the view that significant amounts of fiber Na^+ and water are in a bound state[5, 6].

16.2 DESIGN OF MICROELECTRODES

The main design features of the ion-selective microelectrodes are illustrated in Fig. 16.1. Only the active tip and part of the nonactive shaft are shown for each electrode. The three cation-selective glass microelectrodes (H^+, Na^+, and K^+) have one construction detail in common, that is, a sensitive glass microcapillary inside a nonsensitive glass microcapillary. The active tips of the Na^+ and H^+ electrodes are effectively isolated by fusing the outer and inner microcapillaries at the junction where the outer microcapillary ends. The active tip of the K^+ electrode is isolated by filling the space between the outer and inner microcapillaries with beeswax. It is not difficult to manipulate molten wax down to the broken tip of the outer microcapillary so that it does not run out — provided the two microcapillaries fit snugly at the junction. After these electrodes are filled with an appropriate conducting solution, their tips are sealed and the electrode is virtually complete. A more detailed description of microelectrode construction can be found in a recent publication[1].

The Cl^- microelectrode consists of a 20 μ Pt wire inside the lumen of a glass microcapillary. A small Pt tip is localized by means of a glass-Pt seal. The Pt tip is covered first with solid Ag and then with solid AgCl. The main steps in the construction of this microelectrode are illustrated in Fig. 16.2. First, the Pt thread is placed inside a length of stock tubing and anchored to one end of the tubing. The latter is placed

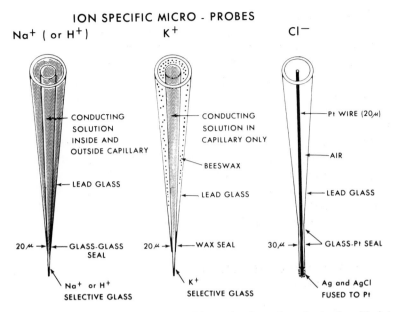

ION SPECIFIC MICRO - PROBES

Na⁺ (or H⁺)

- CONDUCTING SOLUTION INSIDE AND OUTSIDE CAPILLARY
- LEAD GLASS

20µ → GLASS-GLASS SEAL

Na⁺ or H⁺ SELECTIVE GLASS

K⁺

- CONDUCTING SOLUTION IN CAPILLARY ONLY
- BEESWAX
- LEAD GLASS

20µ → WAX SEAL

K⁺ SELECTIVE GLASS

Cl⁻

- Pt WIRE (20µ)
- AIR
- LEAD GLASS

30µ → GLASS-Pt SEAL

Ag and AgCl FUSED TO Pt

Fig. 16.1. Diagram of shaft and tips of ion-selective microelectrodes. Na⁺ (and H⁺) microelectrode has glass-glass seal between inner sensitive and outer resistant capillaries. K⁺ microelectrode has wax between two capillaries. Cl⁻ microelectrode is conventional Ag–AgCl electrode; Ag and AgCl are fused to tip of a 20µ Pt wire.

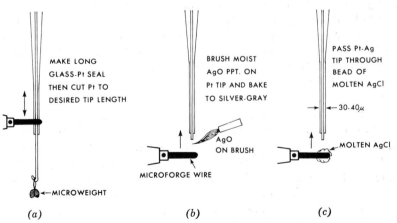

MAKE LONG GLASS-Pt SEAL THEN CUT Pt TO DESIRED TIP LENGTH

—MICROWEIGHT

BRUSH MOIST AgO PPT. ON Pt TIP AND BAKE TO SILVER-GRAY

AgO ON BRUSH

MICROFORGE WIRE

PASS Pt-Ag TIP THROUGH BEAD OF MOLTEN AgCl

← 30-40µ

MOLTEN AgCl

(a)   (b)   (c)

Fig. 16.2. Illustration of main steps in construction of Ag–AgCl microelectrode: (a) glass capillary is fused to Pt wire over 1–1.5 cm length; (b) AgO is applied to Pt tip and reduced to Ag by heating; (c) Ag-coated Pt tip is immersed in bead of molten AgCl.

in a microelectrode puller and drawn down to a microcapillary. In about 50% of the attempts, the microcapillary will break exactly where the Pt thread fills the lumen of the microcapillary. Sometimes a water-tight glass-Pt seal may occur at this junction, but usually one must be produced by carefully heating the junction with the microforge wire. In order to eliminate the wide taper of these microelectrodes, we routinely collapse the glass on the Pt thread for a distance of 1.0–1.5 cm from the tip. The excess Pt holding the weight (Fig. 16.1a) is then cut away so that only a small tip of Pt protrudes from the end of the microcapillary. The Pt tip is coated with solid Ag by painting the tip with a moist AgO precipitate and then heating the tip (with the microforge) until the AgO is converted to Ag. This chemical transformation is easily visualized because AgO is black and Ag is silver-gray. Finally AgCl is fused over the Pt–Ag by immersing the tip briefly in a bead of molten AgCl (Fig. 16.2c).

16.3 ELECTRICAL PERFORMANCE OF Na$^+$ AND K$^+$ MICRO-ELECTRODES

The cation-selective microelectrodes routinely demonstrate ideal electrical behavior, that is, at 24°C their potential changes about 59 mV per decalog change in cation activity. After adequate soaking and aging, they all become sufficiently stable so that the mV reading of a given microelectrode in a given standard may not vary by more than 1 mV over a 10-hr period. For the past three years, the cation-selective glasses have been obtained from Corning Glass Works, and the Na$^+$ and K$^+$ microelectrodes have been giving Na/K selectivity ratios of 100/1 and 1/10, respectively. Such selectivities are quite satisfactory for intracellular use because in the cytoplasm the Na$^+$ concentration is low and the K$^+$ concentration is high.

   The working life span of the cation microelectrodes varies from days to months. Apart from breakage, the limiting feature of the K$^+$ microelectrode is its insulation. As long as the latter remains intact, the K$^+$ microelectrode usually shows increased selectivity to K$^+$ and decreased electrical resistance with age. These desirable features are probably related to hydration of the K$^+$ glass membrane[10]. The main limitation of the Na$^+$ microelectrode is a gradual decrease in response time, even when the glass microcapillary has a very thin wall (not measurable under light microscopy). Generally, the responsiveness of a Na$^+$ microelectrode is considered unsatisfactory when it cannot achieve 90% of its expected voltage after 10 secs in a given standard. In the last year we discovered that sluggish Na$^+$ microelectrodes can be rejuvenated by heating the tip of the microelectrode with the microforge wire. A half-hour treatment usually produces a desirable response time without altering the electrical stability or the ionic selectivity of the electrode.

One serious criticism which has been raised against the K$^+$ microelectrode is that it responds to NH$_4^+$ ions and may well respond in a similar manner to the charged amino groups in the myoplasm[11]. Recently we re-examined the degree to which organic compounds containing charged amino groups can interfere with the ability of the K$^+$ microelectrode to predict the [K$^+$] in solution. Concentrated solutions of NH$_3$, lysine, arginine, albumin and protamine were made up in a 0.253 M KCl solution at pH 7.4. The solutions were serially diluted with 0.253 KCl at pH 7.4. At each dilution, the solution was tested with K$^+$, Na$^+$, and Cl$^-$ microelectrodes and a sample was taken for the chemical determination of the K$^+$, Na$^+$, and Cl$^-$ concentrations. The K$^+$ microelectrode results of this experiment are shown in Fig. 16.3. The dashed horizontal line indicates the measured [K$^+$] in the pure KCl solution. The other curves indicate the calculated [K$^+$] in the same KCl solution in the presence of the

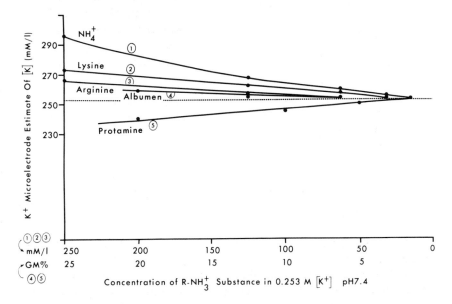

Fig. 16.3. Effect of NH$_4^+$ and R–NH$_3^+$ compounds on prediction of [K$^+$] by means of K$^+$ microelectrode in 0.253 M KCl solution. Dashed horizontal line defines behavior of microelectrode in pure KCl solution (0.253 M). When NH$_4^+$, lysine, arginine, or albumin is added to KCl solution, calculated [K$^+$] increases with increasing [R–NH$_3^+$]. Calculated [K$^+$] appears to decrease when protamine is added.

different amino compounds at varying concentrations. These curves show that amino compounds, particularly in high concentrations, do interfere with the performance of the K^+ microelectrode. The magnitude of interference is as follows. When the concentration of NH_4^+, lysine, and arginine is 50 mM, the K^+ microelectrode predicts a falsely high $[K^+]$ of 2.0, 1.2, and 0%, respectively. When 10 gm% albumin or protamine is present, the $[K^+]$ appears to be 1.2% too high in the former case and 2.4% too low in the latter case.

After obtaining these results, the barnacle muscle was analyzed for free amino-acid content. The only unusual finding was a rather large amount of arginine, equivalent to a concentration of 30 mM. The results depicted in Fig. 16.3 show, however, that arginine does not alter the voltage of the microelectrode even at a concentration of 50 mM.

The muscle fiber contains about 10 gm% of myosin, which has about the same lysine, arginine, and histidine content as albumin. Yet 10 gm% of albumin produces a falsely high $[K^+]$ of only 1.2% (Fig. 16.3). This result indicates that it is unlikely that the extra amino groups of actin or myosin can interfere significantly with the ability of the K^+ microelectrode to predict the K^+ activity in the myoplasm of a living barnacle muscle.

The magnitudes of deviation shown in Fig. 16.3 are such that they may well be caused by changes in junction potential at the reference electrode or by changes in the activity coefficient of the solutions. It is quite possible that these changes may not occur because of an NH_4^+ ion effect from the charged amino groups on the organic compounds. This conclusion is supported by the finding that protamine (which contains a large number of extra amino groups) failed to elevate the potential of the K^+ microelectrode (Fig. 16.3).*

16.4 ELECTRICAL PERFORMANCE OF THE Cl^- MICROELECTRODE

The Cl^- microelectrode usually always behaves ideally (59 mV voltage change per unit log change in Cl^- activity). In addition, the Cl^- microelectrode is electrically stable and responds rapidly to a solution change. The main weakness of this microelectrode is that the fused AgCl may flake off during a particularly difficult impalement. Such an occurrence alters the electrical characteristics of the electrode and makes the preceding standardization of the electrode meaningless. To guard against this mis-

---

*Glycine was found in a very large concentration (> 100 mM), but this amino acid is neutral and therefore not analogous to NH_4^+.

hap, the Cl⁻ microelectrode must be standardized immediately before and after each impalement.

Ions that interfere most seriously with the Cl⁻ microelectrode are those of other halides. This fact cannot be ignored, especially when one is working with a marine animal which may have significant Br⁻ and I⁻ concentration in its cytoplasm. Figure 16.4 illustrates the effect of small Br⁻ concentrations on the response of the microelectrode to Cl⁻. Bromide concentrations as low as 0.05 mM in the presence of 50 mM of Cl⁻ can alter the potential of the electrode significantly. Notice that a small amount of Br⁻ produces a falsely high prediction of chloride activity. In order to obtain a rough idea of how serious the presence of Br⁻ in

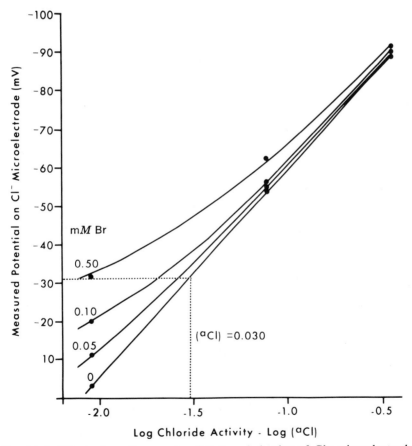

Fig. 16.4. Effect of small amounts of Br⁻ on behavior of Cl⁻ microelectrode. Dashed lines indicate that mV reading is similar for solution containing 30 mM Cl⁻ and for one containing 26 mM Cl⁻ + 0.05 mM Br⁻.

the myoplasm might be in the Cl^- determination let us suppose that the Br^-/Cl^- ratio in myoplasm is identical to sea water. This ratio is about 0.0033 (gram-weight basis) for sea water with a 35% salinity[12]. Thus, when the myoplasmic Cl^- activity (a_{Cl}) is 30 mM, the myoplasmic Br^- activity (a_{Br}) is about 0.06 mM. The dashed lines in Fig. 16.4 show that when the microelectrode reads 30 mM a_{Cl} in the presence of 0.05 mM Br^-, the real Cl^- activity is 26 mM. Our estimate of intracellular Br^- is probably high because the muscle fibers are equilibrated in a Ringer's solution which contains lower quantities of Br^- than sea water does. The Br^- correction becomes even smaller when one remembers that the calibrating solutions also contain traces of Br^- ion. In the experiments to be reported, we have not corrected for a Br^- error, which means that our calculated intracellular a_{Cl} is probably 2–4 mM too high.

16.5 MICROELECTRODE IMPALEMENT OF SINGLE FIBERS

Dissection of single intact viable muscle fibers from the depressor muscles of *Balanus nubilus* have been described elsewhere[5, 6, 13]. Each muscle fiber has its own tendonous insertion which can be easily isolated from neighboring tendons, and the origin of a given fiber to the stony basis need not be disturbed during isolation. Figure 16.5 illustrates how we cannulate the tendon of a single fiber and suspend the fiber vertically in an artificial bath. Such a fiber may elongate 10–30% of its in-vivo length, depending on the size of the stony fragment (not shown) which the fiber suspends at its lower end. This figure also illustrates how the ion-selective microelectrode is fed through the cannula and into the muscle fiber. The microelectrode is advanced downward into the fiber until the active tip is at least 1 cm from the puncture site. Thus measurements are taken at a location which is surrounded by undamaged membrane. The other microelectrode in this diagram is an open-tipped KCl-filled microelectrode for the measurement of membrane potential. Figure 16.6 is a photograph of a typical single muscle fiber impaled with a cation-selective microelectrode. This photograph gives a better idea of the microelectrode size relative to the diameter of the fiber. It also illustrates absence of structural distortion around the microelectrode.

16.6 INTERNAL MORPHOLOGY OF THE FIBER

A typical longitudinal section of a dissected muscle fiber is shown in Fig. 16.7. The fiber interior is occupied primarily by myofibrils packed closely together. Fragments of sarcoplasmic reticulum can be seen between the myofibrils, but it does not appear to be as extensive and well-

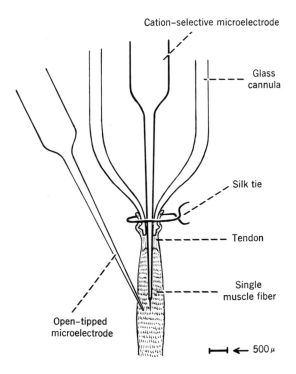

Cation–selective microelectrode

Glass
cannula

Silk tie

Tendon

Single
muscle fiber

Open–tipped
microelectrode

← 500 μ

Fig. 16.5. Diagram showing cannulation of tendon of single muscle fiber. Cation-selective microelectrode is guided through cannula before it impales fiber.

organized as in the frog muscle[14]. Mitochondria are not abundant in this muscle and the nuclei lie mainly at the periphery of the fiber. From this picture, it seems reasonable to suppose that the tip of an ion-selective microelectrode is surrounded exclusively by myofibrils.

A typical cross section of the surface of a dissected fiber is shown in Fig. 16.8. Notice first that the fiber surface is covered with cellular and noncellular elements. Notice also that the fiber surface invaginates frequently into the fiber interior and that these invaginations are filled with the noncellular surface elements. The relatively large invaginations give rise to or communicate with smaller tubules which appear to ramify deeply into the myoplasm. An example of a large extracellular invagination which penetrates deeply into the fiber is shown in Fig. 16.9. These cross sections indicate that the extracellular compartment of an isolated single fiber is quite substantial and cannot be ignored in subsequent calculations.

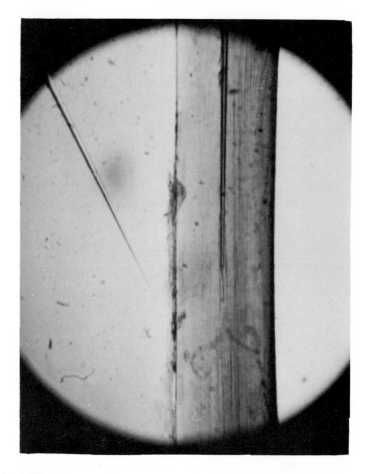

Fig. 16.6. Photograph of single muscle fiber impaled with cation-selective micro-electrode. (See Fig. 16.5 for orientation.) Note lack of structural distortion around impaled microelectrode.

16.7   COMPARTMENTALIZATION VERSUS BINDING IN THE FIBER

The single muscle fiber has several fluid-filled membrane-enclosed com-partments, each of which may have a different ionic composition. These compartments may be listed as the myoplasm, the extra-cellular space, sarcoplasmic reticulum, nuclei, and mitochondria. The ions in each of these compartments may be either free in solution or bound out of solu-tion. Ion binding may be to soluble organic compounds or to insoluble

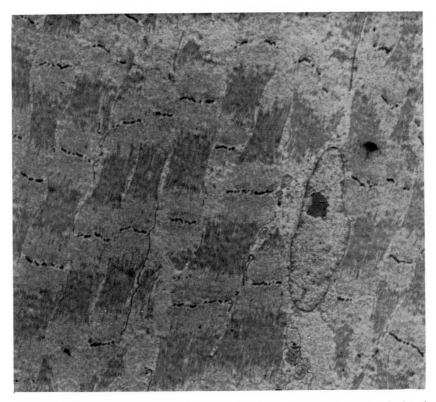

Fig. 16.7. Electronmicrograph of muscle fiber from *Balanus nubilus* in longitudinal section (fixed in 1% gluteraldehyde and 1% OsO_4, stained in lead citrate).
Myofibrils occupy most of muscle volume. A few membranous tubules between
myofibrils and nucleus can be seen in this section. ($700 \times$).

formed structures. In the myoplasm, for example, the myosin in the myofilaments may bind significant quantities of Na^+ and K^+ ions [15] or just
Na^+ ions [5, 6].

As a working approximation, we postulate that most of the monovalent
ions are in the free ionic state and are located mainly in the myoplasm
and the extracellular spaces. Thus the distribution of K^+, Na^+, Cl^-
in the single fiber can be defined by the following equations:

$$V[K] = \alpha_1 V(K)_1 + \alpha_2 V(K)_2 + \beta_K \qquad (16.1)$$

$$V[Na] = \alpha_1 V(Na)_1 + \alpha_2 V(Na)_2 + \beta_{Na} \qquad (16.2)$$

$$V[Cl] = \alpha_1 V(Cl)_1 + \alpha_2 V(Cl)_2 + \beta_{Cl}, \qquad (16.3)$$

where V is the total fiber water; $V[\ ]$ is the total ion content; α_1 and α_2
are the fractions of fiber water acting as solvent in compartments 1 and

Fig. 16.8. Periphery of muscle fiber in cross section, showing cellular and un-
cellular composition of fiber surface. Note large trabeculae arising from surface
and penetrating deep into fiber. Tubules can be seen branching from trabeculae
and ramifying into myoplasm (700×).

2; $(\ )_1$ and $(\ )_2$ are the free ion concentrations in compartments 1 and 2;
and β_K, β_{Na}, and β_{Cl} are the quantities of fiber K^+, Na^+, and Cl^- which are
not in compartments 1 and 2. In this chapter, compartment 1 represents
the extracellular space and compartment 2 represents the myoplasm.

The terms [K], [Na], [Cl], and V can be determined by chemical analysis; $(K)_2$, $(Na)_2$ and $(Cl)_2$ can be measured directly with the ion-selective
microelectrodes. If the fiber has been adequately equilibrated with the
bathing solution, then $(K)_1 = [K]_0$, $(Na)_1 = [Na]_0$, and $(Cl)_1 = [Cl]_0$,
(where $[\ ]_0$ is ion concentration in bathing solution). Thus five unknowns
remain in the three equations: α_1, α_2, β_K, β_{Na}, and β_{Cl}.

As in a previous publication[5], we assume here that $\beta_K = 0$. This
assumption makes α large and β_{Na} and β_{Cl} small. Put another way, the
assumption assigns maximum water and electrolytes to the fluid compartments. Clearly, if $\beta_K \neq 0$ then α will become proportionally smaller and

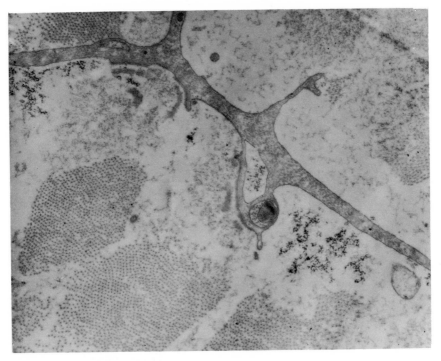

Fig. 16.9. Large extracellular trabecula or cleft located deep inside single fiber well away from periphery (1200×).

β_{Na} and β_{Cl} will become proportionally larger — depending on the magnitude of β_K. In (16.1) the term $\alpha_1 V(K)_1$ can be neglected because $(K)_1 = [K]_0 \ll (K)_2$ and $\alpha_1 \ll \alpha_2$. This equation reduces to $\alpha_2 = [K]/(K)_2$ and can be solved for α_2. Thus the determination of β_{Na} and β_{Cl} is reduced to the problem of estimating α_1, the fraction of fiber water in the extracellular space.

16.8 DETERMINATION OF EXTRACELLULAR SPACE

Because compartment 1 communicates freely with the bath, the ionic concentrations in compartment 1 should adjust to any ionic changes in the bathing solution. The results shown in Fig. 16.10 are from an experiment in which the muscle fiber was equilibrated for 25 min in a Na^+-free sucrose Ringer solution. During this period, the fiber K^+ concentration remained constant, but the fiber Na^+ concentration decreased significantly Yet no change in the myoplasmic Na^+ activity (a_{Na}) was observed. As

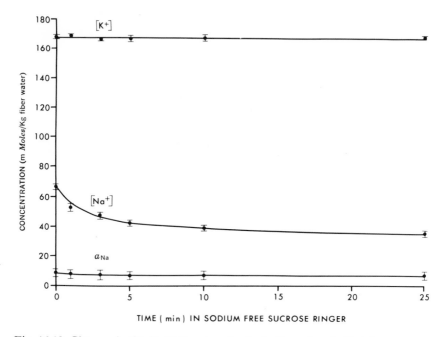

Fig. 16.10. Changes in $[Na^+]$, $[K^+]$ and a_{Na} in single fiber when in Na^+-free sucrose Ringer solution. $[Na^+]$ and $[K^+]$ were obtained by flame-photometric analysis of single fibers; a_{Na} was measured directly by Na^+ microelectrode.

the fiber volume and the fiber water content were unaltered, then the myoplasmic Na^+ content $\alpha_2 V(Na)_2$ remained constant. It is probable, therefore, that the Na^+ loss in this experiment was confined to compartment 1. Assuming $(Na)_1$ dropped from 450 mM to 0 mM in 25 min, we calculated α_1 to be 6.7% of the fiber water. A similar experiment was performed using Li^+ rather than sucrose as a substitute for Na^+ in the Ringer solution. The data from this experiment (Table 16.1), item 5) lead to a calculation of α_1 equal to 7.5% of fiber water.

As the extracellular compartment contains a significant quantity of Cl^- as well as Na^+, it should be equally valid to estimate the size of compartment 1 from a Cl^- washout experiment. The curves in Fig. 16.11 summarize the changes in $[Cl^-]$ and $(a_{Cl})_2$ when the fiber is equilibrated in a 50% chloride (methanesulfonate) Ringer solution. Notice that myoplasmic chloride (a_{Cl}) remained nearly constant for 100 min, even though the total fiber chloride content was reduced significantly. If it is assumed that all the Cl^- was lost from compartment 1 and $(Cl)_1$ dropped from 540 to 270 mM in 30 mins, then α_1 can be calculated to 7.8% of fiber water.

Table 16.1 Estimates of the Extracellular Space on Single Muscle Fibers By Different Methods

Type of Experiment	Myoplasmic Activity		Total Fiber Content		Extracellular Space (% Fiber Water)
	At Start	After 30 min $(a_{Cl})_i$	At Start	After 30 min $[Cl]_i$	
1. Washout into 50% Cl Ringer	0.028 ± 0.0019 (12)	0.0255 ± 0.0012 (9)	0.075 ± 0.0025 (29)	0.051 ± 0016 (20)	7.8
2. Washout into 50% Cl Ringer $[K]_0[Cl]_0$ constant	0.031 ± 0.0004	0.031 ± 0.0008 (5)	0.075 ± 0.0012 (14)	0.056 ± 0.0023 (16)	7.0
3. Washout into 25% Cl Ringer $[K]_0[Cl]_0$ constant	0.029 ± 0.0025 (8)	0.029 ± 0.0030 (8)	0.075 ± 0.0012 (14)	0.044 ± 0.0007 (8)	7.6
	$(a_{Na})_i$		$[Na]_i$		
4. Washout into zero Na^+ Ringer (sucrose)	0.0082 ± 0.0030	0.0058 ± 0.0028 (5)	0.066 ± 0.0021 (20)	0.034 ± 0.0023 (11)	6.7
5. Washout into zero Na^+ Ringer (Li^+)	0.0060 ± 0.0013	0.0045 ± 0.0010 (11)	0.080 ± 0.0035 (20)	0.044 ± 0.0031 (12)	7.5
6. Inulin space (analytic method)	From inulin saturation curve				6·5
7. Inulin space (C^{14} method)	From C^{14} inulin saturation curve				6·5

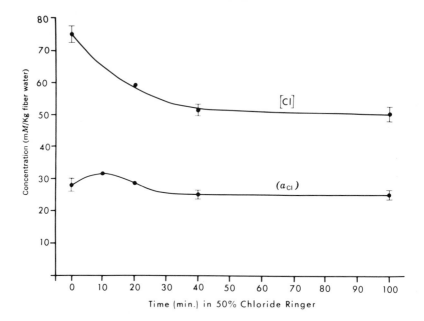

Fig. 16.11. Changes in [Cl⁻] and a_{Cl} in single fiber when in 50% Cl⁻ (methane-sulfonate) Ringer solution. [Cl⁻] was determined on single fibers by Cotlove method and a_{Cl} was measured directly by Cl⁻ microelectrode.

This chloride washout experiment can be made more sophisticated by increasing bath $[K]_0$ as bath $[Cl^-]_0$ is reduced so that the product $[K^+]_0 [Cl^-]_0$ remains constant. This maneuver should maintain Donnan equilibrium across the fiber membrane and minimize outflux of Cl⁻ from the myoplasm. The results of these experiments are shown in Fig. 16.12. The curves on the left are for fibers equilibrated in 50% Cl⁻ Ringer solution and the curves on the right are for fibers equilibrated in 25% Cl⁻ Ringer solution. In all cases the myoplasmic Cl⁻ (a_{Cl}) remained constant. Making the usual assumptions about compartment 1, we calculated α_1 to be 7.0% and 7.6% of fiber water from the left and right hand curves, respectively.

The classical method of estimating the extracellular space of a tissue is the determination of the space occupied by a substance, such as inulin, which does not penetrate the cell membrane. Figure 16.13 shows typical inulin space saturation curves for dissected single muscle fibers. The upper curve was obtained by an analytic method[16]: the fibers were equilibrated in a Ringer solution containing 300 mg% inulin; at varying times, the fibers were removed and colorimetrically analyzed for inulin. The lower curve was obtained by the radioactive method: the fibers were

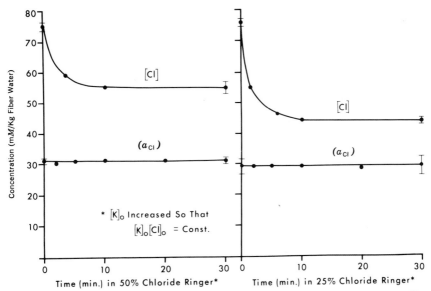

Fig. 16.12. Loss of Cl^- from single fiber when in 50% (left) and 25% (right) Cl^- Ringer solutions in which $[K^+][Cl^-]$ is held constant. In both experiments, myoplasmic Cl^- (a_{Cl}) remained constant.

equilibrated in a Ringer solution containing 2 $\mu c/ml$ C^{14} inulin; at varying times, the fibers were removed and analyzed for C^{14} inulin by means of a liquid scintillation counter. Notice that both curves do not start at zero and do not reach a plateau. To obtain the correct inulin space in the fiber the straight part of each curve beyond 20 min must be extrapolated to zero time. In addition, the apparent inulin reading at zero time must be subtracted from the extrapolated value. These maneuvers yield an inulin space equal to 6.5% of fiber water in both experiments. Table 16.1 lists the pertinent data from all these experiments together with the calculated values of α_1. It can now be said with confidence that the extracellular compartment contains between 6.5 and 7.5% of the fiber water. This estimate is quite compatible with the electronmicroscopic evidence (Figs. 16.8 and 16.9).

16.9 DETERMINATION OF ION AND WATER BINDING

Table 16.2 lists the calculated values of α_2, β_{Na}, and β_{Cl}, together with the experimental data used in the calculation. These data were taken from recent experiments in which the three microelectrode measurements and

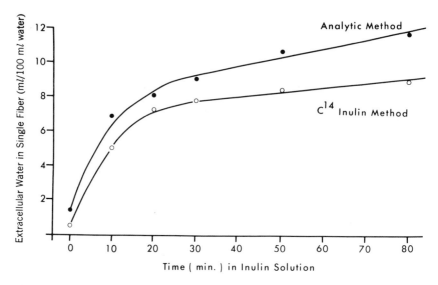

Fig. 16.13. Inulin space saturation curves for single fibers. For upper curve, fibers were equilibrated in Ringer containing 300 mg% inulin; for lower curve, fibers were equilibrated in Ringer containing 2 μc/ml C^{14}-inulin. (See text for more details.)

the analytic determinations of [K$^+$], [Na$^+$], and [Cl$^-$] were made on fibers from the same barnacle. Each of the (K$^+$)$_2$, (Na$^+$)$_2$, and (Cl$^-$)$_2$ values is the mean from at least eight fibers and each of the [K$^+$], [Na$^+$], and [Cl$^-$] values is the mean from at least 16 fibers. The numerical values are in good agreement with values from other experiments conducted over the past three years.

Notice first that the fraction of fiber water in the myoplasm (α_2) is 0.676 which means only 68% of the fiber water is used as solvent for ions in the myoplasm. As $\alpha_1 + \alpha_2 = 0.74$, it follows that 26% of the fiber water is not used as a solvent in compartments 1 and 2. Some of this water may be distributed among the smaller compartments (nuclei, mitochondria, etc.) and some of it may be structured around protein[17].

Table 16.2 shows that about 49% of the total fiber Na$^+$ and about 16% of the total fiber Cl$^-$ is not in the fluid of the two compartments. If the extracellular space is excluded on the grounds that it does not really belong to the fiber interior, then the percentage of internal fiber Na$^+$ and Cl$^-$ which is not measureable by the microelectrodes becomes 83 and 31%, respectively (Table 16.2). It is hard to imagine from the electron-micrographs where such large quantities of ions, particularly Na$^+$, might be located. If one supposed a third fluid compartment containing ionic

Table 16.2   Calculation of α_2, β_{Na}, and β_{Cl} for Single Muscle Fiber in Barnacle Ringer (24°C)

Fraction of Fiber Water in Myoplasm (α_2)	Fraction of Fiber Na$^+$ "Bound"	Fraction of Fiber Cl$^-$ "Bound"
$\dfrac{[K]}{(K)_2} = 0.676$	$\dfrac{\beta_{Na}}{V[Na]} = 0.486$	$\dfrac{\beta_{Cl}}{V[Cl]} = 0.157$
	$\dfrac{\beta_{Na}}{V[Na] - \alpha_1 V(Na)_1} = 0.830$	$\dfrac{\beta_{Cl}}{V[Cl] - \alpha_1 V(Cl)_1} = 0.314$

Calculations based on following values:

$[K] = 160$ mM/kg H_2O $(K)_1 = 8$ mM/kg H_2O $(K)_2 = 235$ mM/kg H_2O

$[Na] = 70$ mM/kg H_2O $(Na)_1 = 450$ mM/kg H_2O $(Na)_2 = 10$ mM/kg H_2O

$[Cl] = 70$ mM/kg H_2O $(Cl)_2 = 540$ mM/kg H_2O $(Cl)_2 = 35$ mM/kg H_2O

$$\alpha_1 = 0.065$$

Na$^+$ at a concentration equal to $(K)_2$ in the myoplasm (for osmotic balance), then the size of this compartment would need about 13% of the fiber water. As about 25% of the fiber water has not been accounted for, there is certainly enough available for the hypothetical third compartment. The only organelles which might contain Na$^+$ in a high concentration and which may be regarded as separate from the myoplasm are the nuclei, mitochondria, and sarcoplasmic reticulum. The mitochondria can be eliminated because they have been shown to concentrate K$^+$ and not Na$^+$ ions[18]. The nuclei can probably be eliminated because they, too, have shown to concentrate K$^+$ in preference to Na$^+$[19]. Finally, the sarcoplasmic reticulum is just not large or extensive enough in the barnacle muscle (Fig. 16.7) to contain 13% of the fiber water or to account for 10% of the fiber volume. For that matter, none of the organelles (even when taken together) appears to have sufficient volume to account for 10% of the fiber volume. For these and other reasons, we have postulated [5, 6] that most of the fiber Na$^+$ exists in the myoplasm, but much of it is bound out of solution – probably to the contractile myofilaments in the myofibrils.

16.10 COMPOSITION AND VOLUME OF COMPARTMENTS DURING OSMOTIC STRESS

If the fiber is made to swell in hypotonic Ringer and made to shrink in hypertonic Ringer, one can expect large volume changes in compartment 2 but not in compartment 1. Furthermore, the ionic concentrations in compartment 2 should depend primarily on the compartment's water content, whereas the ionic concentrations in compartment 1 should depend primarily on the Ringers solution. Finally, it is difficult to predict what changes might occur in the unknown or bound Na^+, Cl^-, and water fractions.

Figure 16.14 shows the changes in total fiber K^+, myoplasmic K^+, and total fiber water when the single fiber is in a Ringer solution made hypotonic by adding an equal volume of water to normal Ringer solution. As expected, the fiber volume increases and the $[K^+]$ decreases. Notice, however, that $(K)_2$ decreases faster than $[K]$, which means that water enters the myoplasm faster than in the fiber as a whole. This observation is reflected in the calculated increase in α_2 from 0.68 to 0.85 in 30 min. The dashed curve in Fig. 16.14 is calculated from the water results, using

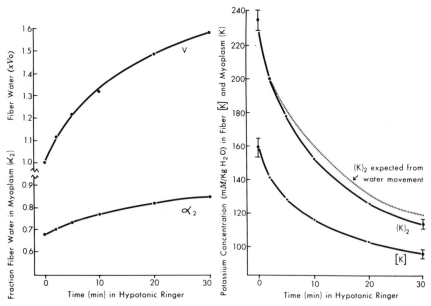

Fig. 16.14. Changes in K^+ concentration in whole fiber $[K^+]$ and in myoplasm (K^+), and changes in total fiber water V and fraction of total fiber water in myoplasm α_2 when fiber was immersed in hypotonic Ringer solution. Vertical bars at an experimental point are 2 (S.E.) long. (See text for more details.)

the assumption that the myoplasmic K^+ content remained constant. As this curve closely matches the experimental curve (obtained from the K^+ microelectrode), it can be concluded that an insignificant number of K^+ ions shifted out of the myoplasmic compartment.

In Fig. 16.15, comparable data are shown for fibers immersed for 30 min in hypertonic Ringer solution. The solution was made hypertonic by adding an osmotically equivalent amount of sucrose to normal Ringer solution. As expected, the fiber volume decreases and the [K] increases. Notice again that $(K)_2$ changes faster than [K], which means α_2 is changing. As illustrated in the Figure, α_2 decreased from 0.68 to 0.52 in 30 min. The dashed curve for $(K)_2$ has the same significance as in Fig. 16.14. However, here it demonstrates that a larger quantity of K^+ moved out of the fiber with water—yet not enough to prevent $(K)_2$ from increasing. In summary, when the fiber was in hypotonic Ringer for 30 min (Fig. 16.14), myoplasmic water increased $2\times$ and the myoplasmic K^+ content remained unchanged; when the fiber was in hypertonic Ringer for 30 min (Fig. 16.15), myoplasmic water decreased $2.5\times$ and the myoplasmic K^+ content decreased $0.25\times$.

Figure 16.16 illustrates the changes in the Na^+ and Cl^- content in the whole fiber and in the myoplasm when the fiber is in the hypotonic and hypertonic Ringer solutions. The curves to the right belong to the

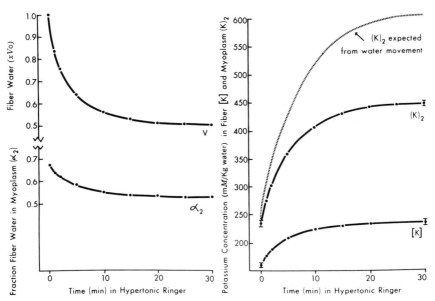

Fig. 16.15. Changes in $[K^+]$, $(K^+)_2$, V, and α_2 when single fiber was immersed in hypertonic Ringer solution. (See text and Fig. 16.14 for more details.)

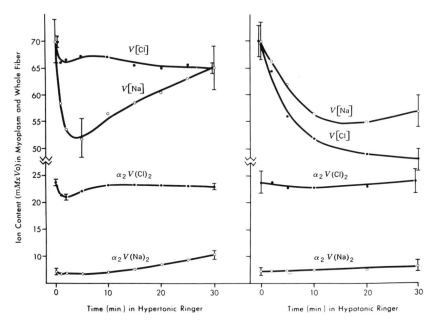

Fig. 16.16. Changes in Na^+ and Cl^- content of whole fiber (upper two curves) and of myoplasm (lower two curves) when fiber was in hypertonic (left) and hypotonic (right) Ringer solution.

hypotonic experiment. Notice that the myoplasmic Na^+ and Cl^- content (lower two curves) remains constant even though significant quantities of Na^+ and Cl^- ions are lost from the fiber (upper two curves). Most likely these ions are lost from the extracellular compartment as it adjusts its Na^+ and Cl^- concentrations to those in the bathing solution (Na^+ and Cl^- are one-half normal). In contrast, the hypertonic results are rather unexpected (Fig. 16.16, left), especially when one remembers that the Na^+ and Cl^- concentrations were not changed in this bathing solution. The total fiber Cl^- (top curve) remained relatively constant, but the total fiber Na^+ content decreased rapidly in the first 10 min and then increased. In 30 min, total fiber Na^+ was nearly back to its starting value. As myoplasmic Na^+ and Cl^- hardly changed at all (lower two curves), the large transitory change in fiber Na^+ must have occurred either in the extracellular space or in the bound compartment. The former is more probable. This transitory loss of Na^+ from the extracellular space may be related to a transitory flooding of the extracellular space with K^+ ions as the latter leave the myoplasm with water.

These experiments illustrate how different the conclusions would be if only the chemical analyses for $[K^+]$ and $[Na^+]$ were performed without the microelectrode measurements. One would have concluded that large changes in the Na^+ and Cl^- content had taken place in the myoplasm. With the microelectrode measurements, however, we conclude that the Na^+ and Cl^- content of the myoplasm remains relatively constant in spite of large changes in myoplasmic volume (Figs. 16.14 and 16.15) and in total fiber Na^+ and Cl^- content. Obviously, the concentrations of Na^+ and Cl^- in the myoplasm did not remain constant in these experiments. As did $(K^+)_2$, both $(Na^+)_2$ and $(Cl^-)_2$ increased during the hypertonic experiment and decreased during the hypotonic experiment.

It was stated earlier that the size of the extracellular space should not be affected by the hypotonic and hypertonic bath solutions. This assumption may not be valid in view of our observing the unexpected transitory changes in fiber Na^+. Also, Girardier et al. observed transient swelling in the transverse tubular system of the crayfish muscle when it was placed in a hypertonic medium[20]. To be certain that the extracellular volume remains relatively constant under osmotic stress, we measured the inulin space on single fibers when they were placed in hypotonic and hypertonic Ringer solutions. The experimental method was as follows. Three groups of dissected fibers were equilibrated in normal Ringer solution containing 2 μc/ml C^{14} inulin. Forty minutes later, one group of fibers was transferred to hypotonic Ringer containing 2 μc/ml C^{14} inulin and another group was transferred to hypertonic containing 2 μc/ml C^{14} inulin. At times 0, 2, 5, 10, 20, and 30 min, fibers from all three Ringer solutions were removed and analyzed for inulin content. The results of this analysis are shown in Table 16.3. Clearly, the inulin space did not vary significantly either in hypotonic or hypertonic Ringer solution. These findings support the assumption that the volume of compartment 1 (α_1) is unaffected by osmotic stresses because it communicates freely with the bathing solution.

Table 16.3   Changes in Inulin Space of Single Fibers after Transfer from Barnacle Ringer to Hypotonic or Hypertonic Ringer (24°C)

Time in New Solution (min)	Inulin Space (ml) Expressed as Fraction of V_0		
	Control (B.R.)	Hypotonic Ringer	Hypertonic Ringer
0	0.065	0.066	0.064
2	0.058	0.070	0.058
5	0.070	0.059	0.059
10	0.071	0.056	0.060
20	0.071	0.060	0.058
30	0.075	0.060	0.080
MEAN	0.068 ± 0.0024	0.062 ± 0.0021	0.064 ± 0.0035

Now that α_1 has been shown to remain relatively constant, the changes in the unknown or bound Na^+, Cl^- and water content of the fiber can be calculated using (16.1–16.3). Table 16.4 lists the values for bound Na^+, Cl^-, and water in the fiber after 30 min in hypotonic and hypertonic Ringer solution. The quantities are expressed as a function of V_0, the initial fiber water content. Notice first that more than 65% of the bound water was retained after 30 min in hypertonic Ringer, even though 50% of the fiber water (Fig. 16.15) and more than 60% of the myoplasmic water ($0.263V_0$ versus $0.676V_0$ at start) was lost. In addition, no increase in bound water occurred in hypotonic Ringer even though myoplasmic water doubled. Observe next that relatively minor changes in bound Na and bound Cl occurred, which indicates most of the unlocated Na^+ and Cl^- remained immobile in spite of large changes in fiber water. Clearly, these unlocated quantities of Na^+, Cl^-, and water cannot be imagined as existing in a fluid compartment in osmotic equilibrium either with the myoplasm or with the bath. These results support the postulate that significant quantities of Na^+ and water (possibly also Cl^-) are bound [5, 6].

Finally, Fig. 16.17 summarizes the changes in the water, Na^+, and Cl^- content of each compartment (extracellular, myoplasmic, and bound) when the fiber is in hypotonic and hypertonic Ringer solutions. The water that enters the fiber during hypotonicity is received by the myoplasm almost exclusively; the water that leaves the fiber during hypertonicity is lost from the myoplasm almost exclusively. The largest losses of Na^+ and Cl^- when the fiber is in hypotonic Ringer come almost exclusively from the extracellular space; the Na^+ and Cl^- content in the other two compartments remains unaltered. Probably also the large transient change in Na^+ content when the fiber is in hypertonic Ringer is confined to the extracellular space. There is an indication, however, that some Na^+ may

Table 16.4   Calculation of α_2, β_{Na}, and β_{Cl} for Single Muscle Fiber After 30 min in Hypotonic or Hypertonic Ringer (24°C)

	Barnacle Ringer		After 30 min in Hypotonic Ringer		After 30 min in Hypertonic Ringer	
Fraction fiber water in myoplasm α_2	0.676		0.845		0.525	
"Bound Na" β_{Na}	34	V_0	34	V_0	26	V_0
"Bound Cl" β_{Cl}	11	V_0	7	V_0	7	V_0
"Bound water" $(1-\alpha_1-\alpha_2)V$	$0.259\,V_0$		$0.180\,V_0$		$0.172\,V_0$	
Myoplasmic water $\alpha_2 V$	$0.676\,V_0$		$1.335\,V_0$		$0.263\,V_0$	

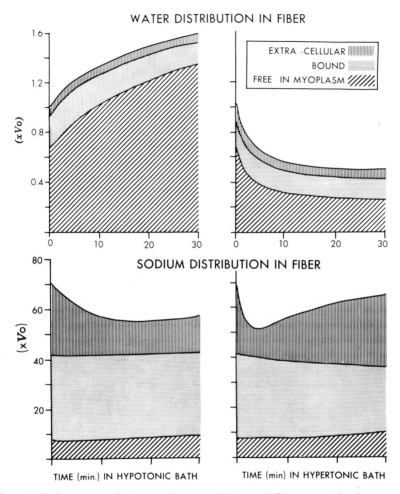

Fig. 16.17. Summary of changes in water, Na$^+$, and Cl$^-$ content in three compartments (extracellular, bound, and myoplasmic) of single muscle fiber when latter is in hypotonic Ringer solution (left) and in hypertonic Ringer solution (right).

be shifting from the bound reservoir to the myoplasm when water and K$^+$ are leaving the fiber in response to the hypertonic stress (see also Table 16.4). Clearly illustrated in Fig. 16.17 is the fact that the fluid myoplasm contains relatively small fractions of the fiber's total Na$^+$ and Cl$^-$ content, particularly Na$^+$, even though this compartment contains most (68%) of the fiber water. Undoubtedly, the K$^+$ ion must be the predominant free cation in the myoplasm (Table 16.1; Figs. 16.14 and 16.15).

16.11 FINAL COMMENT

The conclusions from these experiments are consistent with the general idea that significant quantities of cellular monovalent cations are not free in solution[21, 22], but they do not support the view that K^+ is the cation which is predominantly bound out of the free solution[21–24]. Our micro-electrode measurements strongly indicate that Na^+ ions and not K^+ ions are preferred by intracellular binding sites. This does not mean, however, that no K^+ ions are bound. For example, Robertson[25] obtained some evidence from the lobster muscle that about 26% of fiber K^+ may be bound compared with about 82% of fiber Na^+. Future experiments may well reveal that some fiber K^+ is in fact bound ($\beta_K \neq 0$; equation 16.1), but such a finding will serve to increase, not decrease, our present estimates of Na^+ and water binding. For example, if we assume 50% of the fiber K^+ is bound and recalculate α_2 and β_{Na} from equations 16.1–16.3 (using data in Table 16.2), then $\alpha_2 = 0.34$ (instead of 0.68) and $\beta_{Na} = 0.55$ (instead of 0.49). This calculation illustrates the important relationship between K^+ binding (β_K) and water binding ($1 - \alpha_1 - \alpha_2$) in our equations. As β_K is increased from 0 to 50%, bound water increases from 26% (Table 16.4) to 60%. Thus the microelectrode measurements do not place restrictions on the size of the fluid myoplasmic compartment. Nevertheless, they do indicate that the fluid myoplasm, whatever its volume, contains K^+ ions in a high concentration and Na^+ ions in a low concentration.

REFERENCES

[1] J. A. M. Hinke, in *Glass Electrodes for Hydrogen and Other Cations* (G. Eisenman, ed.), Dekker, New York, 1967.

[2] P. C. Caldwell, *J. Physiol. (London)*, **142**, 22 (1958).

[3] J. A. M. Hinke, *J. Physiol. (London)*, **156**, 314 (1961).

[4] J. A. M. Hinke, *Nature*, **184**, 1257 (1959).

[5] S. G. A. McLaughlin and J. A. M. Hinke, *Can. J. Physiol. Pharmacol.*, **44**, 837 (1966).

[6] J. A. M. Hinke and S. G. A. McLaughlin, *Can. J. Physiol. Pharmacol.*, **45**, 655 (1967).

[7] A. A. Lev and E. P. Buzhinsky, *Cytology (USSR)*, **3**, 614 (1961).

[8] A. A. Lev, *Nature*, **201**, 1132 (1964).

[9] P. G. Kostyuk and Z. A. Sorokina, *Symposia Čsav* (A. Kleinzeller and A. Kotyk, eds.), Czechoslovak. Acad. Sci., 1961.

[10] G. Eisenman, personal communication.

[11] G. N. Ling, *Ann. N. Y. Acad. Sci.*, **137**, 837 (1966).

[12] H. W. Harvey, *The Chemistry and Fertility of Sea Waters*, University Press, Cambridge, 1957.

[13] G. Hoyle and T. Smyth, Jr., *Comp. Biochem. Physiol.*, **10**, 291 (1963).

[14] L. D. Peachey, *J. Cell Biol.*, **25**, 209 (1965).

[15] H. A. Saroff, *Arch. Biochem. Biophys.*, **71**, 194 (1957).

[16] G. Ross and R. Mokotoff, *J. Biol. Chem.*, **190**, 659 (1951).

[17] G. N. Ling, *Ann. N. Y. Acad. Sci.*, **125**, 401 (1965).

[18] R. E. Thiers, E. S. Reynolds and B. L. Vallee, *J. Biol. Chem.*, **235**, 2130 (1960).

[19] S. Itoh and I. L. Schwartz, *Am. J. Physiol.*, **188**, 490 (1957).

[20] L. Girardier, J. P. Reuben, P. W. Brandt and H. Grundfest, *J. Gen. Physiol.*, **47**, 189 (1963).

[21] A. S. Troschin, *Das Problem der Zellpermeabilität.*, Fischer, Jena, 1958.

[22] G. N. Ling, *A Physical Theory of the Living State*, Blaisdell, New York, 1962.

[23] F. H. Shaw and S. E. Simon, *Australian J. Exp. Biol.*, **33**, 153 (1955).

[24] S. E. Simon, F. H. Shaw, S. Bennett and M. Muller, *J. Gen Physiol.*, **40**, 753 (1957).

[25] J. D. Robertson, *J. Exp. Biol.*, **38**, 707 (1961).

CHAPTER 17

Liquid Ion-Exchanger Microelectrodes

FRANK W. ORME

Department of Physiology–Anatomy, University of California at
Berkeley

The author acknowledges support from U.S. Dept. of Public Health.

17.1 INTRODUCTION

One of the most significant aspects of cellular physiology is the measurement of intracellular ion activities. The importance of ions in nerve activity[1, 2], muscle contraction[3], enzyme regulation[4], embryonic development[5], control of osmotic pressure[6], and other physiological processes is well known and because these processes respond to the activities of ions rather than to their concentrations, reliable methods of estimating ion activities are highly desirable. The method which has thus far shown the most promise is potentiometric measurement with ion-specific electrodes. In recent years the construction of ion-specific electrodes has undergone three major advances which promise to give new directions to the intracellular problem.

1. The development of cation specific glasses, particularly for potassium and sodium[7, 8].
2. The production of crystal and precipitate electrodes for various anions[9–11].
3. The discovery of liquid ion exchangers capable of incorporation into electrodes for both cations and anions[12, 13].

Of these methods only the cation glass electrode has been adapted to intracellular use[14, 15]. This Chapter concerns some first attempts to construct ion-sensitive microelectrodes from liquid ion exchangers.

376

17.1.1 Liquid Ion-Exchanger Electrodes

Liquid ion-exchanger electrodes are not new. In 1906 Cremer[16], in his classic paper on the glass electrode, discussed electrodes made from phenol and nitrobenzene saturated with picric acid. These studies were later advanced by a number of workers, including Loeb and Beutner[17], Osterhout[18], and Bonhoeffer et al.[19]. Much of this early work has been reviewed by McClendon[20], Beutner[21, 22], and Wilbrandt[23]. Although the early investigators were principally interested in developing models of the cell membrane, they discovered a number of principles of great importance to those wishing to use liquid membranes for analytical purposes. For instance, they showed that the incorporation of an acidic material into the exchanger produced a cation electrode, while basic materials formed anion electrodes. They investigated the concentration dependence of the potential and found that it was roughly that predicted by the Nernst equation. They were also able to demonstrate that the membranes responded more readily to some ions than to others, thus revealing properties of ion specificity. In general, however, these early electrodes suffered from three defects.

1. The relative transference of anion and cation, or permselectivity, was very poor. While many of these electrodes approached Nernstian behavior at low concentrations, because of the poor permselectivity they became increasingly insensitive at higher concentrations. A similar phenomenon has been noted by Sollner[24] in electrodes made from solid ion-exchanger membranes which are very porous.

2. The membranes had considerable water permeability. Water flow can cause large deviations from theoretical behavior[25, 26].

3. The ion specificity between ions of like charge was generally low, as shown by low biionic potentials. There were some notable exceptions to this, however. Osterhout[18] reported a biionic potential of 67 mV for a nitrobenzene membrane separating KCl from NaCl. This potential corresponds to an apparent selectivity of potassium over sodium of more than ten to one.

As the investigations progressed, certain liquid ion exchangers were found to perform much better than others. Dupeyrat[27] found that a liquid membrane composed of $5 \times 10^{-5} M$ lauryl trimethyl bromide in nitrobenzene between potassium bromide solutions gave potentials very close to those calculated from the Nernst equation. Surprisingly, in view of Osterhout's report, this membrane could not differentiate between sodium and potassium[28]. Dupeyrat found other exchangers with high sodium-potassium selectivity, but no data were given on the concentration potentials. Bonner and Lunney[29] studied several types of liquid cation (dinonylnapthalenesulfonate) and anion (Kemamine Q–1902–C, National Dairy Products Co., Aliquat 336, General Mills Co.) exchangers; they were able, under favorable conditions, to obtain potentials about 98% of those predicted by the Nernst equation. They attributed the deviations to water transfer and the solubility of the exchanger in the aqueous phase. An exceptional liquid ion-exchanger electrode was made by Sollner and Shean[13] with solutions of lauryl trialkylmethyl amine (Amberlite LA-2, Rohm & Haas Co.). Membranes of this material had excellent perm-selectivity and low water transfer, even when the surrounding solutions were quite concentrated. Concentration potentials approached to within a few tenths of millivolts of the Nernst values, and considerable selectivity between different anions was shown.

At the present time three cation electrodes (for calcium, divalent, and cupric ions) and three anion electrodes (for perchlorate, nitrate, and chloride ions) using liquid ion exchangers that are available commercially. One should not take the designations of these electrodes too seriously. They are based upon what the manufacturer feels they will be most useful for; but the "nitrate electrode" is more sensitive to perchlorate than to nitrate, and the calcium electrode gives a greater response to zinc than to calcium. Their uses will depend to a large degree upon the ionic compositions of the solutions into which they are placed. Properties of the commercial electrodes have been described by a number of authors [30–36].

The microelectrodes described here incorporate three different liquid ion exchangers. Although the exact formulations of these exchangers are still not available, their general compositions are as follows[37]:

Electrode	Orion Code No.	Type of Exchanger
Calcium	92–20–02	High-molecular-weight organophosphoric acid
Divalent cation	92–32–02	High-molecular-weight organophosphoric acid
Chloride	92–17–02	High-molecular-weight quaternary ammonium compound in decanol

17.2   METHOD

17.2.1   Construction of Liquid Ion-Exchanger Microelectrodes

In principle liquid ion-exchanger microelectrodes are extremely easy to make. A Pyrex capillary micropipette is pulled by hand or by one of several types of pipette pullers[38, 39]. The tip is then filled with the proper exchanger and connected to an electrometer through a reference electrode.

Several methods of filling the tip are available. If the tip is simply dipped into the exchanger it will gradually fill by capillary action. This process is extremely slow, however, and may be greatly speeded by applying a vacuum to the shank of the capillary. For large electrodes (tips greater than 5 μ) the latter method is quite suitable, 5–20 min being sufficient for the filling process. For smaller electrodes a more convenient method of filling is as follows:

1. The capillaries are first filled with a low-molecular-weight organic solvent by boiling at room temperature under reduced pressure[40]. The solvent must be one capable of dissolving the ion exchanger. For the Orion exchangers methanol works well. For other types of exchangers we have used solvents such as acetone and chloroform.

2. Most of the low-molecular-weight solvent is then removed from the shank and replaced with ion-exchange fluid.

3. The tip is placed in ion-exchange fluid, and several days are allowed for the exchanger to replace the solvent by diffusion. Routinely the electrodes are left in the ion-exchange fluid for about a week, but one day seems to be adequate.

Electrical connection to the electrode may be made in one of several ways (Fig. 17.1). The simplest method is to place a wire directly into the exchanger (Type B, Fig. 17.1). Unless the wire is a reversible electrode, such a connection is subject to polarization—but the currents drawn are usually so small that this factor is not significant. Chloride electrodes so constructed, using a silver-silver chloride electrode, work very well.

A second method of construction is to place a salt solution directly above the exchanger (Type C, Fig. 17.1). Although liquid ion exchangers generally have lower densities than salt solutions, there is no problem of flotation because the exchanger is held in the tip by capillary forces. The salt solution should contain ions which react both with the exchanger and with the reversible electrode (usually Ag/AgCl) that dips into it.

Electrical connection may be made through an internal microelectrode (Type A, Fig. 17.1). A conventional salt-filled microelectrode is inserted into the exchanger-filled capillary to within a few microns of the tip, leaving a thin liquid membrane between the inner electrode and the test solution. The advantage of this construction is that the resistance may be considerably lowered.

No insulation of the electrodes is necessary. Insulation is provided by the glass shank of the electrode itself. It is important that the shank be free of exchanger left over from the filling operation and that it be kept

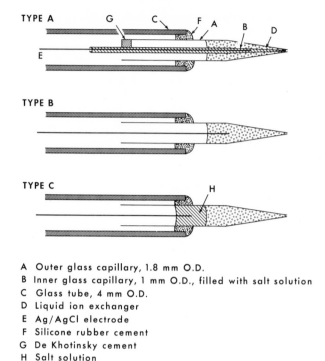

A   Outer glass capillary, 1.8 mm O.D.
B   Inner glass capillary, 1 mm O.D., filled with salt solution
C   Glass tube, 4 mm O.D.
D   Liquid ion exchanger
E   Ag/AgCl electrode
F   Silicone rubber cement
G   De Khotinsky cement
H   Salt solution

Fig. 17.1. Schematic diagrams of liquid ion-exchanger microelectrodes.

clean and dry. Although we have not siliconized the shanks of our electrodes, in some situations this might help to reduce surface conduction caused by absorbed water.

17.3   RESULTS AND DISCUSSION

17.3.1   Microelectrodes for Calcium and Magnesium

Successful calcium electrodes have been made from all three designs shown in Fig. 17.1, but the most promising construction seems to be the one embodying an internal microelectrode (Fig. 17.1, Type A). Electrodes of this construction with tip diameters of 1–2 μ have resistances of about $10^{11}\,\Omega$. Initially the internal capillary was filled with 1 M CaCl$_2$ but recently, in an attempt to lower the resistance, we have been using a mixture of 1 M CaCl$_2$ and 3 M KCl. To insure internal equilibrium, the salt solution should be saturated with AgCl and with the calcium exchanger.

Response of the calcium electrodes was evaluated in a cell without liquid junction:*

$$\text{Ca microelectrode } (CaCl_2 (1\ M) | AgCl | Ag)$$

The voltage of the cell is directly proportional to the logarithm of the mean activity of calcium chloride. A plot of these quantities (Fig. 17.2) shows fairly good agreement with theory, the points falling close to a line with the theoretical slope of 88.7 mV per log activity (25°C). However, the scatter of these points is rather large, which indicates poor reproducibility.

If the Ag/AgCl electrode is replaced with a standard reference electrode with liquid junction, there is an approximately linear relationship between voltage and log concentration in the range of 10^{-5} to $1\ M$

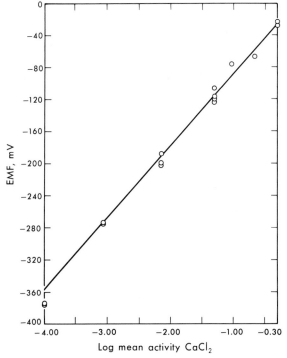

Fig. 17.2. Voltage response of calcium microelectrode – Ag/AgCl electrode pair at 25°C in unbuffered $CaCl_2$ solutions. Straight line drawn through mean of points for 100 mM $CaCl_2$ has theoretical slope of 87.6 mV per log activity.

*Unless otherwise noted all measurements have been made at 25 ± 1°C with a Cary Model #31 vibrating reed electrometer.

(Fig. 17.3). The slope of the curve is approximately 20–22 mV per log concentration. If it is assumed that the liquid junction potential is constant and that the activity coefficient of calcium is equal to the mean activity coefficient of calcium chloride, the slope becomes 23 to 25 mV per log activity—somewhat lower than that of 26–27 mV found by Shatkay[35] for the commercial Orion calcium electrode.

The response time of the calcium electrodes has been quite slow, 20–30 min being required to reach a steady potential (Fig. 17.4). Workers using this material have also been plagued by the pickup of low-frequency (<5 cps) noise with an amplitude of 1–2 mV. This noise, which has so far resisted all screening attempts, has made accurate measurements very difficult. Another defect is that if the electrodes are stored with their tips in salt solutions for a period of a week or two, they often cease to function.

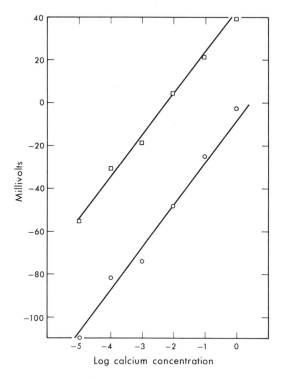

Fig. 17.3. Voltage response of microelectrode containing calcium ion exchanger (lower curve) and divalent ion exchanger (upper curve). Circuit is completed with reference electrode with liquid junction and lines have slopes of approximately 20 mV/log concentration. Readings were made in unbuffered $CaCl_2$ solutions at 23°C (lower curve) and 25°C (upper curve).

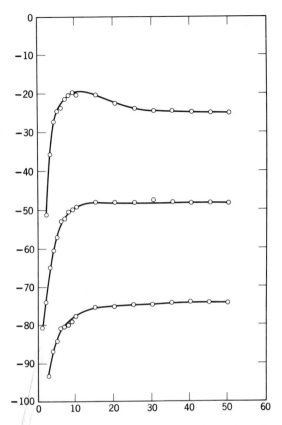

Fig. 17.4. Rate of approach to equilibrium of calcium microelectrode at 25°C. Electrode was equilibrated in 1 M CaCl$_2$ and then placed in 100 (top curve), 10 (middle curve), and 1 (bottom curve) mM CaCl$_2$ solutions in that order. Ordinate shown emf's in mV; abscissa, time in minutes.

For this reason we now rinse the tips with distilled water after use and store them dry.

It is possible that some of these troubles are caused by the formation of a film of water on the inside wall of the tip. Many glasses are markedly hygroscopic[41], and after prolonged immersion in aqueous solutions one would expect some water to creep up the tip wall. Support for this idea comes from the report of Bonner and Lunney[29], who constructed electrodes in which the contact between the exchanger and the salt solution took place at the mouth of a narrow glass capillary (1.5-mm diameter). They found that if the capillaries became wet, the reproducibility became very poor and the electrodes became quite noisy. The problem was solved by siliconizing the capillaries. In this connection it may be noted that Corning has found it necessary to siliconize the porous glass membrane used in their calcium electrode. We are currently undertaking experiments with five different silicone compounds in the hope of finding a way to siliconize microcapillaries.

The pH response of a calcium microelectrode in 1 mM $CaCl_2$ is shown in Fig. 17.5. For comparison a curve for a commercial calcium electrode is also shown. At this concentration of calcium the voltage is fairly constant to about pH 6, which is in good agreement with the claims of the manufacturer[42]. The pronounced dip in the curve around pH 4 is apparently caused by an impurity which has now been removed from some batches of exchanger[43]. In the pH curves published by Ross[12] for what is presumably a slightly different calcium exchanger, the humps are absent — which extends the useful range of the electrode almost by another pH unit.

The voltage, E, of an electrode for divalent cations may be described fairly accurately by the following equation[12,44]:

$$E = E_0 + \frac{RT}{2F} \ln\left(a_i + \sum_j K_{ij} a_j^{2/z_j}\right), \qquad (17.1)$$

where E_0 represents the voltage under standard conditions, R is the gas constant, T is the absolute temperature, F is Faraday's constant, a_i is the

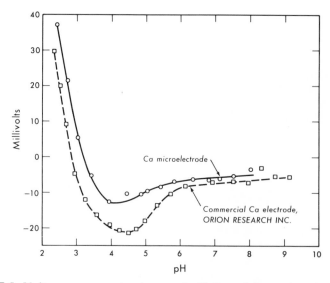

Fig. 17.5. Voltage response to change of pH for calcium microelectrode contrasted with that of commercial calcium electrode (25°C). Calcium was constant at 1 mM.

activity of ion i, and z_j is the valence of ion j. The measurement error in percentage of cation activity is given by:

$$\text{error} = 100 \sum_j K_{ij} \frac{a_j^{2/z_j}}{a_i}. \qquad (17.2)$$

The microelectrode results give a K_{CaH} of about 2×10^5, and if only Ca and H are present in the system the error will be:

$$(100)\,(2 \times 10^5)\,\frac{a_{\text{H}}^2}{a_{\text{Ca}}}$$

It is easy to compute that for 10^{-3}, 10^{-4} and $10^{-5}\,M$ calcium solutions, 99% accuracy requires that the pH be kept above 5.15, 5.66, and 6.15, respectively. The hump in the curve, which is not described by the equation, is, of course, ignored.

Because of the poor performance of these electrodes, we have not yet attempted to measure the interference of other cations. Values given by the manufacturer for the commercial electrode are [42] $K_{\text{CaMg}} = 1.4 \times 10^{-2}$ and $K_{\text{CaNa}} = 3.4 \times 10^{-4}$. Inorganic cation interferences for commercial calcium electrodes are discussed by Glauser et al.[33] and by Shatkay[35].

In biological materials one must also be concerned with interferences from organic ions. Little work has been done on this problem, but Arnold et al.[36] have measured calcium activity in serum with a calcium electrode. In a single experiment we have measured the binding of calcium to serum albumin with a commercial calcium electrode over the pH range of 5.56–11.58. The results are close to those obtained by Carr[45] with a sulfonated polystyrene membrane electrode. These experiments give an indication that the protein error of calcium electrodes is small, but much more work remains to be done.

We have made a few microelectrodes containing the Orion divalent cation exchanger. A concentration curve for one of them is shown in Fig. 17.3. Its slope is almost identical to that of the calcium electrode. According to the manufacturer[46], this electrode is equally sensitive to calcium and magnesium. It is less sensitive to hydrogen ion than the calcium electrode, but it is about ten times more sensitive to sodium.

17.3.2 Microelectrodes for Chloride

Chloride microelectrodes of Type B (Fig. 17.1) with an internal Ag/AgCl electrode have a resistance of approximately $5 \times 10^9\,\Omega$. As these electrodes perform quite well, no attempts have been made to lower the resistance further; however, this might be done by using an internal microelectrode (Type A, Fig. 17.1) or by putting a small Ag/AgCl electrode well down into the tip. With regard to the latter, platinum wire can be sharpened electrolytically to a tip diameter of 1 μ or less [47, 48] The sharpened tip could then be coated with silver and chloridized to produce the desired electrode. Silver wire can also be sharpened electrolytically (in HNO_3), but the tips produced are of a larger diameter and tend to be brittle.

The concentration dependence of the potential of an electrode of a 1 μ tip is shown in Fig. 17.6. The electrode responds almost ideally to chloride in concentrations from 10^{-5} to 2 M. An additional test of the chloride response was made in a cell containing two different types of chloride electrodes:

chloride microelectrodes |X Cl (M) | AgCl | Ag.

The voltage of this cell should be independent of chloride concentration throughout the range that both electrodes respond ideally to chloride.

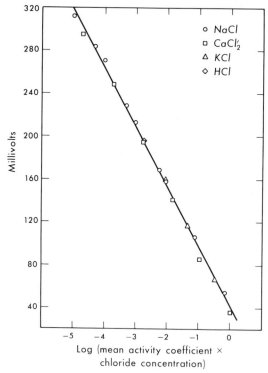

Fig. 17.6. Voltage response of chloride microelectrode at 25°C. Circuit is completed with liquid junction and straight line has theoretical Nernstian slope of − 59.1 mV per log activity.

Three microelectrodes were studied in this manner and over the concentration range of $2 \times 10^{-4} M$ to $1 M$ the average deviations were within 1 mV of the mean voltage. The electrode seems to function at much higher concentrations than the $0.1 M$ claimed by the manufacturer[49].

The chloride response seems to be indifferent to the nature of the cation. Data from four (Na, K, Ca, H) different chloride salts are plotted in Fig. 17.6. We have also used the electrodes successfully in Li, NH_4, and Cs chlorides. The electrode is not affected by tris buffer as long as the ionic strength of the buffer is taken into account when assigning activity coefficients.

Compared with the calcium microelectrodes, the chloride electrodes respond rapidly; a stable voltage is attained in about 2–4 min after changing chloride solutions (Fig. 17.7). They are also less noisy and more reproducible than calcium electrodes. Voltage drifts corresponding to a change in E_0 of (17.1) are seen, however, and for the greatest accuracy frequent calibration is necessary.

It is possible that the positive groups of the chloride ion exchanger react with the negative sites of the glass in the tip wall, forming a waterproof seal. The chloride electrodes seem to last much longer than the

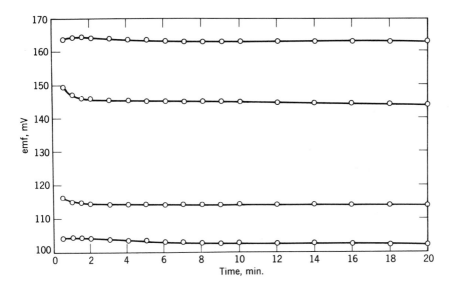

Fig. 17.7. Approach to equilibrium of chloride microelectrode with 2-μ tip. Electrode was equilibrated in 200 mM chloride and then placed in 100 (bottom curve), 55 (second curve), 20 (third curve), and 10 (top curve) mM solutions in that order (25°C). Ordinate shows emf's in mV; abscissa, time in minutes.

calcium electrodes when continuously immersed in aqueous solutions. Eventually they deteriorate, however, and we now store them dry.

The chloride electrode responds to every anion that we have tested. Some approximate selectivity constants are:

	This Study	*Orion*[49]
$K_{Cl;HS}$	7.5	—
$K_{Cl;\,HPO_4{}^{2-}}$	0.97	—
$K_{Cl;\,OH}$	0.72	1
$K_{Cl;\,acetate}$	0.26	0.32
$K_{Cl;HCO^-}$	0.15	0.19
$K_{Cl;\,H_2PO_4{}^-}$	0.092	—

Except for bisulfide, the response to these ions, based on measurements at different concentrations seems to be Nernstian (there is some uncertainty because of the difficulty of assigning single ion activity coefficients). In bisulfide solutions the electrodes drift for hours and the curve has a slope approximately 40 mV/log activity instead of the theoretical 59 mV/log activity. Evidently drastic chemical changes take place in the exchanger when the electrode is exposed to bisulfide ions. After immersion in bisulfide, an electrode may take 5–10 h to return to its original potential in a chloride solution.

At first glance the response to monovalent phosphate ion also appears to be non-Nernstian. The response was tested in serial dilutions of NaH_2PO_4, and the slope obtained was about 45 mV/log concentration. However, taking into account activity coefficients and the amounts of $HPO_4{}^{2-}$ present, the slope of the curve does not appear to be far from that predicted by the Nernst equation. The phosphate response was also tested at different pH's in mixtures of H_3PO_4, NaH_2PO_4, and K_2HPO_4 at a constant phosphate concentration of 10 mM. The theoretical curve based upon activities calculated by the extended form of the Debye-Huckel equation[50] and the Henderson-Hasselbach equation gives a fairly good fit to the experimental data (Fig. 17.8).

The pH curve for a microelectrode in 10^{-3} M chloride is shown in Fig. 17.9. At this level of chloride, interference from hydroxide ion is negligible below pH 8.2. On the acid end of the scale, the electrode seems to respond properly at pH's at least as low as 0.7.

We have attempted to measure chloride binding to serum albumin with the use of a chloride microelectrode; preliminary results are in the range found by Carr[45]. There is some evidence of protein interference with the electrode, particularly at high pH. We are planning to study this problem more thoroughly using the microelectrode-Ag/AgCl cell.

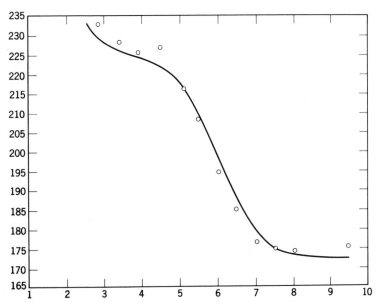

Fig. 17.8. Voltage response of chloride microelectrode in 10 mM phosphate as pH changes (26°C). As pH rises the $HPO_4^{2-}/H_2PO_4^-$ ratio increases; different response of the electrode to these ions is shown by the drop in voltage. Curve is theoretical, based on measured values of selectivity constants for HPO_4^{2-} and $H_2PO_4^-$. Ordinate shows voltage in mV; abscissa, pH values.

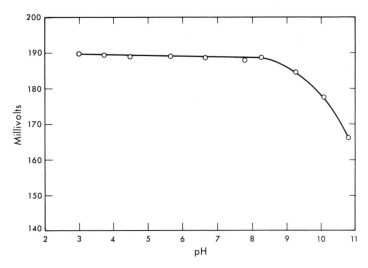

Fig. 17.9. pH response of chloride microelectrode at constant chloride level of 1 mM (25°C).

17.4 BIOLOGICAL APPLICATIONS

In our work with microelectrodes we have not gone much further than studying the interferences of inorganic ions. We can make some tentative statements about their suitability for biological measurements, subject to revision as further information is obtained. Taking the calcium electrode first, we can see that its intracellular use will be very limited. There are few reliable estimates of intracellular calcium concentration, but those that we have show that it must be extremely low. Early microinjection experiments showed that the free calcium ion in amoeba cytoplasm must be below 2 mM because injection at this concentration caused coagulation of the protoplasm[51]. Nanninga[52] calculated — on the basis of dissociation constants for calcium complexes with ATP, creatine phosphate, and myosin — that free muscle calcium is about 1 mM. However, he neglected calcium binding to other muscle fractions, in particular to the sarcoplasmic reticulum, and the true value must be considerably lower than his estimate. Just how low it probably is was shown by Hagiwara and Nakajima[53] in experiments on barnacle muscles. By microinjecting $CaCl_2$ solutions, using EDTA as a calcium buffer, they were able to show (on the basis of physiological responses) that free muscle calcium must be between 8×10^{-8} and $8 \times 10^{-7} M$. Hodgkin and Keynes[54] did measurements of the electrophoretic mobility of calcium in the squid axon. They concluded that at least 98% of the intracellular calcium was complexed, giving a free concentration of less than $10^{-5} M$. The muscle concentration of $10^{-7} M$ is already beyond the range of resolution of present calcium electrodes.

If we make some rough estimates of the activities of other cations in muscle and nerve, we can calculate the errors such activities would produce. In both tissues, potassium, sodium, and hydrogen have been measured with glass electrodes[55-63]. In various types of muscle and nerve the ranges of activities reported for these ions are:

	Nerve	*Muscle*
K	0.154–0.238	0.084–0.193
Na	0.021–0.049	0.007–0.016
H	4.0×10^{-8}–$1.0 \times 10^{-7} M$	7.9×10^{-8}–1.3×10^{-7}

There is much less information on which to base an estimate of intracellular magnesium activity. According to Naninga's calculations[52], in muscle about 70% of the magnesium will be complexed by ATP creatine phosphate, and myosin — leaving about 3 mM free. This may be taken as an upper limit; the actual free concentration is probably much less.

Taking the highest values found in muscle and nerve we can calculate that even if the intracellular calcium is as high as $10^{-5} M$, the electrode reading will give a value six times too high. The limiting factor is the magnesium activity, and even if it is much less than 3 mM there is enough potassium present to give a 200% error. Hydrogen interference, on the other hand, is less than 0.1%.

It is interesting to calculate the maximum values of the selectivity constants for practical intracellular calcium electrodes. Suppose that there are no other important interferring substances and that we can tolerate a 3% error (1% for each ion). Then, using the same values for K, Na, and H, the selectivity constants must be smaller than these values:

	Cells with $10^{-5} M Ca$	*Cells with $10^{-7} M Ca$*
$K_{Ca;H}$	5.9×10^6	5.9×10^4
$K_{Ca;K}$	6.3×10^{-6}	6.3×10^{-8}
$K_{Ca;Mg}$	3.3×10^{-5}	3.3×10^{-7}

Except for hydrogen, the present electrode does not approach this performance.

Despite these drawbacks, if the stability and reproducibility of the present electrode can be improved, there are several situations in which it could be useful for intracellular measurements.

1. It can be used to determine an upper limit for calcium, which in our present ignorance could be quite useful. The present electrode is capable of pushing this limit down to $10^{-4} M$.

2. If simultaneous measurements are made of the other cations, the electrode can be corrected for their interference. This can be done accurately only in the range in which the measured cation and the interferring ion contribute about equally to the voltage. With this principle as a criterion, it can be shown that the maximum selectivity constants given above can be increased about 20 times. The upper limit of calcium can be lowered to $10^{-5} M$, but magnesium interference will prohibit accurate measurements at that level.

3. Some cell compartments (e.g., the vacuoles of plant cells) may have calcium activities at levels that can be measured with the electrode.

4. Calcium activities could be measured in experiments in which the intracellular calcium is raised above its normal level by immersing the cell in a high calcium medium. If the electrode readings were combined with chemical analysis, it would be possible to do an intracellular calcium titration curve.

The divalent cation electrode can be used to measure intracellular magnesium if the activity is in the millimolar range. The intracellular potassium would produce a 10–20% error at that level, but if potassium were measured separately with a glass electrode, a correction could be made.

The usefulness of the chloride electrode is difficult to assess because of the lack of data on the interferences of many anions and the disagreement over their concentrations. According to Conway[64], muscle chloride is about 5 mM, bicarbonate is 1 mM, and phosphate is around 10 mM. For a pH of 7, the bicarbonate would give a 3% error, and the singly and doubly charged species of phosphate, errors of 5 and 25%, respectively. However, much of the phosphate and some of the bicarbonate may be complexed with calcium and magnesium. Also, Conway's estimate for phosphate may be too high. Nanninga[52] gives a value of only 1 mM total orthophosphate. The phosphate errors, then, may be considerably lower than those calculated. On the other hand, there is at least 20 mM of creatine phosphate in muscle[52, 64] and according to Nanninga about 80% of it is in an uncomplexed form. The concentration of hexose monophosphates is below 0.5 mM according to Nanninga, but Conway places it at 15 mM. These ions may or may not add a significant error to chloride measurements.

In nerve fibers the chloride concentrations are considerably higher than in muscle, 20–40 mM being a common figure[65,66]. The squid axon has a chloride concentration above 100 mM[67, 68]. In these tissues the chloride-phosphate ratio is much higher than in muscle, and phosphate is probably not a significant source of error. The squid axon contains about 250 mM of sulphonic acid, however, and the axons of other species have equivalent amounts of dicarboxylic acids[1, 66]. The response to these ions must be measured before the feasibility of measuring chloride activity with the ion-exchanger electrode can be assessed.

17.5 FUTURE POSSIBILITIES

The number of compounds potentially available for liquid ion-exchange electrodes is almost limitless. For description of some of them, the reader is referred to reviews of liquid ion exchange[69, 70] and solvent extraction[71–74]. Among the types of compounds that have been used for liquid ion exchange are phosphoric acid esters, alkyl ammonium compounds, β-diketones, dimercapto- compounds, and dithiocarboxylic acids. These compounds may all be considered liquid ion exchangers — although in some the bonding is probably more of a covalent than an ionic nature. Another group of organic extractants (e.g., ethylenediamine and phenanthroline) are uncharged and act by displacing the ionic water of hydration. The ion which is thus solubilized must form an ion-pair with an organic counterion in order to go into solution in the organic phase. The macro-cyclic antibiotics, the biological effects of which have been extensively studied[75–77], are apparently of this nature. Although not strictly ion exchangers, liquid ion-sensitive membranes have been made from some of them[78, 79]. One compound in particular, valinomycin, shows great promise as a potassium microelectrode.

REFERENCES

[1] A. L. Hodgkin, *Proc. Roy. Soc.*, **148B**, 1 (1958).

[2] A. L. Hodgkin, *Proc. 23rd Intern. Congr. Physiol. Sci.*, (1965).

[3] A. Sandow, *Pharmacol. Rev.*, **17**, 265 (1965).

[4] H. R. Mahler, in *Mineral Metabolism*, vol. 1, Part B (C. L. Comar and F. Bronner, eds.), Academic, New York, 1961.

[5] L. G. Barth, *Biol. Bull.* (Wood's Hole), **129**, 471 (1965).

[6] J. R. Robinson, *Physiol. Rev.*, **40**, 112 (1965).

[7] G. Eisenman, *Advan. Anal. Chem. Instr.*, **4**, 213 (1965).

[8] G. Eisenman, in *Glass Electrodes for Hydrogen and Other Cations* (G. Eisenman, ed.), Dekker, New York, 1967.

[9] E. Pungor, J. Havas, and K. Toth, *Acta Chim. Hung.*, **41**, 239 (1964).

[10] E. Pungor, K. Toth, and J. Havas, *Acta Chim. Acad. Sci. Hung.*, **48**, 17 (1966).

[11] M. Frant, and J. W. Ross, Jr., *Science*, **154**, 1553 (1966).

[12] J. W. Ross, *Science*, **156**, 1378 (1967).

[13] K. Sollner, and G. M. Shean, *J. Am. Chem. Soc.*, **86**, 1901 (1964).

[14] J. A. M. Hinke, in *Glass Electrodes for Hydrogen and Other Cations* (G. Eisenman, ed.), Dekker, New York, 1967,

[15] R. N. Khuri, in *Glass Electrodes for Hydrogen and Other Cations* (G. Eisenman, ed.), Dekker, New York, 1967.

[16] M. Cremer, *Z. Biol.*, **47**, 562 (1906).

[17] R. Beutner, *Am. J. Physiol.*, **31**, 343 (1913).

[18] W. J. V. Osterhout, *Cold Spring Harbor Symp. Quant. Biol.*, **8**, 51 (1940).

[19] K. F. Bonhoeffer, M. Kahlweit, and H. Strehlow, *Z. Phys. Chem.*, **1**, 21 (1954).

[20] J. F. McClendon, *Physical Chemistry of Vital Phenomena*, Ch. 10, Princeton University Press, 1917.

[21] R. Beutner, *Physical Chemistry of Living Tissues and Life Processes*, Williams and Wilkins, Baltimore, 1933.

[22] R. Beutner, in *Medical Physics*, vol. 1 (O. Glasser, ed.) Year Book Publishers, Chicago, 1944.

[23] W. Wilbrandt, *Ergeb. Physiol. Biol. Chem. Exp. Pharmakol.*, **40**, 204 (1938).

[24] K. Sollner, *Svensk Kem. Tidskr.*, **7**, 267 (1958).

[25] N. Lakshminarayanaiah, *Chem. Rev.*, **65**, 491 (1965).

[26] G. Scatchard, *J. Am. Chem. Soc.*, **75**, 2883 (1953).

[27] M. Dupeyrat, *J. Chim. Phys.*, **61**, 306 (1964).

[28] M. Dupeyrat, *J. Chim. Phys.*, **61**, 323 (1964).

[29] O. D. Bonner, and D. C. Lunney, *J. Phys. Chem.*, **70**, 1140 (1966).

[30] M. E. Thompson, *Science*, **153**, 866 (1966).

[31] M. E. Thompson, and J. W. Ross, Jr., *Science*, **154**, 1643 (1966).

[32] E. W. Moore, J. W. Ross, J. Riseman, and W. L. Hughes, Jr., *Fed. Proc.* **25**, 508 (1966).

[33] S. C. Glauser, E. Ifkovits, E. M. Glauser, and R. W. Sevy, *Proc. Soc. Exp. Biol.*, **124**, 131 (1967).

[34] J. A. King, and A. K. Mulherji, *Naturwissenschaften*, **53**, 702 (1966).

[35] A. Shatkay, *Anal. Chem.*, **39**, 1056 (1967).

[36] D. E. Arnold, M. J. Stansell, and H. H. Malvin, USAF School of Air Force Medicine, SAM-TR-67-43 (1967).

[37] M. Frant, personal communication (1967).

[38] D. W. Kennard, in *Electronic Apparatus for Biological Research* (P. E. K. Donaldson, ed.), Academic, New York, 1958.

[39] K. Frank, and M. C. Becker, in *Physical Techniques in Biological Research*, vol. 5, Part A (W. L. Nastuk, ed.), Academic, New York, 1964.

[40] I. Tasaki, E. H. Polley, and F. Orego, *J. Neurophysiol.*, **17**, 454 (1954).

[41] D. Hubbard, *J. Res. Nat. Bur. Std.*, **36**, 511 (1946).

[42] Orion Research, Inc.: *Instruction Manual, Calcium Activity Electrode*, Cambridge, Mass., 1966.

[43] M. Frant, personal communication (1966).

[44] A. H. Truesdell, and C. L. Christ, in *Glass Electrodes for Hydrogen and Other Cations* (G. Eisenman, ed.), Dekker, New York, 1967.

[45] C. W. Carr, in *Electrochemistry in Biology and Medicine* (T. Shedlovsky, ed.), Wiley, New York, 1955.

[46] Orion Research, Inc.: *Instruction Manual, Divalent Cation Activity Electrode*, Cambridge, Mass., 1966.

[47] M. L. Wohlbarsht, E. F. MacNichol, Jr., and H. G. Wagner, *Science*, **132**, 1309 (1960).

[48] C. Guld, *Med. Electron. Biol. Eng.*, **2**, 317 (1964).

[49] Orion Research, Inc.: *Chloride Ion Activity Electrode, Data Sheet*, Cambridge, Mass., 1967.

[50] R. A. Robinson, and R. H. Stokes, *Electrolyte Solutions*, Butterworths, London, 1959.

[51] R. Chambers, and E. L. Chambers, in *Explorations into the Nature of the Living Cell*, Ch. 11, Harvard University Press, Cambridge, Mass., 1961.

[52] L. B. Nanninga, *Biochim. Biophys. Acta*, **54**, 338 (1961).

[53] S. Hagiwara, and S. Nakajima, *J. Gen. Physiol.*, **49**, 807 (1966).

[54] A. L. Hodgkin, and R. D. Keynes, *J. Physiol.*, **138**, 253 (1957).

[55] Z. A. Sorokina, *Bull. Exp. Biol. Med.*, **58**, 1398 (1964).

[56] S. G. A. McLaughlin, and J. A. M. Hinke, *Can. J. Physiol. Pharmacol.*, **44**, 837 (1966).

[57] A. A. Lev, *Nature*, **201**, 1132 (1964).

[58] J. A. M. Hinke, *J. Physiol.*, **156**, 314 (1961).

[59] P. C. Caldwell, *J. Physiol.*, **126**, 169 (1954).

[60] P. C. Caldwell, *J. Physiol.*, **142**, 22 (1958).

[61] C. S. Spyropoulos, *J. Neurochem.*, **5**, 185 (1960).

[62] P. G. Kostyuk, and Z. A. Sorokina, in *Symposium on Membrane Transport and Metabolism* (A. Kleinzeller and A. Kotyk, eds.), Academic, New York, 1961.

[63] J. A. M. Hinke, *Nature*, **184**, 1257 (1959).

[64] E. J. Conway, *Physiol. Rev.*, **37**, 84 (1957).

[65] G. Wallin, *Nature*, **212**, 521 (1966).

[66] O. Schmitt, and N. Geschwind, *Progr. Biophys. Biophys. Chem.*, **8**, 166 (1957).

[67] R. D. Keynes, *J. Physiol.*, **169**, 690 (1963).

[68] J. Villegas, L. Villegas, and R. Villegas, *J. Gen. Physiol.*, **49**, 1 (1965).

[69] C. F. Coleman, C. A. Blake, Jr., and K. B. Brown, *Talanta*, **9**, 297 (1962).

[70] E. Hogfeldt, in *Ion Exchange*, vol. 1 (J. A. Marinsky, ed.), Dekker, New York, 1967.

[71] G. H. Morrison, and H. Freiser, *Solvent Extraction in Analytical Chemistry*, Wiley, New York, 1957.

[72] J. Stary, *Solvent Extraction of Metal Chelates*, Pergamon, London, 1964.

[73] E. A. Martell, and M. Calvin, *Chemistry of the Metal Chelate Compounds*, Prentice-Hall, Englewood Cliffs, N.J., 1952.

[74] R. M. Diamond, and D. G. Tuck, *Progr. Inorg. Chem.*, **2**, 109 (1960).

[75] B. C. Pressman, *Proc. Nat. Acad. Sci. U.S.*, **53**, 1076 (1965).

[76] R. S. Cockrell, E. J. Harris, and B. C. Pressman, *Biochemistry*, **5**, 2326 (1966).

[77] S. N. Graven, H. A. Lardy, and S. Estrada, *Biochemistry*, **6**, 365 (1967).

[78] P. Mueller, and D. O. Rudin, *Biochem. Biophys. Res. Commun.*, **26**, 398 (1967).

[79] Z. Stefanac, and W. Simon, *Microchem. J.*, **12**, 125 (1967).

CHAPTER 18

The pO_2 in Isolated Muscle Measured with an Intracellular Electrode

WILLIAM J. WHALEN

St. Vincent Charity Hospital, Cleveland

The research reported here was supported in part by U.S. Public Health Service Grant HE 5390.

18.1. INTRODUCTION

When an investigator uses an isolated tissue, he is usually concerned with the question of whether the tissue is adequately oxygenated. Even if he is not concerned, his critics almost invariably are. To answer this and other questions (e.g., see Ref. [1]), we constructed a micro oxygen electrode having a tip of 1–2 μ. This electrode has been used to measure the pO_2 in the core of the isolated cat papillary muscle during rest and activity, and while exposed to various concentrations of oxygen.

18.2 CONSTRUCTION AND CALIBRATION OF THE OXYGEN ELECTRODE

Although the details of the construction and testing of the electrode have been published[2], some recapitulation seems appropriate here. A 4- to 5-in. length of Pyrex glass capillary tube (about 0.9 mm o.d. and 0.4 mm i.o.) was filled to half its length with a molten alloy of Wood's metal and gold. The alloy was prepared on a hot plate at 300°C by mixing 5 g of Wood's metal (melting point 73–75°C) with 0.3–0.5 g of precipitated gold powder. It appeared to us that a superior alloy was formed if the heating was continued for several hours. The molten alloy was then drawn into the glass capillary tube by suction, applied through means of a syringe connected to vinyl tubing which fitted snugly over the glass capillary tube. The capillary tube was placed in a pipette puller with the top edge of the alloy at the top of the heating element and pulled out to a long, tapering point. In a few instances the metal was continuous down to within a few microns of the tip, which suited our aim. Most often, however, it was necessary to reheat the alloy at the base end of the capillary, insert the metal wire contact, and gently rewarm the alloy near the tip to force the alloy toward the tip. Electrodes were chosen which had a tip of less than $2\,\mu$ and which had a recess of 10–$30\,\mu$ at the tip. The electrode was placed in gold-plating solution for several minutes, and gold was plated on the metal at the base of the recess. The electrode was then allowed to stand in distilled water for several hours to remove the plating solution, and was subsequently placed in 95% ethanol for 2 h or more. Finally the electrode tip was put in collodion for 5–10 min. Collodion has been shown to improve the performance of recessed oxygen electrodes[3]. The tip end of a completed electrode is shown in Fig. 18.1.

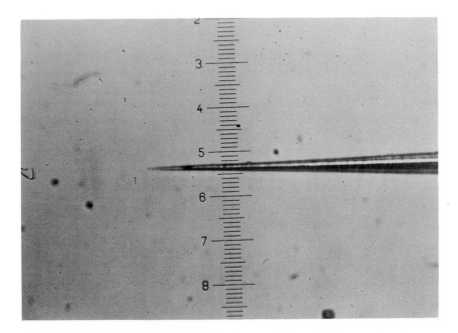

Fig. 18.1. Microphotograph of tip end of an electrode. Each small division on scale represents $2.5\,\mu$. Gold at base of recess shows as very dark region.

The oxygen electrode (properly, "cathode")[3] was used with a separate Ag–AgCl reference electrode. Usually 0·8 V was the polarizing voltage.

The electrodes were calibrated and tested, as previously described[2], in saline, 10% albumin in saline, and in nonrespiring brain tissue. The curves were nearly independent of the calibrating solutions. Furthermore, stirring of the solution did not affect the oxygen current. The electrodes have a rapid response time (1 sec or less), give a good current-voltage plateau, remain stable for several weeks, and consume less than $2 \times 10^{-6} \mu l$ O_2 per minute. After preliminary aging they show little or no "poisoning" [3] in tissue[2, 4].

Figure 18.2 shows the experimental set-up we used. The voltmeter circuit was not used in these experiments because it was not essential to know that the electrode was inside a cell. In other previous experiments, we have used the membrane potential to indicate a cell penetration. We did make some early attempts to measure membrane potentials in the heart muscle, but without much success.

Physiological solutions equilibrated with various gas mixtures could be rapidly passed through the chamber. Mammalian Ringer's solution (calcium reduced to half) at a temperature of 29–30°C was used in all experiments. All gas mixtures contained 2% CO_2, which maintained the pH at 7.2–7.4.

The muscle was stretched to its resting length and the threshold voltage for contractions was determined. Then the diameter of the muscle was measured with a micrometer. With the cat papillary muscle exposed to 25% O_2, the electrode was inserted by definite steps into the tissue (using a micromanipulator) until the pO₂ showed no further fall or began to rise slightly, indicating that the electrode was in the center of the muscle or had passed through the center. If the pO₂ fell to zero or close to it, a

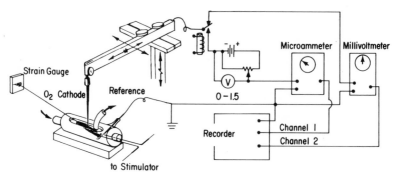

Fig. 18.2. Sketch of muscle chamber and associated electronic circuitry.

solution saturated with 98% O_2 was run into the chamber and the core of the muscle relocated as above. As a check on the zero level, solutions equilibrated with gases containing no oxygen (98% helium or nitrogen) were run into the chamber, followed by an oxygenated solution to recheck the previous level of the core pO_2 and to observe the time course of the rise in pO_2. Usually two or three values were averaged.

In the second phase of the experiments (using many of the same muscles), the muscle was stimulated to contract at 6, 15, 30, and 60 times per minute in solutions containing either 25% or 98% O_2. Stimulation at each frequency was continued until the core pO_2 reached a plateau, then the next higher frequency was begun. If the core pO_2 reached zero, no higher frequency was used.

18.3   MEASUREMENT OF THE INTRACELLULAR pO₂

A tracing of the core pO_2 in a thin, resting muscle in response to sudden changes in external pO_2 is shown in Fig. 18.3. The return of the core pO_2 to the prehypoxic level was in this instance more prolonged than usual.

Data for the core pO_2 for all the muscles studied in the first phase are shown in Fig. 18.4. Of particular interest was the finding that in 98% O_2,

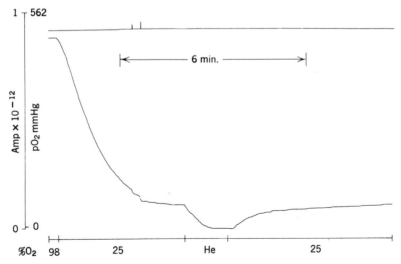

Fig. 18.3. Muscle tension (upper trace) and core pO_2 (lower trace) of muscle 0.7 mm in diameter exposed to solutions containing 98% O_2, 25% O_2 or helium. Two spontaneous contractions occurred during first period of exposure to 25% O_2. Note fall in pO_2 with each contraction.

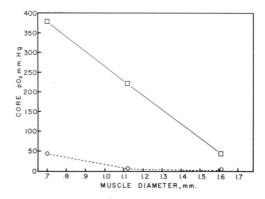

Fig. 18.4. Plot of core pO₂ and diameter of muscles exposed to 98% O₂ (solid line) and 25% O₂ (dashed line). Each point represents mean of 11 or more muscles.

none of the 11 largest muscles which averaged 1.6 mm in diameter (range, 1.4–1.8) indicated a pO₂ that reached zero. The lowest value in this group was 11 mm Hg. Muscles in the range of diameters most commonly used by investigators (0.9–1.3 mm, mean of 1.1) were amply supplied with O₂ in 98% O₂. In 25% O₂, 7 of the 30 muscles in the latter group had a core pO₂ of zero. The mean pO₂ of this group was 6.4 ± 2 mm Hg. These data indicate that if the muscles are not stimulated, even relatively thick muscles (up to 1.8 mm) can be used with impunity at 98% O₂.

During the course of the above study we observed that if the period of exposure to nitrogen or helium was prolonged, there was a latent period before the core pO₂ began to rise following readmission of oxygenated solution. This phenomenon is illustrated in Fig. 18.5. After a short hypoxic period the pO₂ rose almost immediately after the solution was changed to one equilibrated with 25% O₂. After the second, longer exposure there was a latency of more than 30 sec. We interpret the latency as owing to the acceleration of respiration caused by an accumulation of ADP during the hypoxic period. When we used thin muscles or solutions high in oxygen, the latency was far less evident.

In the second phase of the experiments the muscles were stimulated with square wave pulses while recording muscle tension and core pO₂. Figure 18.6 (especially *a*) should help to clarify the procedure as well as to show a typical result. As seen previously in Fig. 18.3, the pO₂ falls very soon after the contraction starts. Commonly, as shown in Fig. 18.6, there was an early brief rise in the current as the contraction began — which confounded the interpretation of the trace. It is possible that the transient

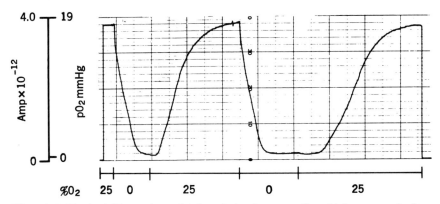

Fig. 18.5. Typical illustration of delayed rise in core pO$_2$ which occurred after prolonged period of hypoxia. Chart speed 1 in/min (dark lines). Muscle diameter 1.1 mm.

rise is caused by the action potential, as it often occurred (as in Fig. 18.6) in the absence of oxygen. It was not a result of stimulation artifact for in control experiments without muscle, pulses at two to three times threshold were not detectable on the oxygen current trace.

Of particular interest to me was the finding that the contractile force did not immediately fall when the core pO$_2$ reached zero; yet there was often a long latency before the core pO$_2$ began to rise (Fig. 18.6a). This finding suggests that isolated heart muscle can contract an oxygen debt, in contrast to the generally accepted view regarding the intact heart. It may be that isolated heart muscle has a larger store of ATP; this aspect of the work requires further study.

It became evident early in this study that the force of the contractions was an important factor in determining the "tolerable" rate of contraction. For example, the first few hours after excision of the muscle, when the force was maximal, core pO$_2$ in a muscle 1-mm thick would usually reach zero after 1–3 min of contraction at a rate of 30/min. Many hours later when the force had declined 50% or more, the core pO$_2$ would seldom reach zero even at a rate of 60–90 contractions per minute. Consequently, only values obtained 6 h or less after excision are given in this report. Even so, some caution is necessary as the solution used contained somewhat less than the optimum amount of calcium.

Figure 18.7 shows a plot of the data obtained from one experiment in which a muscle 0.85 mm in diameter was used. The core pO$_2$ plateaued (in 98% O$_2$) at 86 mm Hg at a contraction rate of 14/min and fell to zero when the rate was increased to 30/min. The diameter of the 15 muscles exposed to 98% O$_2$ ranged from 0.7 to 1.6 mm. Only 2 of the 15 muscles

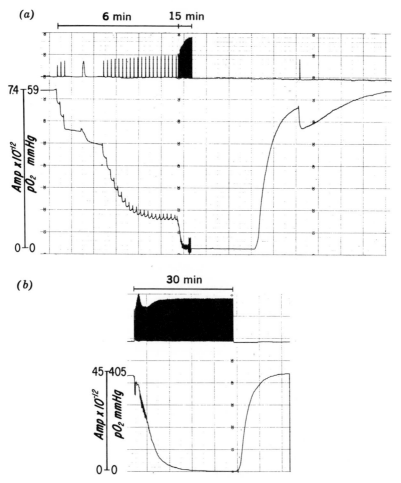

Fig. 18.6. Typical changes in core pO_2 during contraction of muscle (diameter 0.85 mm) exposed to (a) 25% O_2 and (b) 98% O_2. Chart speed was 1 in/min except for fourth contraction in (a) when it was 8 in/min. Note delayed recovery in (a) after stimulation. Muscle diameter 0.85 mm.

studied in 98% O_2 could not tolerate a contraction rate of 15/min. Surprisingly, one of the smaller muscles in this group (1.0-mm diameter) did not tolerate a contraction rate of 6/min for more than 3 min. Twelve of the 15 muscles were also studied during exposure to 25% O_2. All of the 4 muscles 0.85 mm and below in diameter could be stimulated at 6/min without oxygen depletion whereas more than half of the larger muscles became hypoxic in 1 or 2 min.

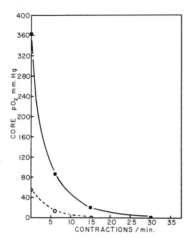

Fig. 18.7. Plot of core pO_2 and frequency of contraction during exposure to solutions equilibrated with 98% O_2 (solid line) and 25% O_2 (dashed line). Muscle diameter 0.85 mm.

It is clear from the second phase of the experiments that intermittent periods of contraction and rest provide a safety factor for investigators using isolated heart muscles. In 98% O_2, muscles as large as 1.6 mm in diameter can contract at a rate of 30/min for at least 1 min without becoming hypoxic. A rest period of 5 min should be ample for at least 95% recovery. If prolonged periods of stimulation are used, contraction at 15/min can be tolerated by most muscles — certainly those below 0.9 mm in diameter.

In most of the earlier work on cardiac energetics in which cat papillary muscle was used, the stimulation rate was 30 or more per minute (e.g., Ref. [5]). In the light of the present findings it appears that some of the earlier work in this field should be reinvestigated.

REFERENCES

[1] W. J. Whalen, *Am. J. Physiol.*, **211**, 862 (1966).
[2] W. J. Whalen, J. Riley and P. Nair. *J. Appl. Physiol.*, in press.
[3] P. W. Davies, in *Physical Techniques in Biological Research*, vol. 4 (W. L. Nastuk, ed.), Academic, New York, 1962.
[4] W. J. Whalen and P. Nair, *Circulation Res.*, **21**, 251 (1967).
[5] W. J. Whalen, *J. Physiol. (London)*, **157**, 1 (1961).

CHAPTER 19

Techniques of Intracellular Microinjection

TUSHAR K. CHOWDHURY

Department of Physiology, George Washington University

I wish to acknowledge the generous help of my former advisor, Professor Fred M. Snell, of the State University of New York in Buffalo. My sincere thanks are due to my colleague at George Washington University, Professor Charles S. Tidball, Chairman of the Department of Physiology, for his valuable suggestions and criticisms during the preparation of the manuscript. It is also a pleasure to acknowledge the assistance of Dr. Gunter F. Bahr of the Armed Forces Institute of Pathology, who permitted me to use the fluorescence microscope equipment in his laboratory.

19.1 INTRODUCTION

Glass micropipettes have opened a new era in the field of intracellular microinjection. Traditionally there have been two different approaches to introducing substances into living cells. The first approach utilizes a controlled extraneous pressure at one end of the micropipette to force material out of the tip and hereafter will be called the "pressure technique." The second approach makes use of an applied electromotive force to send current through the micropipette so that some of the current-carrying ions are ejected from the tip of the micropipette. This approach is commonly known as the "electrophoretic technique." Regarding the usefulness of these two standard techniques, certain limitations—especially with regard to use in biological systems—should be pointed out. For example, the "pressure technique," although useful in injecting both electrolytes as well as nonelectrolytes into a cell, is not practical when the tip of the micropipette is less than 0.5 μ. Yet, to insure the viability of most cellular systems, the outside diameter of the tip of the micropipette should be as small as possible. On the other hand, the

404

"electrophoretic technique," which can be utilized with fine-tipped micropipettes, is restricted to injecting electrolytes or ionizable substances.

In addition to the above-noted techniques, a third approach has been developed by the author in which a high-frequency vibration is imposed on the micropipette to permit the extrusion of materials from the tip of the micropipette; this approach will be termed the "vibration technique." As there are several adequate review articles on the conventional approaches already in print[1–3], only a brief discussion of them will be undertaken in this Chapter. The vibration technique will be described in greater detail.

19.2 MICROINJECTION BY PRESSURE TECHNIQUES

The pioneer in the development of "pressure techniques" for micro-injection is Robert W. Chambers. In his approach the extraneous pressure is transmitted to the column of the injection fluid at the tip or proximal end of the micropipette through a mercury or air medium in the distal portion of the shaft of the micropipette. When the micropipettes have tip diameters in the range of $5-10\,\mu$, significant quantities of the injection material can be forced out of the pipette by using moderate pressure. The main feature in the design of various other devices for this type of microinjection is the fabrication of the micropipette holder. A simplified form of the micropipette holder designed by Grundfest et al.[4] is shown in Fig. 19.1. Air-free mineral oil fills the entire system except the tip and a small portion of the shaft of the micropipette, which are both filled with an injection fluid to which chlorophenol red (0.05–0.10%) has been added. The meniscus at the oil-water interface in the shaft is visualized sufficiently easily so that its excursion can be measured with a calibrated ocular micrometer to indicate the amount of solution injected into the cell. A

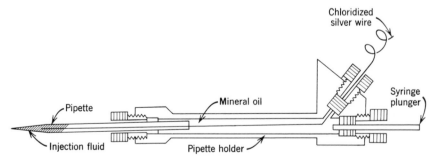

Fig. 19.1. Micropipette holder devised by Grundfest et al.[4].

special feature of this particular device is that a chloridized silver wire extends from a side arm at the distal end of the micropipette and passes through the lumen up to the conical portion of the pipette. Thus use of an electrolyte in the injection fluid enables one to record the intracellular electrical potential utilizing the same micropipette, during and following the injection of the solution under study. Operation of the plunger of the syringe for expelling precise quantities of injection fluid from the micropipette requires considerable practice.

A special type of pressure device for microinjection has been devised by De Fonbrune[5] in which the pressure on the column of the injection fluid is achieved by way of thermal expansion. For this technique a micropipette having a long shaft (approximately 5 cm) is sealed at the distal or the wide end. The shaft is then inserted into a metal tube which can be heated electrically. By alternate heating and cooling, the shaft of the micropipette is filled with some oil or other less volatile liquid. The remaining portion of the micropipette, that is, the tapered region, is filled with the desired injection fluid. The pipette system is then mounted on a micromanipulator and it is ready for injection. When the temperature of the metal jacket is raised, the volume of the trapped oil expands and pushes a portion of the injection fluid out of the tip of the pipette. An injection can be controlled more efficiently by this technique than in the previously described ones. Although the injection can be controlled, subsequent cooling of the metal jacket would cause a withdrawal of material from the cell. Therefore, in practice, the micropipette is withdrawn soon after the injection is made. For this reason, unlike the pressure device of Grundfest et al.[4], the micropipette is not suitable for a simultaneous recording of the intracellular electrical potential. Moreover, because of the alternate heating and cooling filling procedure required by this system, only micropipettes of large tip diameter are suitable; this restricts the type of cell in which they can be used without producing irreversible damage.

There are at least five other microinjection devices developed in various laboratories, all based on the principle of ejection by application of external pressure. The various refinements are concerned with control of external pressure or the uniformity of the injection rate[6, 7]. A detailed description of the earlier devices can be found in McClung's *Handbook of Microscopical Techniques*[8].

The ability of all these pressure techniques to inject microquantities of fluids into a single cell depends not only on accurate control of the applied pressure but also on the viscosity of the injection fluid and, moreso, on the tip diameter of the micropipette. The pressure techniques are quite useful in injecting electrolytes as well as nonelectrolytes into cells or tissues, particularly when the latter systems are relatively large (e.g., muscle and nerve fibers). For injection into such cellular systems, micropipettes having outside diameter down to 10 μ can be used without much difficulty. However, where the cellular system warrants the use of fine-tipped micropipettes, that is, less than 0.5 μ in outside tip diameter, the magnitude of pressure required to move fluid out becomes so large that injection would appear to be only hypothetically possible. This becomes evident from the following numerical considerations. Thus suppose one needs to inject a total of 1 nanol of a solution of the desired substance into a nerve axon within a period of 1 sec. Assuming that the flow of solution out of the cylindrically shaped tip of the micropipette (the region between the planes C and D in Fig. 19.2)* occurs in bulk, the pressure difference across the micropipette required to yield the desired flow would be given by the Poisseuille equation of flow for incompressible fluid:

$$P_1^+ - P_2 = \left(\frac{8}{\pi}\right)\left(\frac{L}{R^4}\right) \eta \text{(note of volume flow)} \qquad (19.1)$$

where L is the length of the narrow region of the micropipette (between C and D), R is the radius of the tip, and η is the viscosity of the injection fluid. Taking a very conservative value of 100 μ for L and 0.01 dynes-sec cm^{-2} for the viscosity of the injection fluid, the pressure difference

*For a conventional micropipette, the cone angle subtended by the two opposite sides of the wall at the tip is about 1°; therefore, for the present calculation, the region between C and D can be considered to have a uniform radius R.

†The radius of the shaft of the pipette is much larger than that at the tip; hence the pressure drop between A and C can be neglected in relation to that between C and D.

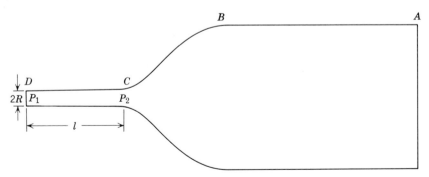

Fig. 19.2. Fluid flow through micropipette. P_1, pressure at D; P_2, pressure at C.

required to be applied across the micropipette would be 2.6 atm. In this computation, the value for R has been taken as 1μ; this means that the outside diameter of the micropipette tip is close to 3μ. With a micropipette having an outside tip diameter of 0.5μ (i.e., with R of about 0.16μ), the pressure required for the above microinjection is 3900 atm — which is not at all practical to employ. In actuality, especially with a conventional micropipette, the length of the narrow region is about 1 mm, in which case the magnitude of the pressure required is even larger.

19.3 MICROINJECTION BY ELECTROPHORETIC TECHNIQUES

Electrophoresis involves the migration of charged solute particles in an electrical field. A rather arbitrary distinction is often made in this definition on the basis of the size of the migrating particles. Thus the term "electrophoresis" is usually reserved for migration of macromolecules and colloid particles, while the migration of the smaller charged particles or ions is generaly referred to as "ionophoresis." Throughout this Chapter the general term "electrophoresis" will be used regardless of the size of the particles.

When particles carrying net electrical charges* on their surface are suspended or dissolved in a conducting medium and subjected to the influence of an external electrical field, they tend to move toward the electrodes of opposite polarities. Thus the positively charged particles move toward the cathode and the negatively charged ones toward the anode as long as the external electrical field is maintained. The rate of movement of these particles will depend, among several other things, on the surface charge density of the particles, the magnitude of the electrical field, and the resistance of the solution medium to the motion of these particles. The latter is determined by the size and shape of the particles as well as the viscosity of the medium. The magnitude of the surface charge depends on the degree of ionization of the various chemical side groups of the particles. It may be altered by changing the pH and/or the ionic strength of the solution. At lower pH, that is, at higher concentration of the hydrogen ions in the solution, the solute particles may acquire more positive charges than at a higher pH. Many substances, because of their zero surface charge at any particular pH or ionic strength

---

*It is important to keep in mind that in free solution any particle carrying a net charge fixed on its surface is accompanied by an equal amount of opposite charge carried by "gegen" ions in the immediate surroundings. Thus in any macro region within the solution, the charges add up to zero.

of the medium, may appear to be unsuitable for electrophoretic study. However, merely by altering the pH of the solution and/or its strength, these solute particles may become electrophoretically mobile.

A large number of experimental methods has been developed in applying the principle of electrophoresis in various fields of research. It is not intended to cover them in this chapter; the interested reader may consult among other references the review of articles by Briggs[9], Stern[2], or Moore[10]. Only a brief description will be given of the techniques used in injecting microquantities of ions or ionizable substances into living cells with the aid of an external electrical field. This approach has also been referred to as "microelectrophoresis"[3], even though Briggs has described a completely different method under the same term. A simple schematic arrangement of the electrical setup for microinjection of cation into a single cell is shown in Fig. 19.3. When the glass micropipette electrode† is made positive with respect to the reference macroelectrode located in the chamber solution, conventional current passes out of the micropipette tip—resulting in an injection of cations into the cell. A portion of the total electrophoretic current is, however, carried by the anions flowing into the pipette through the tip. For an anionic injection, obviously the current has to flow from the cell interior into the microelectrode. As a source of current for electrophoretic microinjection, any variable output dc power supply capable of delivering up to one-tenth of a milliampere at 200 V may be employed. The normal working current, however, is less than 1 μA.

In theory the amount of any ionized particle (say, cation C^+) ejected out of the micropipette electrode can be determined from the equation

$$J_{C^+} \frac{t_+ I}{zF}, \qquad (19.2)$$

where J_{C^+} is the rate of flow of the cation in moles per second, t_+ is the transference number, z is the valence of the cation, I is the electrophoretic current in amperes, and F is the Faraday constant. In practice, however, when an electrical field is applied across a micropipette electrode made from a Pyrex glass capillary, there arises a flow of solvent owing to the phenomenon of electro-osmosis[11]. This solvent flow in turn exerts a drag on all the dissolved or suspended ionized particles. The exact determination of the electro-osmotic flow of the ions is not always

---

†As the term "glass electrode" is commonly used to identify a class of pH electrodes, the glass micropipettes filled with an electrolyte for the purpose of potential measurement have been termed "glass micropipette electrodes" or simply "microelectrodes."

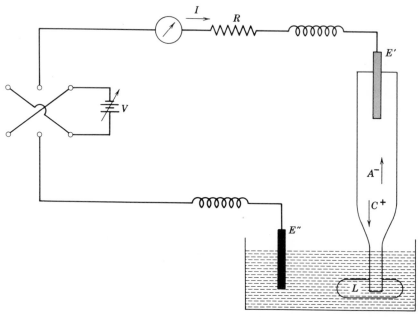

Fig. 19.3. Schematic circuit diagram for electrophoretic microinjection of cation C^+ into living cell, L. E' can be a silver-silver chloride electrode or any other non-polarizable electrode; E'' is the reference electrode. R is series resistance around 100 MΩ and I electrophoretic current. V is a variable voltage source.

possible. However, the magnitude of this electro-osmotic ejection can be comparatively minimized by using a high concentration of the electrolyte (C^+A^-) within the injecting micropipette. Thus, for example, Krnjevic et al.[12] have estimated that in the microinjection of a 3 M solution of acetylcholine, the electro-osmotic ejection is only one-tenth of the electrophoretic ejection.

In a practical situation, the ejection of a substance from the micropipette is governed not only by the electrophoretic current but also by the process of diffusion, especially as the injection fluid within the micropipette is made highly concentrated. This stray diffusion can, however, be minimized by using micropipettes with inside diameter of 0.5 μ or less. An alternate means which has routinely been used in reducing this diffusion is to apply a "retaining current" through the micropipette. This current is made to flow in the direction opposite to that of the electrophoretic current.

In view of the fact that the above-mentioned phenomena can contribute quite significantly and yet in an indeterminate way, it becomes impossible to measure precisely the actual amount of ions injected into the cell.

Only indirect means of calibration for electrophoretic injection have been used to estimate the injected amount.

The major disadvantage of the electrophoretic technique of micro-injection is that when the cellular system warrants the use of a fine-tipped glass micropipette, the injection of anionic substances into the cell becomes very difficult — mainly because of the electrical rectification property of the microelectrode[13]. However, this limitation does not apply in the case of cationic injection.

It is well known to investigators in the field of electrophysiology that the impedance of a fine-tipped glass micropipette electrode is as much as 20 times higher for a positive current inflow than for a positive current outflow through the tip. Such microelectrodes have also been found to possess the cation-exchange properties of a synthetic cation-exchange membrane[14]. An explanation for this cation-exchange and rectification behavior of a fine-tipped glass micropipette electrode may rest on the inherent electrostatic characteristics of the glass surface. The wall of a Pyrex glass micropipette electrode has fixed negative charges on it and thus when the orifice at the tip becomes quite small, the electrical field exerted by the fixed charges on the wall extends far enough within the core of the orifice to favor the passage of cations and hinder the movement of anions.

Another point worth considering here is the fact that, unlike the pressure technique of microinjection, during the period of actual micro-injection the injecting micropipette can not be used in making a simultaneous recording of the electrical activity of the injected cell. The magnitude of the electrophoretic current through the microelectrode is sufficient to generate electrical noise in preventing the recording of the intracellular electrical potential by the same electrode. Even when the injecting electrode is used to record the intracellular potential when the electrophoretic current has ceased, the noise in the recording trace may be quite high unless the retaining current is sufficiently low. Some of the difficulties have been overcome by the use of coaxial or double-barreled microelectrodes[3]. One barrel can be used strictly for sending the electrophoretic current while the second one can be used in monitoring the intracellular potential. Sometimes multibarrel electrodes have been proved to be quite useful for the above-noted purposes. The resulting dimension of the tip of such a composite electrode system obviously is larger than that of a single microelectrode, and this large tip may produce considerable damage to the cell during penetration. Thus the usefulness of the electrophoretic technique for the purpose of intracellular injection of ionized substance depends greatly on the biological system to be studied and the substances to be injected.

19.4. MICROINJECTION BY VIBRATION TECHNIQUE

19.4.1 Theory

It has been observed that when a high-frequency axial vibration is imposed on a fine-tipped glass micropipette electrode having the cone angle of 6 to 14° at the tip, significant quantities of the filling solution are extruded from the tip[13]. The first observation on this phenomenon was made when a micropipette electrode was vibrated at ultrasonic frequencies in air prior to its insertion into the tissue system. With a proper lighting arrangement, the atomized spray of the filling solution coming out of the micropipette was easily visualized under a microscope, by the scattering of light from the atomized fluid particles. This ejection phenomenon is now being utilized in our laboratory to inject microquantities of electrolytes as well as nonelectrolytes into a single living cell. The main factor behind this microinjection technique is the shape of the micropipette, more particularly the cone angle (see Fig. 19.4). When a micropipette with large cone angle is filled with a solution and vibrated only in the direction of the long axis of the micropipette, during the back stroke of the micropipette electrode the mass of solution confined within the infinitesimally small region bounded by the planes at $x + dx$ and x tends to cross the plane at x because of inertia. During the front stroke of the micropipette, the mass of solution confined within the region bounded by the planes at x and $x - dx$ tends to cross the plane x, but from the opposite side. As the number of molecules of solvent and solute particles within the planes $x + dx$ and x is larger than that within the planes x and $x - dx$, there occurs a net transfer of solution toward the tip. This way there is a gradual buildup of a high pressure at the tip within a fraction of a second. Occasionally the tip is found to disintegrate into pieces. The ejection of the fluid takes place with frequencies above 25 kc/sec. The lateral vibration which can produce injury to the injected cell and decrease the control over the injection rate can be eliminated by proper adjustment of the frequency and the amplitude of vibration.

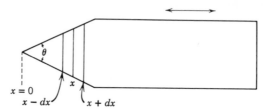

Fig. 19.4. Hydrodynamic rectification in vibrating micropipette. Vibration of micropipette is only in direction of its long axis. θ is cone angle.

Fig. 19.8. Photograph of vibrator assembly coupled to micromanipulator.

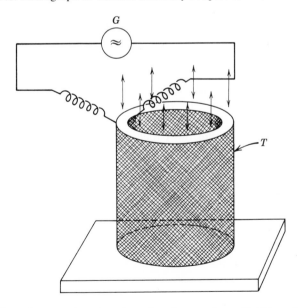

Fig. 19.9. Length expansion of a piezoelectric transducer T. G is ocillator genera-tor; double arrows indicate excursion of transducer held at bottom end.

electrical field. In an actual microinjection, the magnitude of excursion never exceeds 100 Å.

19.4.6   Experimental Results

The first experiments were performed to test whether the vibration technique of microinjection remained efficient when the ejection of solution from the micropipette tip encountered some mechanical resistance. For this purpose, a 1-mm thick layer of carboxymethyl cellulose gel† was used to simulate a biological tissue. The advantage of using such cellulose gel over an agar gel was the consistency of the cellulose gel, which reduced the diffusion rate of salts by a factor of at least ten as compared with the rate of diffusion in agar gels.

The micropipette filled with a solution of either sodium fluoresceinate or euchrysin (0.1%) was inserted into the gel at a depth of about $10\,\mu$ and the micropipette was vibrated at a frequency of 28 kc/sec for a period of a few seconds. The injected fluorescent material was then detected under a high-power fluorescence microscope. In separate gel systems two types of controls were run by inserting the same micropipette without axial vibration, once before and once following the actual injection experiment. In these control experiments the tips were kept within the gel for periods as long as 60 sec. After withdrawal of the micropipette, the gels were studied under the fluorescence microscope and no fluorescent spots were found – indicating that no significant quantity of injection solution came out of the tip of the micropipette by diffusion alone.

Quantitation of the solution injected by this vibration technique has been made to date only by fluorescence microscopy. For this method, injections of the fluorescent dye are performed, using the same micropipette at a fixed frequency of vibration but at different locations in the gel layer and for successively longer durations of the vibrations. As shown in Fig. 19.10, within the capability of the detection system only the fluorescent spots corresponding to a duration of vibration in excess of 2 sec were detectable. Theoretically, it is possible to increase the rate of injection by increasing the frequency of vibration. However, at very high frequencies (above 50 kc/sec) the pressure developed at the tip is so large that the tip disintegrates into small pieces. From the sensitivity of the photographic system and the intensity of the fluorescent spot, the smallest detectable spot corresponds to an ejected solution of $1.2 \pm 0.2 \times 10^{-10}$ ml. This does not mean, however, that there was no ejection of fluid when the duration of vibration was less than 2 sec. This point was

---

†The carboxymethyl cellulose used in this experiment was kindly provided by the Hercules Powder Company.

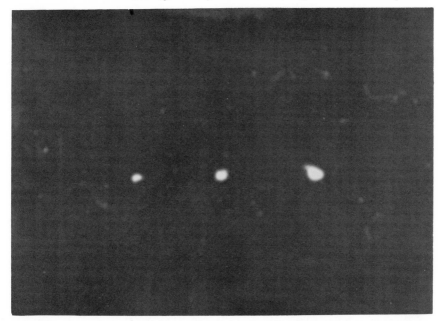

Fig. 19.10. Localization of microinjected spots by fluorescence microscopy. Smallest spot corresponds to vibration period of 2 sec and next two bigger spots correspond to periods of vibration of 5 and 10 sec, respectively. Microelectrode is vibrated at 28 kc/sec at 16 V.

established clearly by the results of some experiments on frog sciatic nerve fibers. One nerve fiber within a bundle was penetrated with a fine-tipped glass micropipette electrode filled with frog Ringer's solution containing 2,4-dinitrophenol (DNP) at a concentration of $10^{-2}\,M$. When a satisfactory resting potential was found to be maintained at a steady level for at least 50 min, the microelectrode was vibrated at 30 kc/sec and 16 V only momentarily. The effect of such vibration on the resting potential of the nerve fiber is shown in Fig. 19.11 (*top*). Within the course of a few minutes, the resting potential started declining. An explanation for the different rates in this decline awaits further investigation. It should be mentioned here that when the frequency is chosen to produce a lateral vibration of the tip of the microelectrode, the onset of the vibration causes the action potential to be developed in the nerve fiber.

In the control experiments that were performed with the second glass micropipette electrode pulled from the same capillary tube as that used for DNP experiments, the micropipette was filled with the same Ringer's solution except for the omission of DNP. The results of a short microinjection of the Ringer's solution alone on the resting potential of a

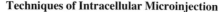

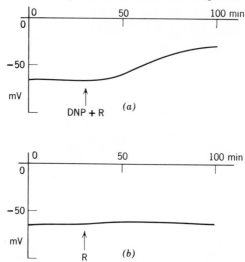

Fig. 19.11. Effect of momentary injection of 2,4-dinitrophenol by vibration technique on resting potential of frog sciatic nerve fiber (*a*) Injection of Ringer solution containing DNP(DNP + R) in frog Ringer's solution made at arrow. (*b*) Injection of Ringer's solution alone (R) made at arrow. Microelectrodes vibrated only momentarily at 30 kc/sec at 16 V.

different nerve fiber is shown in Fig. 19.11 (*bottom*). Here there is no significant effect of the microinjection on the resting potential.

The results of the above two groups of experiments not only indicate that ejection of material takes place even for a period of vibration shorter than 1 sec but also that nerve cells do not undergo any irreversible damage owing to any mechanical effect of vibration or to injection of microquantities of Ringer's solution into them.

In actuality, it may not be possible to know the exact amount of a substance that is injected into the cell by the vibration technique until later when a calibration test is performed with a counterpart of the first microelectrode. Also a better technique of quantitation (other than the fluorescence microscopic technique) would seem to be needed. Perhaps the simultaneous use of a suitable radioisotope within the microelectrode might make it possible to utilize a single microelectrode for both the intracellular injection and the calibration. The specific activity of the radioisotope required for this purpose would obviously have to be quite high. Such a disadvantage of the vibration technique of microinjection is outweighed by the various advantages this technique presents. Thus, for example, any small molecules or ions can be injected into a single living cell even with the utilization of a microelectrode having an outside tip diameter as small as 0.1 μ, thereby insuring minimal damage to the cell. Secondly, the injection micropipette itself can be used to monitor the intracellular electrical potential, thus making it possible to study the transient as well as steady-state effects of injection of a substance. The noise generated during the actual vibration period is not significant. However, because of the flow of fluid through the fine tip of the microelectrode, streaming potential is generated which alters the tip potential of the microelectrode—but only by a magnitude of 0.5 mV or less. This generated potential, however, can easily be differentiated from the tip potential as well as the actual intracellular electrical potential. Moreover, this magnitude is quite insignificant in comparison with the magnitude of the intracellular potential in most living cells. The ease of operation of this injection apparatus is one of the foremost advantages of the vibration technique.

19.5   OTHER USES FOR THE VIBRATION TECHNIQUE

19.5.1   Penetration of Tissue Structure

It is known that any micropipette electrode with an outside tip dimeter above 0.1 μ produces a significant distortion in the tissue or cell before its actual penetration. The indentation in the tissue results in an uneven penetration of the microelectrode, thereby introducing artifact into the actual electrical potential profile within a tissue or a cell[18]. Such indentation of the tissue can be minimized by the imposition of an axial vibration (frequency around 8 kc/sec) on the micropipette electrode[15].

19.5.2   Study of the Streaming Potential

As mentioned in the previous section, associated with the vibration of a micropipette electrode is an alteration in the tip potential. Apparently this extra electrical potential originates within the tip region of the microelectrode because of the extrusion of fluid through the tip. The polarity of this generated potential is consistent with the streaming of electrolyte fluid by the fixed negatively charged wall. Recently, the nature of the streaming potential in microcapillaries has been studied by several investigators[19]. The microelectrode system along with the vibration technique presents its fullest applicability in such investigations.

19.5.3   Filling of Fine-Tipped Micropipettes

The difficulties associated with various procedures of filling fine-tipped microelectrodes have been pointed out in an earlier section. The micropipette first fitted inside the holding screw can be filled up to the tapered region with the aid of a fine hypodermic needle inserted from the wide end. The screw can then be tightened onto the face plate on the vibrator assembly. Imposition of an axial vibration of the micropipette will make the fluid move toward the tip and finally fill the tip with the creation of an empty space at the wide end which can easily be filled with the solution. The main disadvantage in this procedure is that only one micropipette can be filled at one time.

19.5.4   Intracellular Localization of a Microelectrode Tip

In the study of the nature of the electrical potential it often becomes necessary to determine the anatomical site of the tip of the micro-electrode that is measuring the intracellular potential[15, 20]. The approach which is now being followed in this laboratory is to introduce some electron-dense marker substance at a reasonable concentration into the tapered region of the microelectrode. After the intracellular electrical potential is recorded by this microelectrode at any particular position within the cell, an axial vibration is imposed on the microelectrode. Thus a few molecules of the marker substance will come out of the tip. After withdrawal of the microelectrode, the tissue is studied by electron-microscopy, following a serial section of the properly prepared tissue for the localization of the injected marker particles.

These are only some of the applications which are now being tested. It is certain that this new technique of microinjection will soon be adopted in various disciplines of science.

REFERENCES

[1] R. W. Chambers, and M. J. Kopac, in *Handbook of Microscopical Techniques*, 3rd ed. (C. E. McClung, ed.), Hafner, New York, 1950.
[2] K. G. Stern, in *Physical Techniques in Biological Research*, vol. II, Chapter 6 (G. Oster and A. W. Pollister, eds.), Academic, New York, 1956.
[3] D. R. Curtis, in *Physical Techniques in Biological Research*, vol. 5 (W. L. Nastuk, ed.), Academic, New York, 1964.
[4] H. Grundfest, C. Y. Kao, and M. Altamirano, *J. Gen. Physiol.*, **38**, 245 (1954).
[5] P. De Fonbrune, *Recherches et Inventions*, No. 252, (1935).
[6] R. O. Brady, C. S. Spyropoulos, and I. Tasaki, *Am. J. Physiol.*, **194**, 207 (1958).
[7] P. C. Caldwell, A. L. Hodgkin, R. D. Keynes, and T. I. Shaw, *J. Physiol.*, **152**, 561 (1960).
[8] C. E. McClung, in *Handbook of Microscopical Technique*, 3rd ed. (C. E. McClung, ed.), Hafner, New York, 1950.

[9] D. R. Briggs, in *Biophysical Research Methods*, Ch. 9. (F. M. Uber, ed.), Interscience, New York, 1950.

[10] D. H. Moore, in *Technique of Organic Chemistry*, vol. 1 Pt IV (A. Weissberger, ed.), Wiley, New York, 1959.

[11] A. W. Adamson, in *Physical Chemistry of Surfaces*, Interscience, New York, 1960.

[12] K. Krnjevic, J. F. Mitchell, and J. C. Szerb, *J. Physiol.*, London, **165**, 421 (1963).

[13] T. K. Chowdhury, Doctoral dissertation, University Microfilm, Michigan # 65–10, 122, Feb. (1965).

[14] T. K. Chowdhury, unpublished data.

[15] T. K. Chowdhury, and F. M. Snell, *Biochim. Biophys. Acta*, **94**, 461 (1965).

[16] A. L. Byzov, and V. L. Chernyshov, *Biophysics*, **6**, 79 (1961).

[17] C. P. Germano, Engineering Memorandum No. 61–15, Clevite Corporation (1961).

[18] T. K. Chowdhury, and F. M. Snell, *Biochim Biophys. Acta*, **112**, 581 (1966).

[19] C. P. Bean, Research Note from the General Electric Company (1965).

[20] G. Whittembury, *J. Gen. Physiol.*, **47**, 795 (1964).

Author Index

Numbers in **bold** characters show the pages covered by an author's chapter. Numbers in parentheses are reference numbers. Numbers in *italics* show the page on which the complete reference is listed.

Subject Index

Acids, weak, heat of ionization, 19
Acrylamide-acrylic acid mixture, sodium activity coefficient, 345
Alga Nitella, electric potentials between medium and vacuole, 159–162
  ion partition between medium and vacuole, 162–172
  morphology, 154–155
Aluminosilicate minerals, ion exchange, 49, 52
Amplification, ionic, 107–108
Ascaris lumbricoides muscle fibers, anomalous spikes, 207
Atrial tissue, membrane resistance, 239–240

Barnacle *Balanus nubilus* muscle, impalement by microelectrode, 356–357
  intracellular ions activity, 349–375
  morphology, 356–361
  water and ion binding, 365–367
Bases, weak, heat of ionization, 20
Bicarbonate concentration in single nephrons, 296–297
Buffer system, nullifying change of standard potential, 18–19

Caesium exchange, in electrode glass, 37–48
  in glass, $NAS_{10.6-10}$, 41–42
    NAS_{11-18}, 37–48
    $NAS_{17.4-13.4}$, 44–48
    $NAS_{24.7-9.9}$, 44–48
Caesium uptake, by glass, NAS_{11-18}, 50–51
    $NAS_{17.4-13.4}$, 50–51
    $NAS_{24.7-9.9}$, 50–51
Calcium, activity measurements with microelectrode, 376–395
  electrode, liquid exchanger composition, 378
  microelectrode, biological application, 389–392
    effect of pH, 384
    rate of approach to equilibrium, 383

response to calcium, 381–382
Calf heart, electrical constants, 234–235
Calomel electrode, characteristics, 4
  emf standard, 7
  temperature coefficient, nonisothermal, 6–7
  temperature insensitive, 13–17
Carcinus maenas nonmedullated nerve, cytoplasmic resistivity, 300
Cardiac cells, size, 225
Cardiac tissues, electrical constants, 234–235
  resistance measurements, 224–271
    membrane, 232
    myoplasm, 232
Carp retina, intracellular recording, 130–133
  S-compartment, resistance change with light, 151
  spectral response, 131
    of S-potentials, 146
  S-potential recording from receptor side, 147
Cat, heart, electrical constants, 234–235
  motoneuron, inhibitory membrane, excitatory effect, 195
  papillary muscle, oxygen tension measurement, 399–403
Cation activity, in muscle fibers and nerve cells, 322–348
  in single nephrons, 272–298
Cations, divalent, electrode, liquid ion exchanger composition, 378
Cation selective microelectrode, electrical performance, 352–354
  response to charged amino groups, 353
  working life span, 352
Cation sensitive, glass microelectrode, cation specificity mechanism, 84–93
  permselectivity mechanism, 84–93
  microelectrode, construction, 327–331
    fabrication, 280–282

435